Fundamentals of Matrix Algebra

Third Edition, Version 3.1110

Gregory Hartman, Ph.D.

Department of Mathematics and Computer Science

Virginia Military Institute

cc
BY NC

THANKS

This text took a great deal of effort to accomplish and I owe a great many people thanks.

I owe Michelle (and Sydney and Alex) much for their support at home. Michelle puts up with much as I continually read LaTeX manuals, sketch outlines of the text, write exercises, and draw illustrations.

My thanks to the Department of Mathematics and Computer Science at Virginia Military Institute for their support of this project. Lee Dewald and Troy Siemers, my department heads, deserve special thanks for their special encouragement and recognition that this effort has been a worthwhile endeavor.

My thanks to all who informed me of errors in the text or provided ideas for improvement. Special thanks to Michelle Feole and Dan Joseph who each caught a number of errors.

This whole project would have been impossible save for the efforts of the LaTeX community. This text makes use of about 15 different packages, was compiled using MiKTeX, and edited using TeXnicCenter, all of which was provided free of charge. This generosity helped convince me that this text should be made freely available as well.

PREFACE

A Note to Students, Teachers, and other Readers

Thank you for reading this short preface. Allow me to share a few key points about the text so that you may better understand what you will find beyond this page.

This text deals with *matrix* algebra, as opposed to *linear* algebra. Without arguing semantics, I view matrix algebra as a subset of linear algebra, focused primarily on basic concepts and solution techniques. There is little formal development of theory and abstract concepts are avoided. This is akin to the master carpenter teaching his apprentice how to use a hammer, saw and plane before teaching how to make a cabinet.

This book is intended to be read. Each section starts with *"AS YOU READ"* questions that the reader should be able to answer after a careful reading of the section even if all the concepts of the section are not fully understood. I use these questions as a daily reading quiz for my students. The text is written in a conversational manner, hopefully resulting in a text that is easy (and even enjoyable) to read.

Many examples are given to illustrate concepts. When a concept is first learned, I try to demonstrate all the necessary steps so mastery can be obtained. Later, when this concept is now a tool to study another idea, certain steps are glossed over to focus on the new material at hand. I would suggest that technology be employed in a similar fashion.

This text is "open." If it nearly suits your needs as an instructor, but falls short in any way, feel free to make changes. I will readily share the source files (and help you understand them) and you can do with them as you wish. I would find such a process very rewarding on my own end, and I would enjoy seeing this text become better and even eventually grow into a separate linear algebra text. I do ask that the Creative Commons copyright be honored, in that any changes acknowledge this as a source and that it only be used non commercially.

This is the third edition of the *Fundamentals of Matrix Algebra* text. I had not intended a third edition, but it proved necessary given the number of errors found in the second edition and the other opportunities found to improve the text. It varies from the first and second editions in mostly minor ways. I hope this edition is "stable;" I do not want a fourth edition anytime soon.

Finally, I welcome any and all feedback. Please contact me with suggestions, corrections, etc.

Sincerely,
Gregory Hartman

Contents

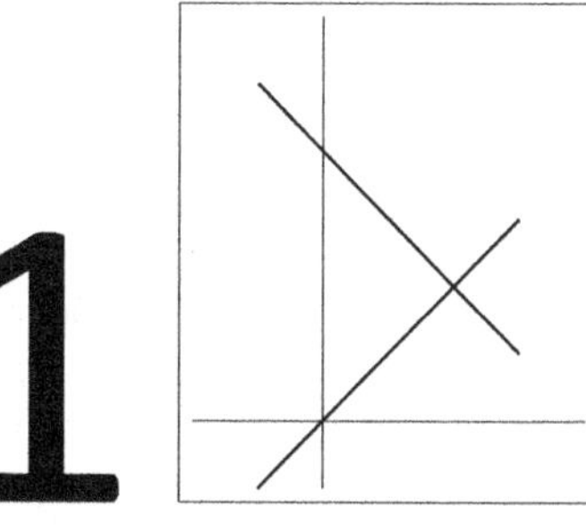
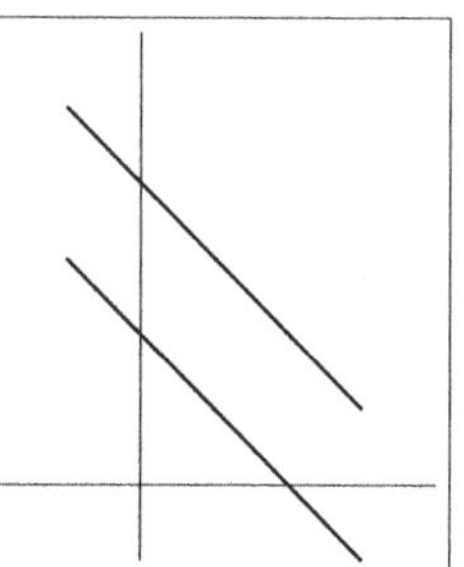
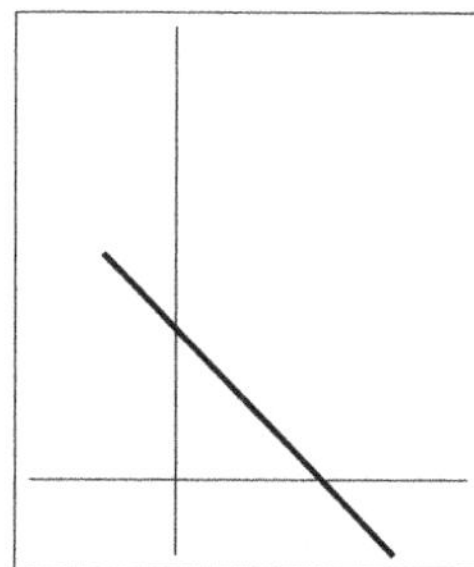

Systems of Linear Equations

You have probably encountered systems of linear equations before; you can probably remember solving systems of equations where you had three equations, three unknowns, and you tried to find the value of the unknowns. In this chapter we will uncover some of the fundamental principles guiding the solution to such problems.

Solving such systems was a bit time consuming, but not terribly difficult. So why bother? We bother because linear equations have many, many, *many* applications, from business to engineering to computer graphics to understanding more mathematics. And not only are there many applications of systems of linear equations, on most occasions where these systems arise we are using far more than three variables. (Engineering applications, for instance, often require thousands of variables.) So getting a good understanding of how to solve these systems effectively is important.

But don't worry; we'll start at the beginning.

1.1 Introduction to Linear Equations

1. What is one of the annoying habits of mathematicians?
2. What is the difference between constants and coefficients?
3. Can a coefficient in a linear equation be 0?

We'll begin this section by examining a problem you probably already know how to solve.

Example 1 Suppose a jar contains red, blue and green marbles. You are told that there are a total of 30 marbles in the jar; there are twice as many red marbles as green ones; the number of blue marbles is the same as the sum of the red and green marbles. How many marbles of each color are there?

SOLUTION We could attempt to solve this with some trial and error, and we'd probably get the correct answer without too much work. However, this won't lend itself towards learning a good technique for solving larger problems, so let's be more mathematical about it.

Let's let r represent the number of red marbles, and let b and g denote the number of blue and green marbles, respectively. We can use the given statements about the marbles in the jar to create some equations.

Since we know there are 30 marbles in the jar, we know that

$$r + b + g = 30. \tag{1.1}$$

Also, we are told that there are twice as many red marbles as green ones, so we know that

$$r = 2g. \tag{1.2}$$

Finally, we know that the number of blue marbles is the same as the sum of the red and green marbles, so we have

$$b = r + g. \tag{1.3}$$

From this stage, there isn't one "right" way of proceeding. Rather, there are many ways to use this information to find the solution. One way is to combine ideas from equations 1.2 and 1.3; in 1.3 replace r with $2g$. This gives us

$$b = 2g + g = 3g. \tag{1.4}$$

We can then combine equations 1.1, 1.2 and 1.4 by replacing r in 1.1 with $2g$ as we did before, and replacing b with $3g$ to get

$$\begin{aligned} r + b + g &= 30 \\ 2g + 3g + g &= 30 \\ 6g &= 30 \\ g &= 5 \end{aligned} \tag{1.5}$$

We can now use equation 1.5 to find r and b; we know from 1.2 that $r = 2g = 10$ and then since $r + b + g = 30$, we easily find that $b = 15$.

Mathematicians often see solutions to given problems and then ask "What if. . .?" It's an annoying habit that we would do well to develop – we should learn to think like a mathematician. What are the right kinds of "what if" questions to ask? Here's another annoying habit of mathematicians: they often ask "wrong" questions. That is, they often ask questions and find that the answer isn't particularly interesting. But asking enough questions often leads to some good "right" questions. So don't be afraid of doing something "wrong;" we mathematicians do it all the time.

So what is a good question to ask after seeing Example 1? Here are two possible questions:

1. Did we really have to call the red balls "r"? Could we call them "q"?
2. What if we had 60 balls at the start instead of 30?

Let's look at the first question. Would the solution to our problem change if we called the red balls q? Of course not. At the end, we'd find that $q = 10$, and we would know that this meant that we had 10 red balls.

Now let's look at the second question. Suppose we had 60 balls, but the other relationships stayed the same. How would the situation and solution change? Let's compare the "orginal" equations to the "new" equations.

Original	New
$r + b + g = 30$	$r + b + g = 60$
$r = 2g$	$r = 2g$
$b = r + g$	$b = r + g$

By examining these equations, we see that nothing has changed except the first equation. It isn't too much of a stretch of the imagination to see that we would solve this new problem exactly the same way that we solved the original one, except that we'd have twice as many of each type of ball.

A conclusion from answering these two questions is this: it doesn't matter what we call our variables, and while changing constants in the equations changes the solution, they don't really change the *method* of how we solve these equations.

In fact, it is a great discovery to realize that all we care about are the *constants* and the *coefficients* of the equations. By systematically handling these, we can solve any set of linear equations in a very nice way. Before we go on, we must first define what a linear equation is.

Definition 1

Linear Equation

A *linear equation* is an equation that can be written in the form

$$a_1x_1 + a_2x_2 + \cdots + a_nx_n = c$$

where the x_i are variables (the unknowns), the a_i are coefficients, and c is a constant.

A *system of linear equations* is a set of linear equations that involve the same variables.

A *solution* to a system of linear equations is a set of values for the variables x_i such that each equation in the system is satisfied.

So in Example 1, when we answered "how many marbles of each color are there?," we were also answering "find a solution to a certain system of linear equations."

The following are examples of linear equations:

$$
\begin{aligned}
2x + 3y - 7z &= 29\\
x_1 + \frac{7}{2}x_2 + x_3 - x_4 + 17x_5 &= \sqrt[3]{-10}\\
y_1 + 14^2y_4 + 4 &= y_2 + 13 - y_1\\
\sqrt{7}r + \pi s + \frac{3t}{5} &= \cos(45^\circ)
\end{aligned}
$$

Notice that the coefficients and constants can be fractions and irrational numbers (like π, $\sqrt[3]{-10}$ and $\cos(45^\circ)$). The variables only come in the form of a_ix_i; that is, just one variable multiplied by a coefficient. (Note that $\frac{3t}{5} = \frac{3}{5}t$, just a variable multiplied by a coefficient.) Also, it doesn't really matter what side of the equation we put the variables and the constants, although most of the time we write them with the variables on the left and the constants on the right.

We would not regard the above collection of equations to constitute a system of equations, since each equation uses differently named variables. An example of a system of linear equations is

$$
\begin{aligned}
x_1 - x_2 + x_3 + x_4 &= 1\\
2x_1 + 3x_2 + x_4 &= 25\\
x_2 + x_3 &= 10
\end{aligned}
$$

It is important to notice that not all equations used all of the variables (it is more accurate to say that the coefficients can be 0, so the last equation could have been written as $0x_1 + x_2 + x_3 + 0x_4 = 10$). Also, just because we have four unknowns does not mean we have to have four equations. We could have had fewer, even just one, and we could have had more.

To get a better feel for what a linear equation is, we point out some examples of what are *not* linear equations.

$$
\begin{aligned}
2xy + z &= 1\\
5x^2 + 2y^5 &= 100\\
\frac{1}{x} + \sqrt{y} + 24z &= 3\\
\sin^2 x_1 + \cos^2 x_2 &= 29\\
2^{x_1} + \ln x_2 &= 13
\end{aligned}
$$

The first example is not a linear equation since the variables x and y are multiplied together. The second is not a linear equation because the variables are raised to powers other than 1; that is also a problem in the third equation (remember that $1/x = x^{-1}$ and $\sqrt{x} = x^{1/2}$). Our variables cannot be the argument of function like sin, cos or ln, nor can our variables be raised as an exponent.

At this stage, we have yet to discuss how to efficiently find a solution to a system of linear equations. That is a goal for the upcoming sections. Right now we focus on identifying linear equations. It is also useful to "limber" up by solving a few systems of equations using any method we have at hand to refresh our memory about the basic process.

Exercises 1.1

In Exercises 1 – 10, state whether or not the given equation is linear.

1. $x + y + z = 10$
2. $xy + yz + xz = 1$
3. $-3x + 9 = 3y - 5z + x - 7$
4. $\sqrt{5}y + \pi x = -1$
5. $(x-1)(x+1) = 0$
6. $\sqrt{x_1^2 + x_2^2} = 25$
7. $x_1 + y + t = 1$
8. $\frac{1}{x} + 9 = 3\cos(y) - 5z$
9. $\cos(15)y + \frac{x}{4} = -1$
10. $2^x + 2^y = 16$

In Exercises 11 – 14, solve the system of linear equations.

11. $\begin{array}{rcrcr} x & + & y & = & -1 \\ 2x & - & 3y & = & 8 \end{array}$

12. $\begin{array}{rcrcr} 2x & - & 3y & = & 3 \\ 3x & + & 6y & = & 8 \end{array}$

13. $\begin{array}{rcrcrcr} x & - & y & + & z & = & 1 \\ 2x & + & 6y & - & z & = & -4 \\ 4x & - & 5y & + & 2z & = & 0 \end{array}$

14. $\begin{array}{rcrcrcr} x & + & y & - & z & = & 1 \\ 2x & + & y & & & = & 2 \\ & & y & + & 2z & = & 0 \end{array}$

15. A farmer looks out his window at his chickens and pigs. He tells his daughter that he sees 62 heads and 190 legs. How many chickens and pigs does the farmer have?

16. A lady buys 20 trinkets at a yard sale. The cost of each trinket is either $0.30 or $0.65. If she spends $8.80, how many of each type of trinket does she buy?

1.2 Using Matrices To Solve Systems of Linear Equations

1. What is remarkable about the definition of a matrix?
2. Vertical lines of numbers in a matrix are called what?
3. In a matrix A, the entry a_{53} refers to which entry?
4. What is an augmented matrix?

In Section 1.1 we solved a linear system using familiar techniques. Later, we commented that in the linear equations we formed, the most important information was the coefficients and the constants; the names of the variables really didn't matter. In Example 1 we had the following three equations:

$$\begin{aligned} r+b+g &= 30 \\ r &= 2g \\ b &= r+g \end{aligned}$$

Let's rewrite these equations so that all variables are on the left of the equal sign and all constants are on the right. Also, for a bit more consistency, let's list the variables in alphabetical order in each equation. Therefore we can write the equations as

$$\begin{array}{ccccccc} b & + & g & + & r & = & 30 \\ & - & 2g & + & r & = & 0 \\ -b & + & g & + & r & = & 0 \end{array}. \tag{1.6}$$

As we mentioned before, there isn't just one "right" way of finding the solution to this system of equations. Here is another way to do it, a way that is a bit different from our method in Section 1.1.

First, lets add the first and last equations together, and write the result as a new third equation. This gives us:

$$\begin{array}{ccccccc} b & + & g & + & r & = & 30 \\ & - & 2g & + & r & = & 0 \\ & & 2g & + & 2r & = & 30 \end{array}.$$

A nice feature of this is that the only equation with a b in it is the first equation.

Now let's multiply the second equation by $-\frac{1}{2}$. This gives

$$\begin{array}{ccccccc} b & + & g & + & r & = & 30 \\ & & g & - & 1/2r & = & 0 \\ & & 2g & + & 2r & = & 30 \end{array}.$$

Let's now do two steps in a row; our goal is to get rid of the g's in the first and third equations. In order to remove the g in the first equation, let's multiply the second equation by -1 and add that to the first equation, replacing the first equation with that sum. To remove the g in the third equation, let's multiply the second equation by -2 and add that to the third equation, replacing the third equation. Our new system of equations now becomes

$$\begin{array}{ccccccc} b & + & & & 3/2r & = & 30 \\ & & g & - & 1/2r & = & 0 \\ & & & & 3r & = & 30 \end{array}.$$

Clearly we can multiply the third equation by $\frac{1}{3}$ and find that $r = 10$; let's make this our new third equation, giving

$$\begin{array}{ccccccc} b & + & & & 3/2r & = & 30 \\ & & g & - & 1/2r & = & 0 \\ & & & & r & = & 10 \end{array}.$$

Now let's get rid of the r's in the first and second equation. To remove the r in the first equation, let's multiply the third equation by $-\frac{3}{2}$ and add the result to the first equation, replacing the first equation with that sum. To remove the r in the second equation, we can multiply the third equation by $\frac{1}{2}$ and add that to the second equation, replacing the second equation with that sum. This gives us:

$$\begin{array}{ccccc} b & & & = & 15 \\ & g & & = & 5 \\ & & r & = & 10 \end{array}.$$

Clearly we have discovered the same result as when we solved this problem in Section 1.1.

Now again revisit the idea that all that really matters are the coefficients and the constants. There is nothing special about the letters b, g and r; we could have used x, y and z or x_1, x_2 and x_3. And even then, since we wrote our equations so carefully, we really didn't need to write the variable names at all as long as we put things "in the right place."

Let's look again at our system of equations in (1.6) and write the coefficients and the constants in a rectangular array. This time we won't ignore the zeros, but rather write them out.

$$\begin{array}{ccccccc} b & + & g & + & r & = & 30 \\ & - & 2g & + & r & = & 0 \\ -b & + & g & + & r & = & 0 \end{array} \quad \Leftrightarrow \quad \begin{bmatrix} 1 & 1 & 1 & 30 \\ 0 & -2 & 1 & 0 \\ -1 & 1 & 1 & 0 \end{bmatrix}$$

Notice how even the equal signs are gone; we don't need them, for we know that the last *column* contains the coefficients.

We have just created a *matrix*. The definition of matrix is remarkable only in how unremarkable it seems.

Definition 2 **Matrix**

A *matrix* is a rectangular array of numbers.

The horizontal lines of numbers form *rows* and the vertical lines of numbers form *columns*. A matrix with m rows and n columns is said to be an $m \times n$ matrix ("an m by n matrix").

The entries of an $m \times n$ matrix are indexed as follows:

$$\begin{bmatrix} a_{11} & a_{12} & a_{13} & \cdots & a_{1n} \\ a_{21} & a_{22} & a_{23} & \cdots & a_{2n} \\ a_{31} & a_{32} & a_{33} & \cdots & a_{3n} \\ \vdots & \vdots & \vdots & \ddots & \vdots \\ a_{m1} & a_{m2} & a_{m3} & \cdots & a_{mn} \end{bmatrix}.$$

That is, a_{32} means "the number in the third row and second column."

In the future, we'll want to create matrices with just the coefficients of a system of linear equations and leave out the constants. Therefore, when we include the constants, we often refer to the resulting matrix as an *augmented matrix*.

We can use augmented matrices to find solutions to linear equations by using essentially the same steps we used above. Every time we used the word "equation" above, substitute the word "row," as we show below. The comments explain how we get from the current set of equations (or matrix) to the one on the next line.

We can use a shorthand to describe matrix operations; let R_1, R_2 represent "row 1" and "row 2," respectively. We can write "add row 1 to row 3, and replace row 3 with that sum" as "$R_1 + R_3 \rightarrow R_3$." The expression "$R_1 \leftrightarrow R_2$" means "interchange row 1 and row 2."

$$\begin{array}{rcrcrcr} b & + & g & + & r & = & 30 \\ & - & 2g & + & r & = & 0 \\ -b & + & g & + & r & = & 0 \end{array} \qquad \begin{bmatrix} 1 & 1 & 1 & 30 \\ 0 & -2 & 1 & 0 \\ -1 & 1 & 1 & 0 \end{bmatrix}$$

Replace equation 3 with the sum of equations 1 and 3

Replace row 3 with the sum of rows 1 and 3. $(R_1 + R_3 \rightarrow R_3)$

$$\begin{array}{rcrcrcr} b & + & g & + & r & = & 30 \\ & - & 2g & + & r & = & 0 \\ & & 2g & + & 2r & = & 30 \end{array} \qquad \begin{bmatrix} 1 & 1 & 1 & 30 \\ 0 & -2 & 1 & 0 \\ 0 & 2 & 2 & 30 \end{bmatrix}$$

Multiply equation 2 by $-\frac{1}{2}$

Multiply row 2 by $-\frac{1}{2}$ $(-\frac{1}{2}R_2 \rightarrow R_2)$

$$\begin{array}{rcrcrcr} b & + & g & + & r & = & 30 \\ & & g & + & -1/2r & = & 0 \\ & & 2g & + & 2r & = & 30 \end{array} \qquad \begin{bmatrix} 1 & 1 & 1 & 30 \\ 0 & 1 & -\frac{1}{2} & 0 \\ 0 & 2 & 2 & 30 \end{bmatrix}$$

Replace equation 1 with the sum of (-1) times equation 2 plus equation 1;
Replace equation 3 with the sum of (-2) times equation 2 plus equation 3

Replace row 1 with the sum of (-1) times row 2 plus row 1 $(-R_2 + R_1 \rightarrow R_1)$;
Replace row 3 with the sum of (-2) times row 2 plus row 3 $(-2R_2 + R_3 \rightarrow R_3)$

$$\begin{array}{rcrcrcr} b & + & & & 3/2r & = & 30 \\ & & g & - & 1/2r & = & 0 \\ & & & & 3r & = & 30 \end{array} \qquad \begin{bmatrix} 1 & 0 & \frac{3}{2} & 30 \\ 0 & 1 & -\frac{1}{2} & 0 \\ 0 & 0 & 3 & 30 \end{bmatrix}$$

Multiply equation 3 by $\frac{1}{3}$

Multiply row 3 by $\frac{1}{3}$ $(\frac{1}{3}R_3 \rightarrow R_3)$

$$\begin{array}{rcrcl} b & + & 3/2r & = & 30 \\ & g & - \ 1/2r & = & 0 \\ & & r & = & 10 \end{array} \qquad \begin{bmatrix} 1 & 0 & \frac{3}{2} & 30 \\ 0 & 1 & -\frac{1}{2} & 0 \\ 0 & 0 & 1 & 10 \end{bmatrix}$$

Replace equation 2 with the sum of $\frac{1}{2}$ times equation 3 plus equation 2;
Replace equation 1 with the sum of $-\frac{3}{2}$ times equation 3 plus equation 1

Replace row 2 with the sum of $\frac{1}{2}$ times row 3 plus row 2 ($\frac{1}{2}R_3 + R_2 \rightarrow R_2$);
Replace row 1 with the sum of $-\frac{3}{2}$ times row 3 plus row 1 ($-\frac{3}{2}R_3 + R_1 \rightarrow R_1$)

$$\begin{array}{rcl} b & = & 15 \\ g & = & 5 \\ r & = & 10 \end{array} \qquad \begin{bmatrix} 1 & 0 & 0 & 15 \\ 0 & 1 & 0 & 5 \\ 0 & 0 & 1 & 10 \end{bmatrix}$$

The final matrix contains the same solution information as we have on the left in the form of equations. Recall that the first column of our matrices held the coefficients of the b variable; the second and third columns held the coefficients of the g and r variables, respectively. Therefore, the first row of the matrix can be interpreted as "$b + 0g + 0r = 15$," or more concisely, "$b = 15$."

Let's practice this manipulation again.

Example 2 Find a solution to the following system of linear equations by simultaneously manipulating the equations and the corresponding augmented matrices.

$$\begin{array}{rcrcrcl} x_1 & + & x_2 & + & x_3 & = & 0 \\ 2x_1 & + & 2x_2 & + & x_3 & = & 0 \\ -1x_1 & + & x_2 & - & 2x_3 & = & 2 \end{array}$$

Solution We'll first convert this system of equations into a matrix, then we'll proceed by manipulating the system of equations (and hence the matrix) to find a solution. Again, there is not just one "right" way of proceeding; we'll choose a method that is pretty efficient, but other methods certainly exist (and may be "better"!). The method use here, though, is a good one, and it is the method that we will be learning in the future.

The given system and its corresponding augmented matrix are seen below.

Original system of equations

$$\begin{array}{rcrcrcl} x_1 & + & x_2 & + & x_3 & = & 0 \\ 2x_1 & + & 2x_2 & + & x_3 & = & 0 \\ -1x_1 & + & x_2 & - & 2x_3 & = & 2 \end{array}$$

Corresponding matrix

$$\begin{bmatrix} 1 & 1 & 1 & 0 \\ 2 & 2 & 1 & 0 \\ -1 & 1 & -2 & 2 \end{bmatrix}$$

We'll proceed by trying to get the x_1 out of the second and third equation.

Replace equation 2 with the sum of (-2) times equation 1 plus equation 2;
Replace equation 3 with the sum of equation 1 and equation 3

$$\begin{array}{rcrcrcl} x_1 & + & x_2 & + & x_3 & = & 0 \\ & & & & -x_3 & = & 0 \\ & & 2x_2 & - & x_3 & = & 2 \end{array}$$

Replace row 2 with the sum of (-2) times row 1 plus row 2 $(-2R_1 + R_2 \to R_2)$;
Replace row 3 with the sum of row 1 and row 3 $(R_1 + R_3 \to R_3)$

$$\begin{bmatrix} 1 & 1 & 1 & 0 \\ 0 & 0 & -1 & 0 \\ 0 & 2 & -1 & 2 \end{bmatrix}$$

Notice that the second equation no longer contains x_2. We'll exchange the order of the equations so that we can follow the convention of solving for the second variable in the second equation.

Interchange equations 2 and 3

$$\begin{array}{rcrcrcl} x_1 & + & x_2 & + & x_3 & = & 0 \\ & & 2x_2 & - & x_3 & = & 2 \\ & & & & -x_3 & = & 0 \end{array}$$

Interchange rows 2 and 3
$R_2 \leftrightarrow R_3$

$$\begin{bmatrix} 1 & 1 & 1 & 0 \\ 0 & 2 & -1 & 2 \\ 0 & 0 & -1 & 0 \end{bmatrix}$$

Multiply equation 2 by $\frac{1}{2}$

$$\begin{array}{rcrcrcl} x_1 & + & x_2 & + & x_3 & = & 0 \\ & & x_2 & - & \frac{1}{2}x_3 & = & 1 \\ & & & & -x_3 & = & 0 \end{array}$$

Multiply row 2 by $\frac{1}{2}$
$(\frac{1}{2}R_2 \to R_2)$

$$\begin{bmatrix} 1 & 1 & 1 & 0 \\ 0 & 1 & -\frac{1}{2} & 1 \\ 0 & 0 & -1 & 0 \end{bmatrix}$$

Multiply equation 3 by -1

$$\begin{array}{rcrcrcl} x_1 & + & x_2 & + & x_3 & = & 0 \\ & & x_2 & - & \frac{1}{2}x_3 & = & 1 \\ & & & & x_3 & = & 0 \end{array}$$

Multiply row 3 by -1
$(-1R_3 \to R_3)$

$$\begin{bmatrix} 1 & 1 & 1 & 0 \\ 0 & 1 & -\frac{1}{2} & 1 \\ 0 & 0 & 1 & 0 \end{bmatrix}$$

Notice that the last equation (and also the last row of the matrix) show that $x_3 = 0$. Knowing this would allow us to simply eliminate the x_3 from the first two equations. However, we will formally do this by manipulating the equations (and rows) as we have previously.

Replace equation 1 with the sum of (-1) times equation 3 plus equation 1;
Replace equation 2 with the sum of $\frac{1}{2}$ times equation 3 plus equation 2

$$\begin{array}{rcrcrcl} x_1 & + & x_2 & & & = & 0 \\ & & x_2 & & & = & 1 \\ & & & & x_3 & = & 0 \end{array}$$

Replace row 1 with the sum of (-1) times row 3 plus row 1 $(-R_3 + R_1 \to R_1)$;
Replace row 2 with the sum of $\frac{1}{2}$ times row 3 plus row 2 $(\frac{1}{2}R_3 + R_2 \to R_2)$

$$\begin{bmatrix} 1 & 1 & 0 & 0 \\ 0 & 1 & 0 & 1 \\ 0 & 0 & 1 & 0 \end{bmatrix}$$

Notice how the second equation shows that $x_2 = 1$. All that remains to do is to solve for x_1.

Replace equation 1 with the sum of (-1) times equation 2 plus equation 1

$$\begin{array}{rcl} x_1 & = & -1 \\ x_2 & = & 1 \\ x_3 & = & 0 \end{array}$$

Replace row 1 with the sum of (-1) times row 2 plus row 1 $(-R_2 + R_1 \rightarrow R_1)$

$$\begin{bmatrix} 1 & 0 & 0 & -1 \\ 0 & 1 & 0 & 1 \\ 0 & 0 & 1 & 0 \end{bmatrix}$$

Obviously the equations on the left tell us that $x_1 = -1$, $x_2 = 1$ and $x_3 = 0$, and notice how the matrix on the right tells us the same information.

Exercises 1.2

In Exercises 1 – 4, convert the given system of linear equations into an augmented matrix.

1. $$\begin{array}{rcrcrcr} 3x & + & 4y & + & 5z & = & 7 \\ -x & + & y & - & 3z & = & 1 \\ 2x & - & 2y & + & 3z & = & 5 \end{array}$$

2. $$\begin{array}{rcrcrcr} 2x & + & 5y & - & 6z & = & 2 \\ 9x & & & - & 8z & = & 10 \\ -2x & + & 4y & + & z & = & -7 \end{array}$$

3. $$\begin{array}{rcl} x_1 + 3x_2 - 4x_3 + 5x_4 & = & 17 \\ -x_1 + 4x_3 + 8x_4 & = & 1 \\ 2x_1 + 3x_2 + 4x_3 + 5x_4 & = & 6 \end{array}$$

4. $$\begin{array}{rcl} 3x_1 - 2x_2 & = & 4 \\ 2x_1 & = & 3 \\ -x_1 + 9x_2 & = & 8 \\ 5x_1 - 7x_2 & = & 13 \end{array}$$

In Exercises 5 – 9, convert the given augmented matrix into a system of linear equations. Use the variables x_1, x_2, etc.

5. $$\begin{bmatrix} 1 & 2 & 3 \\ -1 & 3 & 9 \end{bmatrix}$$

6. $$\begin{bmatrix} -3 & 4 & 7 \\ 0 & 1 & -2 \end{bmatrix}$$

7. $$\begin{bmatrix} 1 & 1 & -1 & -1 & 2 \\ 2 & 1 & 3 & 5 & 7 \end{bmatrix}$$

8. $$\begin{bmatrix} 1 & 0 & 0 & 0 & 2 \\ 0 & 1 & 0 & 0 & -1 \\ 0 & 0 & 1 & 0 & 5 \\ 0 & 0 & 0 & 1 & 3 \end{bmatrix}$$

9. $$\begin{bmatrix} 1 & 0 & 1 & 0 & 7 & 2 \\ 0 & 1 & 3 & 2 & 0 & 5 \end{bmatrix}$$

In Exercises 10 – 15, perform the given row operations on A, where

$$A = \begin{bmatrix} 2 & -1 & 7 \\ 0 & 4 & -2 \\ 5 & 0 & 3 \end{bmatrix}.$$

10. $-1R_1 \rightarrow R_1$

11. $R_2 \leftrightarrow R_3$

12. $R_1 + R_2 \rightarrow R_2$

13. $2R_2 + R_3 \rightarrow R_3$

14. $\frac{1}{2}R_2 \rightarrow R_2$

15. $-\frac{5}{2}R_1 + R_3 \rightarrow R_3$

A matrix A is given below. In Exercises 16 – 20, a matrix B is given. Give the row operation that transforms A into B.

$$A = \begin{bmatrix} 1 & 1 & 1 \\ 1 & 0 & 1 \\ 1 & 2 & 3 \end{bmatrix}$$

16. $B = \begin{bmatrix} 1 & 1 & 1 \\ 2 & 0 & 2 \\ 1 & 2 & 3 \end{bmatrix}$

17. $B = \begin{bmatrix} 1 & 1 & 1 \\ 2 & 1 & 2 \\ 1 & 2 & 3 \end{bmatrix}$

18. $B = \begin{bmatrix} 3 & 5 & 7 \\ 1 & 0 & 1 \\ 1 & 2 & 3 \end{bmatrix}$

19. $B = \begin{bmatrix} 1 & 0 & 1 \\ 1 & 1 & 1 \\ 1 & 2 & 3 \end{bmatrix}$

20. $B = \begin{bmatrix} 1 & 1 & 1 \\ 1 & 0 & 1 \\ 0 & 2 & 2 \end{bmatrix}$

In Exercises 21 – 26, rewrite the system of equations in matrix form. Find the solution to the linear system by simultaneously manipulating the equations and the matrix.

21. $\begin{array}{rcrcr} x & + & y & = & 3 \\ 2x & - & 3y & = & 1 \end{array}$

22. $\begin{array}{rcrcr} 2x & + & 4y & = & 10 \\ -x & + & y & = & 4 \end{array}$

23. $\begin{array}{rcrcr} -2x & + & 3y & = & 2 \\ -x & + & y & = & 1 \end{array}$

24. $\begin{array}{rcrcr} 2x & + & 3y & = & 2 \\ -2x & + & 6y & = & 1 \end{array}$

25. $\begin{array}{rcrcrcr} -5x_1 & & & + & 2x_3 & = & 14 \\ & & x_2 & & & = & 1 \\ -3x_1 & & & + & x_3 & = & 8 \end{array}$

26. $\begin{array}{rcrcrcr} & - & 5x_2 & + & 2x_3 & = & -11 \\ x_1 & & & + & 2x_3 & = & 15 \\ & - & 3x_2 & + & x_3 & = & -8 \end{array}$

1.3 Elementary Row Operations and Gaussian Elimination

AS YOU READ . . .

1. Give two reasons why the Elementary Row Operations are called "Elementary."
2. T/F: Assuming a solution exists, all linear systems of equations can be solved using only elementary row operations.
3. Give one reason why one might not be interested in putting a matrix into reduced row echelon form.
4. Identify the leading 1s in the following matrix:

$$\begin{bmatrix} 1 & 0 & 0 & 1 \\ 0 & 1 & 1 & 0 \\ 0 & 0 & 1 & 1 \\ 0 & 0 & 0 & 0 \end{bmatrix}$$

5. Using the "forward" and "backward" steps of Gaussian elimination creates lots of ________ making computations easier.

In our examples thus far, we have essentially used just three types of manipulations in order to find solutions to our systems of equations. These three manipulations are:

1. Add a scalar multiple of one equation to a second equation, and replace the second equation with that sum

2. Multiply one equation by a nonzero scalar

3. Swap the position of two equations in our list

We saw earlier how we could write all the information of a system of equations in a matrix, so it makes sense that we can perform similar operations on matrices (as we have done before). Again, simply replace the word "equation" above with the word "row."

We didn't justify our ability to manipulate our equations in the above three ways; it seems rather obvious that we should be able to do that. In that sense, these operations are "elementary." These operations are *elementary* in another sense; they are *fundamental* – they form the basis for much of what we will do in matrix algebra. Since these operations are so important, we list them again here in the context of matrices.

Key Idea 1 **Elementary Row Operations**

1. Add a scalar multiple of one row to another row, and replace the latter row with that sum

2. Multiply one row by a nonzero scalar

3. Swap the position of two rows

Given any system of linear equations, we can find a solution (if one exists) by using these three row operations. Elementary row operations give us a new linear system, but the solution to the new system is the same as the old. We can use these operations as much as we want and not change the solution. This brings to mind two good questions:

1. Since we can use these operations as much as we want, how do we know when to stop? (Where are we supposed to "go" with these operations?)

2. Is there an efficient way of using these operations? (How do we get "there" the fastest?)

We'll answer the first question first. Most of the time[1] we will want to take our original matrix and, using the elementary row operations, put it into something called *reduced row echelon form*.[2] This is our "destination," for this form allows us to readily identify whether or not a solution exists, and in the case that it does, what that solution is.

In the previous section, when we manipulated matrices to find solutions, we were unwittingly putting the matrix into reduced row echelon form. However, not all solutions come in such a simple manner as we've seen so far. Putting a matrix into reduced

[1]unless one prefers obfuscation to clarification

[2]Some texts use the term *reduced echelon form* instead.

row echelon form helps us identify all types of solutions. We'll explore the topic of understanding what the reduced row echelon form of a matrix tells us in the following sections; in this section we focus on finding it.

Definition 3 **Reduced Row Echelon Form**

A matrix is in *reduced row echelon form* if its entries satisfy the following conditions.

1. The first nonzero entry in each row is a 1 (called a *leading 1*).
2. Each leading 1 comes in a column to the right of the leading 1s in rows above it.
3. All rows of all 0s come at the bottom of the matrix.
4. If a column contains a leading 1, then all other entries in that column are 0.

A matrix that satisfies the first three conditions is said to be in *row echelon form*.

Example 3 Which of the following matrices is in reduced row echelon form?

a) $\begin{bmatrix} 1 & 0 \\ 0 & 1 \end{bmatrix}$ b) $\begin{bmatrix} 1 & 0 & 1 \\ 0 & 1 & 2 \end{bmatrix}$

c) $\begin{bmatrix} 0 & 0 \\ 0 & 0 \end{bmatrix}$ d) $\begin{bmatrix} 1 & 1 & 0 \\ 0 & 0 & 1 \end{bmatrix}$

e) $\begin{bmatrix} 1 & 0 & 0 & 1 \\ 0 & 0 & 0 & 0 \\ 0 & 0 & 1 & 3 \end{bmatrix}$ f) $\begin{bmatrix} 1 & 2 & 0 & 0 \\ 0 & 0 & 3 & 0 \\ 0 & 0 & 0 & 4 \end{bmatrix}$

g) $\begin{bmatrix} 0 & 1 & 2 & 3 & 0 & 4 \\ 0 & 0 & 0 & 0 & 1 & 5 \\ 0 & 0 & 0 & 0 & 0 & 0 \end{bmatrix}$ h) $\begin{bmatrix} 1 & 1 & 0 \\ 0 & 1 & 0 \\ 0 & 0 & 1 \end{bmatrix}$

Solution The matrices in a), b), c), d) and g) are all in reduced row echelon form. Check to see that each satisfies the necessary conditions. If your instincts were wrong on some of these, correct your thinking accordingly.

The matrix in e) is not in reduced row echelon form since the row of all zeros is not at the bottom. The matrix in f) is not in reduced row echelon form since the first nonzero entries in rows 2 and 3 are not 1. Finally, the matrix in h) is not in reduced row echelon form since the first entry in column 2 is not zero; the second 1 in column 2 is a leading one, hence all other entries in that column should be 0.

We end this example with a preview of what we'll learn in the future. Consider the matrix in b). If this matrix came from the augmented matrix of a system of linear equations, then we can readily recognize that the solution of the system is $x_1 = 1$ and $x_2 = 2$. Again, in previous examples, when we found the solution to a linear system, we were unwittingly putting our matrices into reduced row echelon form.

We began this section discussing how we can manipulate the entries in a matrix with elementary row operations. This led to two questions, "Where do we go?" and "How do we get there quickly?" We've just answered the first question: most of the time we are "going to" reduced row echelon form. We now address the second question.

There is no one "right" way of using these operations to transform a matrix into reduced row echelon form. However, there is a general technique that works very well in that it is very efficient (so we don't waste time on unnecessary steps). This technique is called *Gaussian elimination*. It is named in honor of the great mathematician Karl Friedrich Gauss.

While this technique isn't very difficult to use, it is one of those things that is easier understood by watching it being used than explained as a series of steps. With this in mind, we will go through one more example highlighting important steps and then we'll explain the procedure in detail.

Example 4 Put the augmented matrix of the following system of linear equations into reduced row echelon form.

$$\begin{array}{rcrcrcr} -3x_1 & - & 3x_2 & + & 9x_3 & = & 12 \\ 2x_1 & + & 2x_2 & - & 4x_3 & = & -2 \\ & & -2x_2 & - & 4x_3 & = & -8 \end{array}$$

SOLUTION We start by converting the linear system into an augmented matrix.

$$\begin{bmatrix} \boxed{-3} & -3 & 9 & 12 \\ 2 & 2 & -4 & -2 \\ 0 & -2 & -4 & -8 \end{bmatrix}$$

Our next step is to change the entry in the box to a 1. To do this, let's multiply row 1 by $-\frac{1}{3}$.

$$-\tfrac{1}{3}R_1 \rightarrow R_1 \qquad \begin{bmatrix} 1 & 1 & -3 & -4 \\ 2 & 2 & -4 & -2 \\ 0 & -2 & -4 & -8 \end{bmatrix}$$

We have now created a *leading 1*; that is, the first entry in the first row is a 1. Our next step is to put zeros under this 1. To do this, we'll use the elementary row operation given below.

$$-2R_1 + R_2 \rightarrow R_2 \qquad \begin{bmatrix} 1 & 1 & -3 & -4 \\ 0 & \boxed{0} & 2 & 6 \\ 0 & -2 & -4 & -8 \end{bmatrix}$$

Once this is accomplished, we shift our focus from the leading one down one row, and to the right one column, to the position that is boxed. We again want to put a 1 in this position. We can use any elementary row operations, but we need to restrict ourselves to using only the second row and any rows below it. Probably the simplest thing we can do is interchange rows 2 and 3, and then scale the new second row so that there is a 1 in the desired position.

$$R_2 \leftrightarrow R_3 \qquad \begin{bmatrix} 1 & 1 & -3 & -4 \\ 0 & \boxed{-2} & -4 & -8 \\ 0 & 0 & 2 & 6 \end{bmatrix}$$

$$-\tfrac{1}{2}R_2 \rightarrow R_2 \qquad \begin{bmatrix} 1 & 1 & -3 & -4 \\ 0 & 1 & 2 & 4 \\ 0 & 0 & \boxed{2} & 6 \end{bmatrix}$$

We have now created another leading 1, this time in the second row. Our next desire is to put zeros underneath it, but this has already been accomplished by our previous steps. Therefore we again shift our attention to the right one column and down one row, to the next position put in the box. We want that to be a 1. A simple scaling will accomplish this.

$$\tfrac{1}{2}R_3 \rightarrow R_3 \qquad \begin{bmatrix} 1 & 1 & -3 & -4 \\ 0 & 1 & 2 & 4 \\ 0 & 0 & 1 & 3 \end{bmatrix}$$

This ends what we will refer to as the *forward steps*. Our next task is to use the elementary row operations and go back and put zeros above our leading 1s. This is referred to as the *backward steps*. These steps are given below.

$$\begin{array}{l} 3R_3 + R_1 \rightarrow R_1 \\ -2R_3 + R_2 \rightarrow R_2 \end{array} \qquad \begin{bmatrix} 1 & 1 & 0 & 5 \\ 0 & 1 & 0 & -2 \\ 0 & 0 & 1 & 3 \end{bmatrix}$$

$$-R_2 + R_1 \rightarrow R_1 \qquad \begin{bmatrix} 1 & 0 & 0 & 7 \\ 0 & 1 & 0 & -2 \\ 0 & 0 & 1 & 3 \end{bmatrix}$$

It is now easy to read off the solution as $x_1 = 7$, $x_2 = -2$ and $x_3 = 3$.

We now formally explain the procedure used to find the solution above. As you read through the procedure, follow along with the example above so that the explanation makes more sense.

Forward Steps

1. Working from left to right, consider the first column that isn't all zeros that hasn't already been worked on. Then working from top to bottom, consider the first row that hasn't been worked on.

2. If the entry in the row and column that we are considering is zero, interchange rows with a row below the current row so that that entry is nonzero. If all entries below are zero, we are done with this column; start again at step 1.

3. Multiply the current row by a scalar to make its first entry a 1 (a leading 1).

4. Repeatedly use Elementary Row Operation 1 to put zeros underneath the leading one.

5. Go back to step 1 and work on the new rows and columns until either all rows or columns have been worked on.

If the above steps have been followed properly, then the following should be true about the current state of the matrix:

1. The first nonzero entry in each row is a 1 (a leading 1).

2. Each leading 1 is in a column to the right of the leading 1s above it.

3. All rows of all zeros come at the bottom of the matrix.

Note that this means we have just put a matrix into row echelon form. The next steps finish the conversion into *reduced* row echelon form. These next steps are referred to as the *backward* steps. These are much easier to state.

Backward Steps

1. Starting from the right and working left, use Elementary Row Operation 1 repeatedly to put zeros above each leading 1.

The basic method of Gaussian elimination is this: create leading ones and then use elementary row operations to put zeros above and below these leading ones. We can do this in any order we please, but by following the "Forward Steps" and "Backward Steps," we make use of the presence of zeros to make the overall computations easier. This method is very efficient, so it gets its own name (which we've already been using).

Definition 4

Gaussian Elimination

Gaussian elimination is the technique for finding the reduced row echelon form of a matrix using the above procedure. It can be abbreviated to:

1. Create a leading 1.
2. Use this leading 1 to put zeros underneath it.
3. Repeat the above steps until all possible rows have leading 1s.
4. Put zeros above these leading 1s.

Let's practice some more.

Example 5 Use Gaussian elimination to put the matrix A into reduced row echelon form, where

$$A = \begin{bmatrix} -2 & -4 & -2 & -10 & 0 \\ 2 & 4 & 1 & 9 & -2 \\ 3 & 6 & 1 & 13 & -4 \end{bmatrix}.$$

Solution We start by wanting to make the entry in the first column and first row a 1 (a leading 1). To do this we'll scale the first row by a factor of $-\frac{1}{2}$.

$$-\tfrac{1}{2}R_1 \rightarrow R_1 \qquad \begin{bmatrix} 1 & 2 & 1 & 5 & 0 \\ 2 & 4 & 1 & 9 & -2 \\ 3 & 6 & 1 & 13 & -4 \end{bmatrix}$$

Next we need to put zeros in the column below this newly formed leading 1.

$$\begin{array}{l} -2R_1 + R_2 \rightarrow R_2 \\ -3R_1 + R_3 \rightarrow R_3 \end{array} \qquad \begin{bmatrix} 1 & 2 & 1 & 5 & 0 \\ 0 & \boxed{0} & -1 & -1 & -2 \\ 0 & 0 & -2 & -2 & -4 \end{bmatrix}$$

Our attention now shifts to the right one column and down one row to the position indicated by the box. We want to put a 1 in that position. Our only options are to either scale the current row or to interchange rows with a row below it. However, in this case neither of these options will accomplish our goal. Therefore, we shift our attention to the right one more column.

We want to put a 1 where there is a –1. A simple scaling will accomplish this; once done, we will put a 0 underneath this leading one.

$$-R_2 \rightarrow R_2 \qquad \begin{bmatrix} 1 & 2 & 1 & 5 & 0 \\ 0 & 0 & 1 & 1 & 2 \\ 0 & 0 & -2 & -2 & -4 \end{bmatrix}$$

$$2R_2 + R_3 \rightarrow R_3 \qquad \begin{bmatrix} 1 & 2 & 1 & 5 & 0 \\ 0 & 0 & 1 & 1 & 2 \\ 0 & 0 & 0 & \boxed{0} & 0 \end{bmatrix}$$

Our attention now shifts over one more column and down one row to the position indicated by the box; we wish to make this a 1. Of course, there is no way to do this, so we are done with the forward steps.

Our next goal is to put a 0 above each of the leading 1s (in this case there is only one leading 1 to deal with).

$$-R_2 + R_1 \rightarrow R_1 \qquad \begin{bmatrix} 1 & 2 & 0 & 4 & -2 \\ 0 & 0 & 1 & 1 & 2 \\ 0 & 0 & 0 & 0 & 0 \end{bmatrix}$$

This final matrix is in reduced row echelon form.

Example 6 Put the matrix

$$\begin{bmatrix} 1 & 2 & 1 & 3 \\ 2 & 1 & 1 & 1 \\ 3 & 3 & 2 & 1 \end{bmatrix}$$

into reduced row echelon form.

SOLUTION Here we will show all steps without explaining each one.

$$\begin{array}{l} -2R_1 + R_2 \rightarrow R_2 \\ -3R_1 + R_3 \rightarrow R_3 \end{array} \qquad \begin{bmatrix} 1 & 2 & 1 & 3 \\ 0 & -3 & -1 & -5 \\ 0 & -3 & -1 & -8 \end{bmatrix}$$

$$-\tfrac{1}{3}R_2 \rightarrow R_2 \qquad \begin{bmatrix} 1 & 2 & 1 & 3 \\ 0 & 1 & 1/3 & 5/3 \\ 0 & -3 & -1 & -8 \end{bmatrix}$$

$$3R_2 + R_3 \rightarrow R_3 \qquad \begin{bmatrix} 1 & 2 & 1 & 3 \\ 0 & 1 & 1/3 & 5/3 \\ 0 & 0 & 0 & -3 \end{bmatrix}$$

$$-\tfrac{1}{3}R_3 \rightarrow R_3 \qquad \begin{bmatrix} 1 & 2 & 1 & 3 \\ 0 & 1 & 1/3 & 5/3 \\ 0 & 0 & 0 & 1 \end{bmatrix}$$

$$\begin{array}{l} -3R_3 + R_1 \rightarrow R_1 \\ -\tfrac{5}{3}R_3 + R_2 \rightarrow R_2 \end{array} \qquad \begin{bmatrix} 1 & 2 & 1 & 0 \\ 0 & 1 & 1/3 & 0 \\ 0 & 0 & 0 & 1 \end{bmatrix}$$

$$-2R_2 + R_1 \rightarrow R_1 \qquad \begin{bmatrix} 1 & 0 & 1/3 & 0 \\ 0 & 1 & 1/3 & 0 \\ 0 & 0 & 0 & 1 \end{bmatrix}$$

The last matrix in the above example is in reduced row echelon form. If one thinks of the original matrix as representing the augmented matrix of a system of linear equations, this final result is interesting. What does it mean to have a leading one in the last column? We'll figure this out in the next section.

Example 7 Put the matrix A into reduced row echelon form, where

$$A = \begin{bmatrix} 2 & 1 & -1 & 4 \\ 1 & -1 & 2 & 12 \\ 2 & 2 & -1 & 9 \end{bmatrix}.$$

Solution We'll again show the steps without explanation, although we will stop at the end of the forward steps and make a comment.

$$\tfrac{1}{2}R_1 \rightarrow R_1 \qquad \begin{bmatrix} 1 & 1/2 & -1/2 & 2 \\ 1 & -1 & 2 & 12 \\ 2 & 2 & -1 & 9 \end{bmatrix}$$

$$\begin{array}{c} -R_1 + R_2 \rightarrow R_2 \\ -2R_1 + R_3 \rightarrow R_3 \end{array} \qquad \begin{bmatrix} 1 & 1/2 & -1/2 & 2 \\ 0 & -3/2 & 5/2 & 10 \\ 0 & 1 & 0 & 5 \end{bmatrix}$$

$$-\tfrac{2}{3}R_2 \rightarrow R_2 \qquad \begin{bmatrix} 1 & 1/2 & -1/2 & 2 \\ 0 & 1 & -5/3 & -20/3 \\ 0 & 1 & 0 & 5 \end{bmatrix}$$

$$-R_2 + R_3 \rightarrow R_3 \qquad \begin{bmatrix} 1 & 1/2 & -1/2 & 2 \\ 0 & 1 & -5/3 & -20/3 \\ 0 & 0 & 5/3 & 35/3 \end{bmatrix}$$

$$\tfrac{3}{5}R_3 \rightarrow R_3 \qquad \begin{bmatrix} 1 & 1/2 & -1/2 & 2 \\ 0 & 1 & -5/3 & -20/3 \\ 0 & 0 & 1 & 7 \end{bmatrix}$$

Let's take a break here and think about the state of our linear system at this moment. Converting back to linear equations, we now know

$$\begin{array}{rcl} x_1 + 1/2x_2 - 1/2x_3 & = & 2 \\ x_2 - 5/3x_3 & = & -20/3 \\ x_3 & = & 7 \end{array}.$$

Since we know that $x_3 = 7$, the second equation turns into

$$x_2 - (5/3)(7) = -20/3,$$

telling us that $x_2 = 5$.

Finally, knowing values for x_2 and x_3 lets us substitute in the first equation and find

$$x_1 + (1/2)(5) - (1/2)(7) = 2,$$

so $x_1 = 3$.

This process of substituting known values back into other equations is called *back substitution*. This process is essentially what happens when we perform the backward steps of Gaussian elimination. We make note of this below as we finish out finding the reduced row echelon form of our matrix.

$\frac{5}{3}R_3 + R_2 \rightarrow R_2$
(knowing $x_3 = 7$ allows us to find $x_2 = 5$)

$$\begin{bmatrix} 1 & 1/2 & -1/2 & 2 \\ 0 & 1 & 0 & 5 \\ 0 & 0 & 1 & 7 \end{bmatrix}$$

$\frac{1}{2}R_3 + R_1 \rightarrow R_1$
$-\frac{1}{2}R_2 + R_1 \rightarrow R_1$
(knowing $x_2 = 5$ and $x_3 = 7$ allows us to find $x_1 = 3$)

$$\begin{bmatrix} 1 & 0 & 0 & 3 \\ 0 & 1 & 0 & 5 \\ 0 & 0 & 1 & 7 \end{bmatrix}$$

We did our operations slightly "out of order" in that we didn't put the zeros above our leading 1 in the third column in the same step, highlighting how back substitution works.

In all of our practice, we've only encountered systems of linear equations with exactly one solution. Is this always going to be the case? Could we ever have systems with more than one solution? If so, how many solutions could there be? Could we have systems without a solution? These are some of the questions we'll address in the next section.

Exercises 1.3

In Exercises 1 – 4, state whether or not the given matrices are in reduced row echelon form. If it is not, state why.

1. (a) $\begin{bmatrix} 1 & 0 \\ 0 & 1 \end{bmatrix}$ (b) $\begin{bmatrix} 0 & 1 \\ 1 & 0 \end{bmatrix}$ (c) $\begin{bmatrix} 1 & 1 \\ 1 & 1 \end{bmatrix}$ (d) $\begin{bmatrix} 1 & 0 & 1 \\ 0 & 1 & 2 \end{bmatrix}$

2. (a) $\begin{bmatrix} 1 & 0 & 0 \\ 0 & 0 & 1 \end{bmatrix}$ (b) $\begin{bmatrix} 1 & 0 & 1 \\ 0 & 1 & 1 \end{bmatrix}$ (c) $\begin{bmatrix} 0 & 0 & 0 \\ 1 & 0 & 0 \end{bmatrix}$ (d) $\begin{bmatrix} 0 & 0 & 0 \\ 0 & 0 & 0 \end{bmatrix}$

3. (a) $\begin{bmatrix} 1 & 1 & 1 \\ 0 & 1 & 1 \\ 0 & 0 & 1 \end{bmatrix}$ (b) $\begin{bmatrix} 1 & 0 & 0 \\ 0 & 1 & 0 \\ 0 & 0 & 0 \end{bmatrix}$ (c) $\begin{bmatrix} 1 & 0 & 0 \\ 0 & 0 & 1 \\ 0 & 0 & 0 \end{bmatrix}$ (d) $\begin{bmatrix} 1 & 0 & 0 & -5 \\ 0 & 1 & 0 & 7 \\ 0 & 0 & 1 & 3 \end{bmatrix}$

4. (a) $\begin{bmatrix} 2 & 0 & 0 & 2 \\ 0 & 2 & 0 & 2 \\ 0 & 0 & 2 & 2 \end{bmatrix}$ (b) $\begin{bmatrix} 0 & 1 & 0 & 0 \\ 0 & 0 & 1 & 0 \\ 0 & 0 & 0 & 0 \end{bmatrix}$ (c) $\begin{bmatrix} 0 & 0 & 1 & -5 \\ 0 & 0 & 0 & 0 \\ 0 & 0 & 0 & 0 \end{bmatrix}$ (d) $\begin{bmatrix} 1 & 1 & 0 & 0 & 1 & 1 \\ 0 & 0 & 1 & 0 & 1 & 1 \\ 0 & 0 & 0 & 1 & 0 & 0 \end{bmatrix}$

In Exercises 5 – 22, use Gaussian Elimination to put the given matrix into reduced row echelon form.

5. $\begin{bmatrix} 1 & 2 \\ -3 & -5 \end{bmatrix}$

6. $\begin{bmatrix} 2 & -2 \\ 3 & -2 \end{bmatrix}$

7. $\begin{bmatrix} 4 & 12 \\ -2 & -6 \end{bmatrix}$

8. $\begin{bmatrix} -5 & 7 \\ 10 & 14 \end{bmatrix}$

9. $\begin{bmatrix} -1 & 1 & 4 \\ -2 & 1 & 1 \end{bmatrix}$

10. $\begin{bmatrix} 7 & 2 & 3 \\ 3 & 1 & 2 \end{bmatrix}$

11. $\begin{bmatrix} 3 & -3 & 6 \\ -1 & 1 & -2 \end{bmatrix}$

12. $\begin{bmatrix} 4 & 5 & -6 \\ -12 & -15 & 18 \end{bmatrix}$

13. $\begin{bmatrix} -2 & -4 & -8 \\ -2 & -3 & -5 \\ 2 & 3 & 6 \end{bmatrix}$

14. $\begin{bmatrix} 2 & 1 & 1 \\ 1 & 1 & 1 \\ 2 & 1 & 2 \end{bmatrix}$

15. $\begin{bmatrix} 1 & 2 & 1 \\ 1 & 3 & 1 \\ -1 & -3 & 0 \end{bmatrix}$

16. $\begin{bmatrix} 1 & 2 & 3 \\ 0 & 4 & 5 \\ 1 & 6 & 9 \end{bmatrix}$

17. $\begin{bmatrix} 1 & 1 & 1 & 2 \\ 2 & -1 & -1 & 1 \\ -1 & 1 & 1 & 0 \end{bmatrix}$

18. $\begin{bmatrix} 2 & -1 & 1 & 5 \\ 3 & 1 & 6 & -1 \\ 3 & 0 & 5 & 0 \end{bmatrix}$

19. $\begin{bmatrix} 1 & 1 & -1 & 7 \\ 2 & 1 & 0 & 10 \\ 3 & 2 & -1 & 17 \end{bmatrix}$

20. $\begin{bmatrix} 4 & 1 & 8 & 15 \\ 1 & 1 & 2 & 7 \\ 3 & 1 & 5 & 11 \end{bmatrix}$

21. $\begin{bmatrix} 2 & 2 & 1 & 3 & 1 & 4 \\ 1 & 1 & 1 & 3 & 1 & 4 \end{bmatrix}$

22. $\begin{bmatrix} 1 & -1 & 3 & 1 & -2 & 9 \\ 2 & -2 & 6 & 1 & -2 & 13 \end{bmatrix}$

1.4 Existence and Uniqueness of Solutions

AS YOU READ . . .

1. T/F: It is possible for a linear system to have exactly 5 solutions.
2. T/F: A variable that corresponds to a leading 1 is "free."
3. How can one tell what kind of solution a linear system of equations has?
4. Give an example (different from those given in the text) of a 2 equation, 2 unknown linear system that is not consistent.
5. T/F: A particular solution for a linear system with infinite solutions can be found by arbitrarily picking values for the free variables.

So far, whenever we have solved a system of linear equations, we have always found exactly one solution. This is not always the case; we will find in this section that some systems do not have a solution, and others have more than one.

We start with a very simple example. Consider the following linear system:

$$x - y = 0.$$

There are obviously infinite solutions to this system; as long as $x = y$, we have a solution. We can picture all of these solutions by thinking of the graph of the equation $y = x$ on the traditional x, y coordinate plane.

Let's continue this visual aspect of considering solutions to linear systems. Consider the system

$$\begin{aligned} x + y &= 2 \\ x - y &= 0. \end{aligned}$$

Each of these equations can be viewed as lines in the coordinate plane, and since their slopes are different, we know they will intersect somewhere (see Figure 1.1 (a)). In this example, they intersect at the point $(1, 1)$ – that is, when $x = 1$ and $y = 1$, both equations are satisfied and we have a solution to our linear system. Since this is the only place the two lines intersect, this is the only solution.

Now consider the linear system

$$\begin{aligned} x + y &= 1 \\ 2x + 2y &= 2. \end{aligned}$$

It is clear that while we have two equations, they are essentially the same equation; the second is just a multiple of the first. Therefore, when we graph the two equations, we are graphing the same line twice (see Figure 1.1 (b); the thicker line is used to represent drawing the line twice). In this case, we have an infinite solution set, just as if we only had the one equation $x + y = 1$. We often write the solution as $x = 1 - y$ to demonstrate that y can be any real number, and x is determined once we pick a value for y.

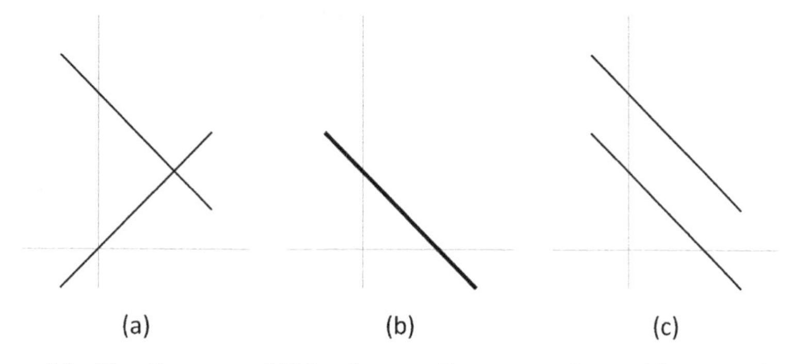

Figure 1.1: The three possibilities for two linear equations with two unknowns.

Finally, consider the linear system

$$\begin{aligned} x + y &= 1 \\ x + y &= 2. \end{aligned}$$

We should immediately spot a problem with this system; if the sum of x and y is 1, how can it also be 2? There is no solution to such a problem; this linear system has no solution. We can visualize this situation in Figure 1.1 (c); the two lines are parallel and never intersect.

If we were to consider a linear system with three equations and two unknowns, we could visualize the solution by graphing the corresponding three lines. We can picture that perhaps all three lines would meet at one point, giving exactly 1 solution; perhaps all three equations describe the same line, giving an infinite number of solutions; perhaps we have different lines, but they do not all meet at the same point, giving no solution. We further visualize similar situations with, say, 20 equations with two variables.

While it becomes harder to visualize when we add variables, no matter how many equations and variables we have, solutions to linear equations always come in one of three forms: exactly one solution, infinite solutions, or no solution. This is a fact that we will not prove here, but it deserves to be stated.

Theorem 1

Solution Forms of Linear Systems

Every linear system of equations has exactly one solution, infinite solutions, or no solution.

This leads us to a definition. Here we don't differentiate between having one solution and infinite solutions, but rather just whether or not a solution exists.

Definition 5

Consistent and Inconsistent Linear Systems

A system of linear equations is *consistent* if it has a solution (perhaps more than one). A linear system is *inconsistent* if it does not have a solution.

How can we tell what kind of solution (if one exists) a given system of linear equations has? The answer to this question lies with properly understanding the reduced row echelon form of a matrix. To discover what the solution is to a linear system, we first put the matrix into reduced row echelon form and then interpret that form properly.

Before we start with a simple example, let us make a note about finding the reduced row echelon form of a matrix.

Technology Note: In the previous section, we learned how to find the reduced row echelon form of a matrix using Gaussian elimination – by hand. We need to know how to do this; understanding the process has benefits. However, actually executing the process by hand for every problem is not usually beneficial. In fact, with large systems, computing the reduced row echelon form by hand is effectively impossible. Our main concern is *what* "the rref" is, not what exact steps were used to arrive there. Therefore, the reader is encouraged to employ some form of technology to find the reduced row echelon form. Computer programs such as *Mathematica*, MATLAB, Maple, and Derive can be used; many handheld calculators (such as Texas Instruments calculators) will perform these calculations very quickly.

As a general rule, when we are learning a new technique, it is best to not use technology to aid us. This helps us learn not only the technique but some of its "inner workings." We can then use technology once we have mastered the technique and are now learning how to use it to solve problems.

From here on out, in our examples, when we need the reduced row echelon form of a matrix, we will not show the steps involved. Rather, we will give the initial matrix, then immediately give the reduced row echelon form of the matrix. We trust that the reader can verify the accuracy of this form by both performing the necessary steps by hand or utilizing some technology to do it for them.

Our first example explores officially a quick example used in the introduction of this section.

Example 8 Find the solution to the linear system

$$\begin{array}{ccccc} x_1 & + & x_2 & = & 1 \\ 2x_1 & + & 2x_2 & = & 2 \end{array}.$$

Solution Create the corresponding augmented matrix, and then put the matrix into reduced row echelon form.

$$\begin{bmatrix} 1 & 1 & 1 \\ 2 & 2 & 2 \end{bmatrix} \quad \overrightarrow{\text{rref}} \quad \begin{bmatrix} 1 & 1 & 1 \\ 0 & 0 & 0 \end{bmatrix}$$

Now convert the reduced matrix back into equations. In this case, we only have one equation,

$$x_1 + x_2 = 1$$

or, equivalently,

$$\begin{aligned} x_1 &= 1 - x_2 \\ x_2 &\text{ is free.} \end{aligned}$$

We have just introduced a new term, the word *free*. It is used to stress that idea that x_2 can take on *any* value; we are "free" to choose any value for x_2. Once this value is chosen, the value of x_1 is determined. We have infinite choices for the value of x_2, so therefore we have infinite solutions.

For example, if we set $x_2 = 0$, then $x_1 = 1$; if we set $x_2 = 5$, then $x_1 = -4$.

Let's try another example, one that uses more variables.

Example 9 Find the solution to the linear system

$$\begin{array}{rcrcrcr} & & x_2 & - & x_3 & = & 3 \\ x_1 & & & + & 2x_3 & = & 2 \\ & & -3x_2 & + & 3x_3 & = & -9 \end{array}.$$

Solution To find the solution, put the corresponding matrix into reduced row echelon form.

$$\begin{bmatrix} 0 & 1 & -1 & 3 \\ 1 & 0 & 2 & 2 \\ 0 & -3 & 3 & -9 \end{bmatrix} \quad \overrightarrow{\text{rref}} \quad \begin{bmatrix} 1 & 0 & 2 & 2 \\ 0 & 1 & -1 & 3 \\ 0 & 0 & 0 & 0 \end{bmatrix}$$

Now convert this reduced matrix back into equations. We have

$$x_1 + 2x_3 = 2$$
$$x_2 - x_3 = 3$$

or, equivalently,

$$x_1 = 2 - 2x_3$$
$$x_2 = 3 + x_3$$
$$x_3 \text{ is free.}$$

These two equations tell us that the values of x_1 and x_2 depend on what x_3 is. As we saw before, there is no restriction on what x_3 must be; it is "free" to take on the value of any real number. Once x_3 is chosen, we have a solution. Since we have infinite choices for the value of x_3, we have infinite solutions.

As examples, $x_1 = 2$, $x_2 = 3$, $x_3 = 0$ is one solution; $x_1 = -2$, $x_2 = 5$, $x_3 = 2$ is another solution. Try plugging these values back into the original equations to verify that these indeed are solutions. (By the way, since infinite solutions exist, this system of equations is consistent.)

In the two previous examples we have used the word "free" to describe certain variables. What exactly is a free variable? How do we recognize which variables are free and which are not?

Look back to the reduced matrix in Example 8. Notice that there is only one leading 1 in that matrix, and that leading 1 corresponded to the x_1 variable. That told us that x_1 was *not* a free variable; since x_2 *did not* correspond to a leading 1, it was a free variable.

Look also at the reduced matrix in Example 9. There were two leading 1s in that matrix; one corresponded to x_1 and the other to x_2. This meant that x_1 and x_2 were not free variables; since there was not a leading 1 that corresponded to x_3, it was a free variable.

We formally define this and a few other terms in this following definition.

Definition 6

Dependent and Independent Variables

Consider the reduced row echelon form of an augmented matrix of a linear system of equations. Then:

a variable that corresponds to a leading 1 is a *basic*, or *dependent*, variable, and

a variable that does not correspond to a leading 1 is a *free*, or *independent*, variable.

One can probably see that "free" and "independent" are relatively synonymous. It follows that if a variable is not independent, it must be dependent; the word "basic" comes from connections to other areas of mathematics that we won't explore here.

These definitions help us understand when a consistent system of linear equations will have infinite solutions. If there are no free variables, then there is exactly one solution; if there are any free variables, there are infinite solutions.

Key Idea 2

Consistent Solution Types

A consistent linear system of equations will have exactly one solution if and only if there is a leading 1 for each variable in the system.

If a consistent linear system of equations has a free variable, it has infinite solutions.

If a consistent linear system has more variables than leading 1s, then the system will have infinite solutions.

A consistent linear system with more variables than equations will always have infinite solutions.

Note: Key Idea 2 applies only to *consistent* systems. If a system is *inconsistent*, then no solution exists and talking about free and basic variables is meaningless.

When a consistent system has only one solution, each equation that comes from the reduced row echelon form of the corresponding augmented matrix will contain exactly one variable. If the consistent system has infinite solutions, then there will be at least one equation coming from the reduced row echelon form that contains more than one variable. The "first" variable will be the basic (or dependent) variable; all others will be free variables.

We have now seen examples of consistent systems with exactly one solution and others with infinite solutions. How will we recognize that a system is inconsistent? Let's find out through an example.

Example 10 Find the solution to the linear system

$$\begin{array}{ccccccc} x_1 & + & x_2 & + & x_3 & = & 1 \\ x_1 & + & 2x_2 & + & x_3 & = & 2 \\ 2x_1 & + & 3x_2 & + & 2x_3 & = & 0 \end{array}.$$

Solution We start by putting the corresponding matrix into reduced row echelon form.

$$\begin{bmatrix} 1 & 1 & 1 & 1 \\ 1 & 2 & 1 & 2 \\ 2 & 3 & 2 & 0 \end{bmatrix} \quad \overrightarrow{\text{rref}} \quad \begin{bmatrix} 1 & 0 & 1 & 0 \\ 0 & 1 & 0 & 0 \\ 0 & 0 & 0 & 1 \end{bmatrix}$$

Now let us take the reduced matrix and write out the corresponding equations. The first two rows give us the equations

$$\begin{aligned} x_1 + x_3 &= 0 \\ x_2 &= 0. \end{aligned}$$

So far, so good. However the last row gives us the equation

$$0x_1 + 0x_2 + 0x_3 = 1$$

or, more concisely, $0 = 1$. Obviously, this is not true; we have reached a contradiction. Therefore, no solution exists; this system is inconsistent.

In previous sections we have only encountered linear systems with unique solutions (exactly one solution). Now we have seen three more examples with different solution types. The first two examples in this section had infinite solutions, and the third had no solution. How can we tell if a system is inconsistent?

A linear system will be inconsistent only when it implies that 0 equals 1. We can tell if a linear system implies this by putting its corresponding augmented matrix into reduced row echelon form. If we have any row where all entries are 0 except for the entry in the last column, then the system implies 0=1. More succinctly, if we have a leading 1 in the last column of an augmented matrix, then the linear system has no solution.

Key Idea 3

Inconsistent Systems of Linear Equations

A system of linear equations is inconsistent if the reduced row echelon form of its corresponding augmented matrix has a leading 1 in the last column.

Example 11 Confirm that the linear system

$$\begin{array}{ccccc} x & + & y & = & 0 \\ 2x & + & 2y & = & 4 \end{array}$$

has no solution.

Solution We can verify that this system has no solution in two ways. First, let's just think about it. If $x+y=0$, then it stands to reason, by multiplying both sides of this equation by 2, that $2x+2y=0$. However, the second equation of our system says that $2x+2y=4$. Since $0 \neq 4$, we have a contradiction and hence our system has no solution. (We cannot possibly pick values for x and y so that $2x+2y$ equals both 0 and 4.)

Now let us confirm this using the prescribed technique from above. The reduced row echelon form of the corresponding augmented matrix is

$$\begin{bmatrix} 1 & 1 & 0 \\ 0 & 0 & 1 \end{bmatrix}.$$

We have a leading 1 in the last column, so therefore the system is inconsistent.

Let's summarize what we have learned up to this point. Consider the reduced row echelon form of the augmented matrix of a system of linear equations.[3] If there is a leading 1 in the last column, the system has no solution. Otherwise, if there is a leading 1 for each variable, then there is exactly one solution; otherwise (i.e., there are free variables) there are infinite solutions.

Systems with exactly one solution or no solution are the easiest to deal with; systems with infinite solutions are a bit harder to deal with. Therefore, we'll do a little more practice. First, a definition: if there are infinite solutions, what do we call one of those infinite solutions?

[3]That sure seems like a mouthful in and of itself. However, it boils down to "look at the reduced form of the usual matrix."

Definition 7 **Particular Solution**

Consider a linear system of equations with infinite solutions. A *particular solution* is one solution out of the infinite set of possible solutions.

The easiest way to find a particular solution is to pick values for the free variables which then determines the values of the dependent variables. Again, more practice is called for.

Example 12 Give the solution to a linear system whose augmented matrix in reduced row echelon form is

$$\begin{bmatrix} 1 & -1 & 0 & 2 & 4 \\ 0 & 0 & 1 & -3 & 7 \\ 0 & 0 & 0 & 0 & 0 \end{bmatrix}$$

and give two particular solutions.

Solution We can essentially ignore the third row; it does not divulge any information about the solution.[4] The first and second rows can be rewritten as the following equations:

$$\begin{aligned} x_1 - x_2 + 2x_4 &= 4 \\ x_3 - 3x_4 &= 7. \end{aligned}$$

Notice how the variables x_1 and x_3 correspond to the leading 1s of the given matrix. Therefore x_1 and x_3 are dependent variables; all other variables (in this case, x_2 and x_4) are free variables.

We generally write our solution with the dependent variables on the left and independent variables and constants on the right. It is also a good practice to acknowledge the fact that our free variables are, in fact, free. So our final solution would look something like

$$\begin{aligned} &x_1 = 4 + x_2 - 2x_4 \\ &x_2 \text{ is free} \\ &x_3 = 7 + 3x_4 \\ &x_4 \text{ is free.} \end{aligned}$$

To find particular solutions, choose values for our free variables. There is no "right" way of doing this; we are "free" to choose whatever we wish.

[4]Then why include it? Rows of zeros sometimes appear "unexpectedly" in matrices after they have been put in reduced row echelon form. When this happens, we do learn *something*; it means that at least one equation was a combination of some of the others.

By setting $x_2 = 0 = x_4$, we have the solution $x_1 = 4$, $x_2 = 0$, $x_3 = 7$, $x_4 = 0$. By setting $x_2 = 1$ and $x_4 = -5$, we have the solution $x_1 = 15$, $x_2 = 1$, $x_3 = -8$, $x_4 = -5$. It is easier to read this when are variables are listed vertically, so we repeat these solutions:

One particular solution is:

$$\begin{aligned} x_1 &= 4 \\ x_2 &= 0 \\ x_3 &= 7 \\ x_4 &= 0. \end{aligned}$$

Another particular solution is:

$$\begin{aligned} x_1 &= 15 \\ x_2 &= 1 \\ x_3 &= -8 \\ x_4 &= -5. \end{aligned}$$

Example 13 Find the solution to a linear system whose augmented matrix in reduced row echelon form is

$$\begin{bmatrix} 1 & 0 & 0 & 2 & 3 \\ 0 & 1 & 0 & 4 & 5 \end{bmatrix}$$

and give two particular solutions.

Solution Converting the two rows into equations we have

$$\begin{aligned} x_1 + 2x_4 &= 3 \\ x_2 + 4x_4 &= 5. \end{aligned}$$

We see that x_1 and x_2 are our dependent variables, for they correspond to the leading 1s. Therefore, x_3 and x_4 are independent variables. This situation feels a little unusual,[5] for x_3 doesn't appear in any of the equations above, but cannot overlook it; it is still a free variable since there is not a leading 1 that corresponds to it. We write our solution as:

$$\begin{aligned} x_1 &= 3 - 2x_4 \\ x_2 &= 5 - 4x_4 \\ x_3 &\text{ is free} \\ x_4 &\text{ is free.} \end{aligned}$$

To find two particular solutions, we pick values for our free variables. Again, there is no "right" way of doing this (in fact, there are . . . infinite ways of doing this) so we give only an example here.

[5] What kind of situation would lead to a column of all zeros? To have such a column, the original matrix needed to have a column of all zeros, meaning that while we acknowledged the existence of a certain variable, we never actually used it in any equation. In practical terms, we could respond by removing the corresponding column from the matrix and just keep in mind that that variable is free. In very large systems, it might be hard to determine whether or not a variable is actually used and one would not worry about it.

When we learn about eigenvectors and eigenvalues, we will see that under certain circumstances this situation arises. In those cases we leave the variable in the system just to remind ourselves that it is there.

One particular solution is:

$$x_1 = 3$$
$$x_2 = 5$$
$$x_3 = 1000$$
$$x_4 = 0.$$

Another particular solution is:

$$x_1 = 3 - 2\pi$$
$$x_2 = 5 - 4\pi$$
$$x_3 = e^2$$
$$x_4 = \pi.$$

(In the second particular solution we picked "unusual" values for x_3 and x_4 just to highlight the fact that we can.)

Example 14 Find the solution to the linear system

$$\begin{array}{ccccccc} x_1 & + & x_2 & + & x_3 & = & 5 \\ x_1 & - & x_2 & + & x_3 & = & 3 \end{array}$$

and give two particular solutions.

Solution The corresponding augmented matrix and its reduced row echelon form are given below.

$$\begin{bmatrix} 1 & 1 & 1 & 5 \\ 1 & -1 & 1 & 3 \end{bmatrix} \quad \overrightarrow{\text{rref}} \quad \begin{bmatrix} 1 & 0 & 1 & 4 \\ 0 & 1 & 0 & 1 \end{bmatrix}$$

Converting these two rows into equations, we have

$$x_1 + x_3 = 4$$
$$x_2 = 1$$

giving us the solution

$$x_1 = 4 - x_3$$
$$x_2 = 1$$
$$x_3 \text{ is free.}$$

Once again, we get a bit of an "unusual" solution; while x_2 is a dependent variable, it does not depend on any free variable; instead, it is always 1. (We can think of it as depending on the value of 1.) By picking two values for x_3, we get two particular solutions.

One particular solution is:

$$x_1 = 4$$
$$x_2 = 1$$
$$x_3 = 0.$$

Another particular solution is:

$$x_1 = 3$$
$$x_2 = 1$$
$$x_3 = 1.$$

The constants and coefficients of a matrix work together to determine whether a given system of linear equations has one, infinite, or no solution. The concept will be fleshed out more in later chapters, but in short, the coefficients determine whether a matrix will have exactly one solution or not. In the "or not" case, the constants determine whether or not infinite solutions or no solution exists. (So if a given linear system has exactly one solution, it will always have exactly one solution even if the constants are changed.) Let's look at an example to get an idea of how the values of constants and coefficients work together to determine the solution type.

Example 15 For what values of k will the given system have exactly one solution, infinite solutions, or no solution?

$$\begin{array}{ccccc} x_1 & + & 2x_2 & = & 3 \\ 3x_1 & + & kx_2 & = & 9 \end{array}$$

Solution We answer this question by forming the augmented matrix and starting the process of putting it into reduced row echelon form. Below we see the augmented matrix and one elementary row operation that starts the Gaussian elimination process.

$$\begin{bmatrix} 1 & 2 & 3 \\ 3 & k & 9 \end{bmatrix} \quad \xrightarrow[-3R_1 + R_2 \to R_2]{} \quad \begin{bmatrix} 1 & 2 & 3 \\ 0 & k-9 & 0 \end{bmatrix}$$

This is as far as we need to go. In looking at the second row, we see that if $k = 9$, then that row contains only zeros and x_2 is a free variable; we have infinite solutions. If $k \neq 9$, then our next step would be to make that second row, second column entry a leading one. We don't particularly care about the solution, only that we would have exactly one as both x_1 and x_2 would correspond to a leading one and hence be dependent variables.

Our final analysis is then this. If $k \neq 9$, there is exactly one solution; if $k = 9$, there are infinite solutions. In this example, it is not possible to have no solutions.

As an extension of the previous example, consider the similar augmented matrix where the constant 9 is replaced with a 10. Performing the same elementary row operation gives

$$\begin{bmatrix} 1 & 2 & 3 \\ 3 & k & 10 \end{bmatrix} \quad \xrightarrow[-3R_1 + R_2 \to R_2]{} \quad \begin{bmatrix} 1 & 2 & 3 \\ 0 & k-9 & 1 \end{bmatrix}.$$

As in the previous example, if $k \neq 9$, we can make the second row, second column entry a leading one and hence we have one solution. However, if $k = 9$, then our last row is $[0\ 0\ 1]$, meaning we have no solution.

We have been studying the solutions to linear systems mostly in an "academic" setting; we have been solving systems for the sake of solving systems. In the next section, we'll look at situations which create linear systems that need solving (i.e., "word problems").

Exercises 1.4

In Exercises 1 – 14, find the solution to the given linear system. If the system has infinite solutions, give 2 particular solutions.

1. $\begin{array}{rcrcr} 2x_1 & + & 4x_2 & = & 2 \\ x_1 & + & 2x_2 & = & 1 \end{array}$

2. $\begin{array}{rcrcr} -x_1 & + & 5x_2 & = & 3 \\ 2x_1 & - & 10x_2 & = & -6 \end{array}$

3. $\begin{array}{rcrcr} x_1 & + & x_2 & = & 3 \\ 2x_1 & + & x_2 & = & 4 \end{array}$

4. $\begin{array}{rcrcr} -3x_1 & + & 7x_2 & = & -7 \\ 2x_1 & - & 8x_2 & = & 8 \end{array}$

5. $\begin{array}{rcrcr} 2x_1 & + & 3x_2 & = & 1 \\ -2x_1 & - & 3x_2 & = & 1 \end{array}$

6. $\begin{array}{rcrcr} x_1 & + & 2x_2 & = & 1 \\ -x_1 & - & 2x_2 & = & 5 \end{array}$

7. $\begin{array}{rcrcrcr} -2x_1 & + & 4x_2 & + & 4x_3 & = & 6 \\ x_1 & - & 3x_2 & + & 2x_3 & = & 1 \end{array}$

8. $\begin{array}{rcrcrcr} -x_1 & + & 2x_2 & + & 2x_3 & = & 2 \\ 2x_1 & + & 5x_2 & + & x_3 & = & 2 \end{array}$

9. $\begin{array}{rcr} -x_1 - x_2 + x_3 + x_4 & = & 0 \\ -2x_1 - 2x_2 + x_3 & = & -1 \end{array}$

10. $\begin{array}{rcr} x_1 + x_2 + 6x_3 + 9x_4 & = & 0 \\ -x_1 - x_3 - 2x_4 & = & -3 \end{array}$

11. $\begin{array}{rcrcrcr} 2x_1 & + & x_2 & + & 2x_3 & = & 0 \\ x_1 & + & x_2 & + & 3x_3 & = & 1 \\ 3x_1 & + & 2x_2 & + & 5x_3 & = & 3 \end{array}$

12. $\begin{array}{rcrcrcr} x_1 & + & 3x_2 & + & 3x_3 & = & 1 \\ 2x_1 & - & x_2 & + & 2x_3 & = & -1 \\ 4x_1 & + & 5x_2 & + & 8x_3 & = & 2 \end{array}$

13. $\begin{array}{rcrcrcr} x_1 & + & 2x_2 & + & 2x_3 & = & 1 \\ 2x_1 & + & x_2 & + & 3x_3 & = & 1 \\ 3x_1 & + & 3x_2 & + & 5x_3 & = & 2 \end{array}$

14. $\begin{array}{rcrcrcr} 2x_1 & + & 4x_2 & + & 6x_3 & = & 2 \\ 1x_1 & + & 2x_2 & + & 3x_3 & = & 1 \\ -3x_1 & - & 6x_2 & - & 9x_3 & = & -3 \end{array}$

In Exercises 15 – 18, state for which values of k the given system will have exactly 1 solution, infinite solutions, or no solution.

15. $\begin{array}{rcrcr} x_1 & + & 2x_2 & = & 1 \\ 2x_1 & + & 4x_2 & = & k \end{array}$

16. $\begin{array}{rcrcr} x_1 & + & 2x_2 & = & 1 \\ x_1 & + & kx_2 & = & 1 \end{array}$

17. $\begin{array}{rcrcr} x_1 & + & 2x_2 & = & 1 \\ x_1 & + & kx_2 & = & 2 \end{array}$

18. $\begin{array}{rcrcr} x_1 & + & 2x_2 & = & 1 \\ x_1 & + & 3x_2 & = & k \end{array}$

1.5 Applications of Linear Systems

AS YOU READ . . .

1. How do most problems appear "in the real world?"
2. The unknowns in a problem are also called what?
3. How many points are needed to determine the coefficients of a 5th degree polynomial?

We've started this chapter by addressing the issue of finding the solution to a system of linear equations. In subsequent sections, we defined matrices to store linear equation information; we described how we can manipulate matrices without changing the solutions; we described how to efficiently manipulate matrices so that a working solution can be easily found.

We shouldn't lose sight of the fact that our work in the previous sections was aimed at finding solutions to systems of linear equations. In this section, we'll learn how to apply what we've learned to actually solve some problems.

Many, many, *many* problems that are addressed by engineers, businesspeople, scientists and mathematicians can be solved by properly setting up systems of linear equations. In this section we highlight only a few of the wide variety of problems that matrix algebra can help us solve.

We start with a simple example.

Example 16 A jar contains 100 blue, green, red and yellow marbles. There are twice as many yellow marbles as blue; there are 10 more blue marbles than red; the sum of the red and yellow marbles is the same as the sum of the blue and green. How many marbles of each color are there?

SOLUTION Let's call the number of blue balls b, and the number of the other balls g, r and y, each representing the obvious. Since we know that we have 100 marbles, we have the equation

$$b + g + r + y = 100.$$

The next sentence in our problem statement allows us to create three more equations.

We are told that there are twice as many yellow marbles as blue. One of the following two equations is correct, based on this statement; which one is it?

$$2y = b \qquad \text{or} \qquad 2b = y$$

The first equation says that if we take the number of yellow marbles, then double it, we'll have the number of blue marbles. That is not what we were told. The second equation states that if we take the number of blue marbles, then double it, we'll have the number of yellow marbles. This *is* what we were told.

The next statement of "there are 10 more blue marbles as red" can be written as either

$$b = r + 10 \qquad \text{or} \qquad r = b + 10.$$

Which is it?

The first equation says that if we take the number of red marbles, then add 10, we'll have the number of blue marbles. This is what we were told. The next equation is wrong; it implies there are more red marbles than blue.

The final statement tells us that the sum of the red and yellow marbles is the same as the sum of the blue and green marbles, giving us the equation

$$r + y = b + g.$$

We have four equations; altogether, they are

$$\begin{aligned} b+g+r+y &= 100 \\ 2b &= y \\ b &= r+10 \\ r+y &= b+g. \end{aligned}$$

We want to write these equations in a standard way, with all the unknowns on the left and the constants on the right. Let us also write them so that the variables appear in the same order in each equation (we'll use alphabetical order to make it simple). We now have

$$\begin{aligned} b+g+r+y &= 100 \\ 2b-y &= 0 \\ b-r &= 10 \\ -b-g+r+y &= 0 \end{aligned}$$

To find the solution, let's form the appropriate augmented matrix and put it into reduced row echelon form. We do so here, without showing the steps.

$$\begin{bmatrix} 1 & 1 & 1 & 1 & 100 \\ 2 & 0 & 0 & -1 & 0 \\ 1 & 0 & -1 & 0 & 10 \\ -1 & -1 & 1 & 1 & 0 \end{bmatrix} \quad \overrightarrow{\text{rref}} \quad \begin{bmatrix} 1 & 0 & 0 & 0 & 20 \\ 0 & 1 & 0 & 0 & 30 \\ 0 & 0 & 1 & 0 & 10 \\ 0 & 0 & 0 & 1 & 40 \end{bmatrix}$$

We interpret from the reduced row echelon form of the matrix that we have 20 blue, 30 green, 10 red and 40 yellow marbles.

Even if you had a bit of difficulty with the previous example, in reality, this type of problem is pretty simple. The unknowns were easy to identify, the equations were pretty straightforward to write (maybe a bit tricky for some), and only the necessary information was given.

Most problems that we face in the world do not approach us in this way; most problems do not approach us in the form of "Here is an equation. Solve it." Rather, most problems come in the form of:

> Here is a problem. I want the solution. To help, here is lots of information. It may be just enough; it may be too much; it may not be enough. You figure out what you need; just give me the solution.

Faced with this type of problem, how do we proceed? Like much of what we've done in the past, there isn't just one "right" way. However, there are a few steps that can guide us. You don't have to follow these steps, "step by step," but if you find that you are having difficulty solving a problem, working through these steps may help.

(Note: while the principles outlined here will help one solve any type of problem, these steps are written specifically for solving problems that involve only linear equations.)

Key Idea 4

Mathematical Problem Solving

1. Understand the problem. What exactly is being asked?
2. Identify the unknowns. What are you trying to find? What units are involved?
3. Give names to your unknowns (these are your *variables*).
4. Use the information given to write as many equations as you can that involve these variables.
5. Use the equations to form an augmented matrix; use Gaussian elimination to put the matrix into reduced row echelon form.
6. Interpret the reduced row echelon form of the matrix to identify the solution.
7. Ensure the solution makes sense in the context of the problem.

Having identified some steps, let us put them into practice with some examples.

Example 17 A concert hall has seating arranged in three sections. As part of a special promotion, guests will recieve two of three prizes. Guests seated in the first and second sections will receive Prize A, guests seated in the second and third sections will receive Prize B, and guests seated in the first and third sections will receive Prize C. Concert promoters told the concert hall managers of their plans, and asked how many seats were in each section. (The promoters want to store prizes for each section separately for easier distribution.) The managers, thinking they were being helpful, told the promoters they would need 105 A prizes, 103 B prizes, and 88 C prizes, and have since been unavailable for further help. How many seats are in each section?

Solution Before we rush in and start making equations, we should be clear about what is being asked. The final sentence asks: "How many seats are in each section?" This tells us what our unknowns should be: we should name our unknowns for the number of seats in each section. Let x_1, x_2 and x_3 denote the number of seats in the first, second and third sections, respectively. This covers the first two steps of our general problem solving technique.

(It is tempting, perhaps, to name our variables for the number of prizes given away. However, when we think more about this, we realize that we already know this – that information is given to us. Rather, we should name our variables for the things we don't know.)

Having our unknowns identified and variables named, we now proceed to forming equations from the information given. Knowing that Prize A goes to guests in the first and second sections and that we'll need 105 of these prizes tells us

$$x_1 + x_2 = 105.$$

Proceeding in a similar fashion, we get two more equations,

$$x_2 + x_3 = 103 \quad \text{and} \quad x_1 + x_3 = 88.$$

Thus our linear system is

$$\begin{aligned} x_1 + x_2 &= 105 \\ x_2 + x_3 &= 103 \\ x_1 + x_3 &= 88 \end{aligned}$$

and the corresponding augmented matrix is

$$\begin{bmatrix} 1 & 1 & 0 & 105 \\ 0 & 1 & 1 & 103 \\ 1 & 0 & 1 & 88 \end{bmatrix}.$$

To solve our system, let's put this matrix into reduced row echelon form.

$$\begin{bmatrix} 1 & 1 & 0 & 105 \\ 0 & 1 & 1 & 103 \\ 1 & 0 & 1 & 88 \end{bmatrix} \quad \overrightarrow{\text{rref}} \quad \begin{bmatrix} 1 & 0 & 0 & 45 \\ 0 & 1 & 0 & 60 \\ 0 & 0 & 1 & 43 \end{bmatrix}$$

We can now read off our solution. The first section has 45 seats, the second has 60 seats, and the third has 43 seats.

Example 18 A lady takes a 2-mile motorized boat trip down the Highwater River, knowing the trip will take 30 minutes. She asks the boat pilot "How fast does this river flow?" He replies "I have no idea, lady. I just drive the boat."

She thinks for a moment, then asks "How long does the return trip take?" He replies "The same; half an hour." She follows up with the statement, "Since both legs take the same time, you must not drive the boat at the same speed."

"Naw," the pilot said. "While I really don't know exactly how fast I go, I do know that since we don't carry any tourists, I drive the boat twice as fast."

The lady walks away satisfied; she knows how fast the river flows.

(How fast *does* it flow?)

Solution This problem forces us to think about what information is given and how to use it to find what we want to know. In fact, to find the solution, we'll find out extra information that we weren't asked for!

We are asked to find how fast the river is moving (step 1). To find this, we should recognize that, in some sense, there are three speeds at work in the boat trips: the speed of the river (which we want to find), the speed of the boat, and the speed that they actually travel at.

We know that each leg of the trip takes half an hour; if it takes half an hour to cover 2 miles, then they must be traveling at 4 mph, each way.

The other two speeds are unknowns, but they are related to the overall speeds. Let's call the speed of the river r and the speed of the boat b. (And we should be careful. From the conversation, we know that the boat travels at two different speeds. So we'll say that b represents the speed of the boat when it travels downstream, so $2b$ represents the speed of the boat when it travels upstream.) Let's let our speed be measured in the units of miles/hour (mph) as we used above (steps 2 and 3).

What is the rate of the people on the boat? When they are travelling downstream, their rate is the sum of the water speed and the boat speed. Since their overall speed is 4 mph, we have the equation $r + b = 4$.

When the boat returns going against the current, its overall speed is the rate of the boat minus the rate of the river (since the river is working against the boat). The overall trip is still taken at 4 mph, so we have the equation $2b - r = 4$. (Recall: the boat is traveling twice as fast as before.)

The corresponding augmented matrix is

$$\begin{bmatrix} 1 & 1 & 4 \\ 2 & -1 & 4 \end{bmatrix}.$$

Note that we decided to let the first column hold the coefficients of b.

Putting this matrix in reduced row echelon form gives us:

$$\begin{bmatrix} 1 & 1 & 4 \\ 2 & -1 & 4 \end{bmatrix} \quad \overrightarrow{\text{rref}} \quad \begin{bmatrix} 1 & 0 & 8/3 \\ 0 & 1 & 4/3 \end{bmatrix}.$$

We finish by interpreting this solution: the speed of the boat (going downstream) is 8/3 mph, or $2.\overline{6}$ mph, and the speed of the river is 4/3 mph, or $1.\overline{3}$ mph. All we really wanted to know was the speed of the river, at about 1.3 mph.

Example 19 Find the equation of the quadratic function that goes through the points $(-1, 6)$, $(1, 2)$ and $(2, 3)$.

Solution This may not seem like a "linear" problem since we are talking about a quadratic function, but closer examination will show that it really is.

We normally write quadratic functions as $y = ax^2 + bx + c$ where a, b and c are the coefficients; in this case, they are our unknowns. We have three points; consider the point $(-1, 6)$. This tells us directly that if $x = -1$, then $y = 6$. Therefore we know that $6 = a(-1)^2 + b(-1) + c$. Writing this in a more standard form, we have the linear equation

$$a - b + c = 6.$$

The second point tells us that $a(1)^2 + b(1) + c = 2$, which we can simplify as $a + b + c = 2$, and the last point tells us $a(2)^2 + b(2) + c = 3$, or $4a + 2b + c = 3$.

Thus our linear system is

$$\begin{aligned} a - b + c &= 6 \\ a + b + c &= 2 \\ 4a + 2b + c &= 3. \end{aligned}$$

Again, to solve our system, we find the reduced row echelon form of the corresponding augmented matrix. We don't show the steps here, just the final result.

$$\begin{bmatrix} 1 & -1 & 1 & 6 \\ 1 & 1 & 1 & 2 \\ 4 & 2 & 1 & 3 \end{bmatrix} \quad \overrightarrow{\text{rref}} \quad \begin{bmatrix} 1 & 0 & 0 & 1 \\ 0 & 1 & 0 & -2 \\ 0 & 0 & 1 & 3 \end{bmatrix}$$

This tells us that $a = 1$, $b = -2$ and $c = 3$, giving us the quadratic function $y = x^2 - 2x + 3$.

One thing interesting about the previous example is that it confirms for us something that we may have known for a while (but didn't know *why* it was true). Why do we need two points to find the equation of the line? Because in the equation of the a line, we have two unknowns, and hence we'll need two equations to find values for these unknowns.

A quadratic has three unknowns (the coefficients of the x^2 term and the x term, and the constant). Therefore we'll need three equations, and therefore we'll need three points.

What happens if we try to find the quadratic function that goes through 3 points that are all on the same line? The fast answer is that you'll get the equation of a line; there isn't a quadratic function that goes through 3 colinear points. Try it and see! (Pick easy points, like $(0, 0)$, $(1, 1)$ and $(2, 2)$. You'll find that the coefficient of the x^2 term is 0.)

Of course, we can do the same type of thing to find polynomials that go through 4, 5, etc., points. In general, if you are given $n + 1$ points, a polynomial that goes through all $n + 1$ points will have degree at most n.

Example 20 A woman has 32 $1, $5 and $10 bills in her purse, giving her a total of $100. How many bills of each denomination does she have?

Solution Let's name our unknowns x, y and z for our ones, fives and tens, respectively (it is tempting to call them o, f and t, but o looks too much like 0). We know that there are a total of 32 bills, so we have the equation

$$x + y + z = 32.$$

We also know that we have $100, so we have the equation

$$x + 5y + 10z = 100.$$

We have three unknowns but only two equations, so we know that we cannot expect a unique solution. Let's try to solve this system anyway and see what we get.

Putting the system into a matrix and then finding the reduced row echelon form, we have

$$\begin{bmatrix} 1 & 1 & 1 & 32 \\ 1 & 5 & 10 & 100 \end{bmatrix} \quad \overrightarrow{\text{rref}} \quad \begin{bmatrix} 1 & 0 & -\frac{5}{4} & 15 \\ 0 & 1 & \frac{9}{4} & 17 \end{bmatrix}.$$

Reading from our reduced matrix, we have the infinite solution set

$$x = 15 + \frac{5}{4}z$$
$$y = 17 - \frac{9}{4}z$$
$$z \text{ is free.}$$

While we do have infinite solutions, most of these solutions really don't make sense in the context of this problem. (Setting $z = \frac{1}{2}$ doesn't make sense, for having half a ten dollar bill doesn't give us \$5. Likewise, having $z = 8$ doesn't make sense, for then we'd have "-1" \$5 bills.) So we must make sure that our choice of z doesn't give us fractions of bills or negative amounts of bills.

To avoid fractions, z must be a multiple of 4 $(-4, 0, 4, 8, \ldots)$. Of course, $z \geq 0$ for a negative number wouldn't make sense. If $z = 0$, then we have 15 one dollar bills and 17 five dollar bills, giving us \$100. If $z = 4$, then we have $x = 20$ and $y = 8$. We already mentioned that $z = 8$ doesn't make sense, nor does any value of z where $z \geq 8$.

So it seems that we have two answers; one with $z = 0$ and one with $z = 4$. Of course, by the statement of the problem, we are led to believe that the lady has at least one \$10 bill, so probably the "best" answer is that we have 20 \$1 bills, 8 \$5 bills and 4 \$10 bills. The real point of this example, though, is to address how infinite solutions may appear in a real world situation, and how suprising things may result.

Example 21 In a football game, teams can score points through touchdowns worth 6 points, extra points (that follow touchdowns) worth 1 point, two point conversions (that also follow touchdowns) worth 2 points and field goals, worth 3 points. You are told that in a football game, the two competing teams scored on 7 occasions, giving a total score of 24 points. Each touchdown was followed by either a successful extra point or two point conversion. In what ways were these points scored?

SOLUTION The question asks how the points were scored; we can interpret this as asking how many touchdowns, extra points, two point conversions and field goals were scored. We'll need to assign variable names to our unknowns; let t represent the number of **t**ouchdowns scored; let x represent the number of e**x**tra points scored, let w represent the number of t**w**o point conversions, and let f represent the number of **f**ield goals scored.

Now we address the issue of writing equations with these variables using the given information. Since we have a total of 7 scoring occasions, we know that

$$t + x + w + f = 7.$$

The total points scored is 24; considering the value of each type of scoring opportunity, we can write the equation

$$6t + x + 2w + 3f = 24.$$

Finally, we know that each touchdown was followed by a successful extra point or two point conversion. This is subtle, but it tells us that the number of touchdowns is equal to the sum of extra points and two point conversions. In other words,

$$t = x + w.$$

To solve our problem, we put these equations into a matrix and put the matrix into reduced row echelon form. Doing so, we find

$$\begin{bmatrix} 1 & 1 & 1 & 1 & 7 \\ 6 & 1 & 2 & 3 & 24 \\ 1 & -1 & -1 & 0 & 0 \end{bmatrix} \quad \overrightarrow{\text{rref}} \quad \begin{bmatrix} 1 & 0 & 0 & 0.5 & 3.5 \\ 0 & 1 & 0 & 1 & 4 \\ 0 & 0 & 1 & -0.5 & -0.5 \end{bmatrix}.$$

Therefore, we know that

$$\begin{aligned} t &= 3.5 - 0.5f \\ x &= 4 - f \\ w &= -0.5 + 0.5f. \end{aligned}$$

We recognize that this means there are "infinite solutions," but of course most of these will not make sense in the context of a real football game. We must apply some logic to make sense of the situation.

Progressing in no particular order, consider the second equation, $x = 4 - f$. In order for us to have a positive number of extra points, we must have $f \leq 4$. (And of course, we need $f \geq 0$, too.) Therefore, right away we know we have a total of only 5 possibilities, where $f = 0, 1, 2, 3$ or 4.

From the first and third equations, we see that if f is an even number, then t and w will both be fractions (for instance, if $f = 0$, then $t = 3.5$) which does not make sense. Therefore, we are down to two possible solutions, $f = 1$ and $f = 3$.

If $f = 1$, we have 3 touchdowns, 3 extra points, no two point conversions, and (of course), 1 field goal. (Check to make sure that gives 24 points!) If $f = 3$, then we 2 touchdowns, 1 extra point, 1 two point conversion, and (of course) 3 field goals. Again, check to make sure this gives us 24 points. Also, we should check each solution to make sure that we have a total of 7 scoring occasions and that each touchdown could be followed by an extra point or a two point conversion.

We have seen a variety of applications of systems of linear equations. We would do well to remind ourselves of the ways in which solutions to linear systems come: there can be exactly one solution, infinite solutions, or no solutions. While we did see a few examples where it seemed like we had only 2 solutions, this was because we were restricting our solutions to "make sense" within a certain context.

We should also remind ourselves that linear equations are immensely important. The examples we considered here ask fundamentally simple questions like "How fast is the water moving?" or "What is the quadratic function that goes through these three points?" or "How were points in a football game scored?" The real "important" situations ask much more difficult questions that often require *thousands* of equations!

(Gauss began the systematic study of solving systems of linear equations while trying to predict the next sighting of a comet; he needed to solve a system of linear equations that had 17 unknowns. Today, this a relatively easy situation to handle with the help of computers, but to do it by hand is a real pain.) Once we understand the fundamentals of solving systems of equations, we can move on to looking at solving bigger systems of equations; this text focuses on getting us to understand the fundamentals.

Exercises 1.5

In Exercises 1 – 5, find the solution of the given problem by:

(a) creating an appropriate system of linear equations

(b) forming the augmented matrix that corresponds to this system

(c) putting the augmented matrix into reduced row echelon form

(d) interpreting the reduced row echelon form of the matrix as a solution

1. A farmer looks out his window at his chickens and pigs. He tells his daughter that he sees 62 heads and 190 legs. How many chickens and pigs does the farmer have?

2. A lady buys 20 trinkets at a yard sale. The cost of each trinket is either \$0.30 or \$0.65. If she spends \$8.80, how many of each type of trinket does she buy?

3. A carpenter can make two sizes of table, grande and venti. The grande table requires 4 table legs and 1 table top; the venti requires 6 table legs and 2 table tops. After doing work, he counts up spare parts in his warehouse and realizes that he has 86 table tops left over, and 300 legs. How many tables of each kind can he build and use up exactly all of his materials?

4. A jar contains 100 marbles. We know there are twice as many green marbles as red; that the number of blue and yellow marbles together is the same as the number of green; and that three times the number of yellow marbles together with the red marbles gives the same numbers as the blue marbles. How many of each color of marble are in the jar?

5. A rescue mission has 85 sandwiches, 65 bags of chips and 210 cookies. They know from experience that men will eat 2 sandwiches, 1 bag of chips and 4 cookies; women will eat 1 sandwich, a bag of chips and 2 cookies; kids will eat half a sandwhich, a bag of chips and 3 cookies. If they want to use all their food up, how many men, women and kids can they feed?

In Exercises 6 – 15, find the polynomial with the smallest degree that goes through the given points.

6. $(1, 3)$ and $(3, 15)$

7. $(-2, 14)$ and $(3, 4)$

8. $(1, 5)$, $(-1, 3)$ and $(3, -1)$

9. $(-4, -3)$, $(0, 1)$ and $(1, 4.5)$

10. $(-1, -8)$, $(1, -2)$ and $(3, 4)$

11. $(-3, 3)$, $(1, 3)$ and $(2, 3)$

12. $(-2, 15)$, $(-1, 4)$, $(1, 0)$ and $(2, -5)$

13. $(-2, -7)$, $(1, 2)$, $(2, 9)$ and $(3, 28)$

14. $(-3, 10)$, $(-1, 2)$, $(1, 2)$ and $(2, 5)$

15. $(0, 1)$, $(-3, -3.5)$, $(-2, -2)$ and $(4, 7)$

16. The general exponential function has the form $f(x) = ae^{bx}$, where a and b are constants and e is Euler's constant (≈ 2.718). We want to find the equation of the exponential function that goes through the points $(1, 2)$ and $(2, 4)$.

(a) Show why we cannot simply subsitute in values for x and y in $y = ae^{bx}$ and solve using the techniques we used for polynomials.

(b) Show how the equality $y = ae^{bx}$ leads us to the linear equation $\ln y = \ln a + bx$.

(c) Use the techniques we developed to solve for the unknowns $\ln a$ and b.

(d) Knowing $\ln a$, find a; find the exponential function $f(x) = ae^{bx}$ that goes through the points $(1, 2)$ and $(2, 4)$.

17. In a football game, 24 points are scored from 8 scoring occasions. The number of successful extra point kicks is equal to the number of successful two point conversions. Find all ways in which the points may have been scored in this game.

18. In a football game, 29 points are scored from 8 scoring occasions. There are 2 more successful extra point kicks than successful two point conversions. Find all ways in which the points may have been scored in this game.

19. In a basketball game, where points are scored either by a 3 point shot, a 2 point shot or a 1 point free throw, 80 points were scored from 30 successful shots. Find all ways in which the points may have been scored in this game.

20. In a basketball game, where points are scored either by a 3 point shot, a 2 point shot or a 1 point free throw, 110 points were scored from 70 successful shots. Find all ways in which the points may have been scored in this game.

21. Describe the equations of the linear functions that go through the point (1,3). Give 2 examples.

22. Describe the equations of the linear functions that go through the point (2,5). Give 2 examples.

23. Describe the equations of the quadratic functions that go through the points $(2, -1)$ and (1,0). Give 2 examples.

24. Describe the equations of the quadratic functions that go through the points $(-1, 3)$ and (2,6). Give 2 examples.

2 Matrix Arithmetic

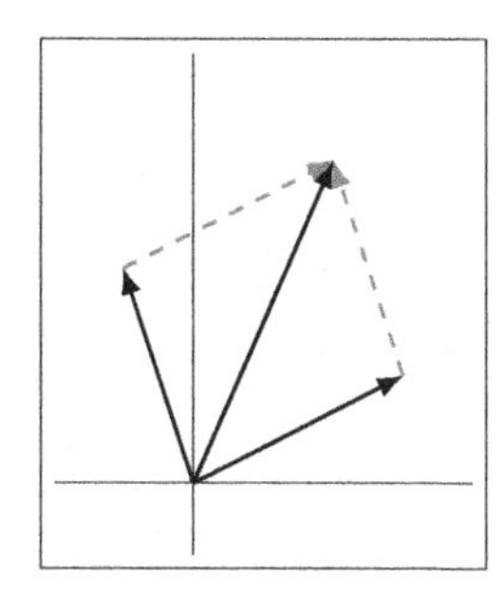

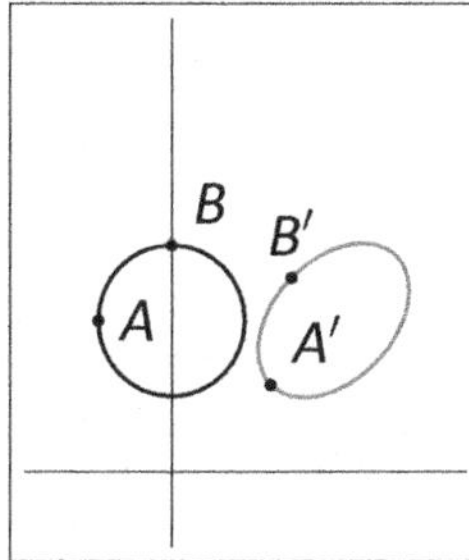

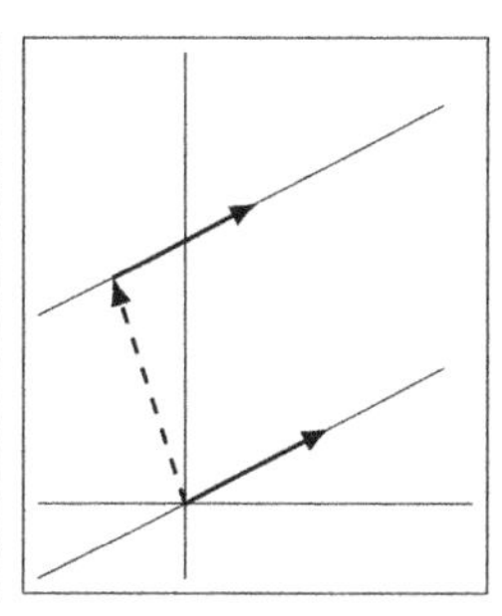

A fundamental topic of mathematics is arithmetic; adding, subtracting, multiplying and dividing numbers. After learning how to do this, most of us went on to learn how to add, subtract, multiply and divide "x". We are comfortable with expressions such as

$$x + 3x - x \cdot x^2 + x^5 \cdot x^{-1}$$

and know that we can "simplify" this to

$$4x - x^3 + x^4.$$

This chapter deals with the idea of doing similar operations, but instead of an unknown number x, we will be using a matrix A. So what exactly does the expression

$$A + 3A - A \cdot A^2 + A^5 \cdot A^{-1}$$

mean? We are going to need to learn to define what matrix addition, scalar multiplication, matrix multiplication and matrix inversion are. We will learn just that, plus some more good stuff, in this chapter.

2.1 Matrix Addition and Scalar Multiplication

1. When are two matrices equal?
2. Write an explanation of how to add matrices as though writing to someone who knows what a matrix is but not much more.
3. T/F: There is only 1 zero matrix.
4. T/F: To multiply a matrix by 2 means to multiply each entry in the matrix by 2.

In the past, when we dealt with expressions that used "x," we didn't just add and multiply x's together for the fun of it, but rather because we were usually given some sort of *equation* that had x in it and we had to "solve for x."

This begs the question, "What does it mean to be equal?" Two numbers are equal, when, . . ., uh, . . ., nevermind. What does it mean for two matrices to be equal? We say that matrices A and B are equal when their corresponding entries are equal. This seems like a very simple definition, but it is rather important, so we give it a box.

Definition 8 **Matrix Equality**

Two $m \times n$ matrices A and B are *equal* if their corresponding entries are equal.

Notice that our more formal definition specifies that if matrices are equal, they have the same dimensions. This should make sense.

Now we move on to describing how to add two matrices together. To start off, take a wild stab: what do you think the following sum is equal to?

$$\begin{bmatrix} 1 & 2 \\ 3 & 4 \end{bmatrix} + \begin{bmatrix} 2 & -1 \\ 5 & 7 \end{bmatrix} = ?$$

If you guessed

$$\begin{bmatrix} 3 & 1 \\ 8 & 11 \end{bmatrix},$$

you guessed correctly. That wasn't so hard, was it?

Let's keep going, hoping that we are starting to get on a roll. Make another wild guess: what do you think the following expression is equal to?

$$3 \cdot \begin{bmatrix} 1 & 2 \\ 3 & 4 \end{bmatrix} = ?$$

If you guessed

$$\begin{bmatrix} 3 & 6 \\ 9 & 12 \end{bmatrix},$$

you guessed correctly!

Even if you guessed wrong both times, you probably have seen enough in these two examples to have a fair idea now what matrix addition and scalar multiplication are all about.

Before we formally define how to perform the above operations, let us first recall that if A is an $m \times n$ matrix, then we can write A as

$$A = \begin{bmatrix} a_{11} & a_{12} & \cdots & a_{1n} \\ a_{21} & a_{22} & \cdots & a_{2n} \\ \vdots & \vdots & \ddots & \vdots \\ a_{m1} & a_{m2} & \cdots & a_{mn} \end{bmatrix}.$$

Secondly, we should define what we mean by the word *scalar*. A scalar is any number that we multiply a matrix by. (In some sense, we use that number to *scale* the matrix.) We are now ready to define our first arithmetic operations.

Definition 9 **Matrix Addition**

Let A and B be $m \times n$ matrices. The *sum* of A and B, denoted $A + B$, is

$$\begin{bmatrix} a_{11}+b_{11} & a_{12}+b_{12} & \cdots & a_{1n}+b_{1n} \\ a_{21}+b_{21} & a_{22}+b_{22} & \cdots & a_{2n}+b_{2n} \\ \vdots & \vdots & \ddots & \vdots \\ a_{m1}+b_{m1} & a_{m2}+b_{m2} & \cdots & a_{mn}+b_{mn} \end{bmatrix}.$$

Definition 10 **Scalar Multiplication**

Let A be an $m \times n$ matrix and let k be a scalar. The *scalar multiplication* of k and A, denoted kA, is

$$\begin{bmatrix} ka_{11} & ka_{12} & \cdots & ka_{1n} \\ ka_{21} & ka_{22} & \cdots & ka_{2n} \\ \vdots & \vdots & \ddots & \vdots \\ ka_{m1} & ka_{m2} & \cdots & ka_{mn} \end{bmatrix}.$$

We are now ready for an example.

Example 22 Let

$$A = \begin{bmatrix} 1 & 2 & 3 \\ -1 & 2 & 1 \\ 5 & 5 & 5 \end{bmatrix}, \quad B = \begin{bmatrix} 2 & 4 & 6 \\ 1 & 2 & 2 \\ -1 & 0 & 4 \end{bmatrix}, \quad C = \begin{bmatrix} 1 & 2 & 3 \\ 9 & 8 & 7 \end{bmatrix}.$$

Simplify the following matrix expressions.

1. $A + B$
2. $B + A$
3. $A - B$
4. $A + C$
5. $-3A + 2B$
6. $A - A$
7. $5A + 5B$
8. $5(A + B)$

Solution

1. $A+B=\begin{bmatrix} 3 & 6 & 9 \\ 0 & 4 & 3 \\ 4 & 5 & 9 \end{bmatrix}$.

2. $B+A=\begin{bmatrix} 3 & 6 & 9 \\ 0 & 4 & 3 \\ 4 & 5 & 9 \end{bmatrix}$.

3. $A-B=\begin{bmatrix} -1 & -2 & -3 \\ -2 & 0 & -1 \\ 6 & 5 & 1 \end{bmatrix}$.

4. $A+C$ is not defined. If we look at our definition of matrix addition, we see that the two matrices need to be the same size. Since A and C have different dimensions, we don't even try to create something as an addition; we simply say that the sum is not defined.

5. $-3A+2B=\begin{bmatrix} 1 & 2 & 3 \\ 5 & -2 & 1 \\ -17 & -15 & -7 \end{bmatrix}$.

6. $A-A=\begin{bmatrix} 0 & 0 & 0 \\ 0 & 0 & 0 \\ 0 & 0 & 0 \end{bmatrix}$.

7. Strictly speaking, this is $\begin{bmatrix} 5 & 10 & 15 \\ -5 & 10 & 5 \\ 25 & 25 & 25 \end{bmatrix}+\begin{bmatrix} 10 & 20 & 30 \\ 5 & 10 & 10 \\ -5 & 0 & 20 \end{bmatrix}=\begin{bmatrix} 15 & 30 & 45 \\ 0 & 20 & 15 \\ 20 & 25 & 45 \end{bmatrix}$.

8. Strictly speaking, this is

$$5\left(\begin{bmatrix} 1 & 2 & 3 \\ -1 & 2 & 1 \\ 5 & 5 & 5 \end{bmatrix}+\begin{bmatrix} 2 & 4 & 6 \\ 1 & 2 & 2 \\ -1 & 0 & 4 \end{bmatrix}\right)=5\cdot\begin{bmatrix} 3 & 6 & 9 \\ 0 & 4 & 3 \\ 4 & 5 & 9 \end{bmatrix}$$
$$=\begin{bmatrix} 15 & 30 & 45 \\ 0 & 20 & 15 \\ 20 & 25 & 45 \end{bmatrix}.$$

Our example raised a few interesting points. Notice how $A+B=B+A$. We probably aren't suprised by this, since we know that when dealing with numbers, $a+b=b+a$. Also, notice that $5A+5B=5(A+B)$. In our example, we were careful to compute each of these expressions following the proper order of operations; knowing these are equal allows us to compute similar expressions in the most convenient way.

Another interesting thing that came from our previous example is that

$$A-A=\begin{bmatrix} 0 & 0 & 0 \\ 0 & 0 & 0 \\ 0 & 0 & 0 \end{bmatrix}.$$

It seems like this should be a special matrix; after all, every entry is 0 and 0 is a special number.

In fact, this is a special matrix. We define $\mathbf{0}$, which we read as "the zero matrix," to be the matrix of all zeros.[1] We should be careful; this previous "definition" is a bit ambiguous, for we have not stated what size the zero matrix should be. Is $\begin{bmatrix} 0 & 0 \\ 0 & 0 \end{bmatrix}$ the zero matrix? How about $\begin{bmatrix} 0 & 0 \end{bmatrix}$?

Let's not get bogged down in semantics. If we ever see $\mathbf{0}$ in an expression, we will usually know right away what size $\mathbf{0}$ should be; it will be the size that allows the expression to make sense. If A is a 3×5 matrix, and we write $A + \mathbf{0}$, we'll simply assume that $\mathbf{0}$ is also a 3×5 matrix. If we are ever in doubt, we can add a subscript; for instance, $\mathbf{0}_{2\times 7}$ is the 2×7 matrix of all zeros.

Since the zero matrix is an important concept, we give it it's own definition box.

Definition 11 **The Zero Matrix**

The $m \times n$ matrix of all zeros, denoted $\mathbf{0}_{m\times n}$, is the *zero matrix*.

When the dimensions of the zero matrix are clear from the context, the subscript is generally omitted.

The following presents some of the properties of matrix addition and scalar multiplication that we discovered above, plus a few more.

Theorem 2 **Properties of Matrix Addition and Scalar Multiplication**

The following equalities hold for all $m \times n$ matrices A, B and C and scalars k.

1. $A + B = B + A$ (Commutative Property)
2. $(A + B) + C = A + (B + C)$ (Associative Property)
3. $k(A + B) = kA + kB$ (Scalar Multiplication Distributive Property)
4. $kA = Ak$
5. $A + \mathbf{0} = \mathbf{0} + A = A$ (Additive Identity)
6. $0A = \mathbf{0}$

Be sure that this last property makes sense; it says that if we multiply any matrix

[1] We use the bold face to distinguish the zero matrix, $\mathbf{0}$, from the number zero, 0.

by the *number* 0, the result is the *zero matrix*, or **0**.

We began this section with the concept of matrix equality. Let's put our matrix addition properties to use and solve a matrix equation.

Example 23 Let

$$A = \begin{bmatrix} 2 & -1 \\ 3 & 6 \end{bmatrix}.$$

Find the matrix X such that

$$2A + 3X = -4A.$$

SOLUTION We can use basic algebra techniques to manipulate this equation for X; first, let's subtract $2A$ from both sides. This gives us

$$3X = -6A.$$

Now divide both sides by 3 to get

$$X = -2A.$$

Now we just need to compute $-2A$; we find that

$$X = \begin{bmatrix} -4 & 2 \\ -6 & -12 \end{bmatrix}.$$

Our matrix properties identified **0** as the Additive Identity; i.e., if you add **0** to any matrix A, you simply get A. This is similar in notion to the fact that for all numbers a, $a + 0 = a$. A *Multiplicative Identity* would be a matrix I where $I \times A = A$ for all matrices A. (What would such a matrix look like? A matrix of all 1s, perhaps?) However, in order for this to make sense, we'll need to learn to multiply matrices together, which we'll do in the next section.

Exercises 2.1

Matrices A and B are given below. In Exercises 1 – 6, simplify the given expression.

$$A = \begin{bmatrix} 1 & -1 \\ 7 & 4 \end{bmatrix} \quad B = \begin{bmatrix} -3 & 2 \\ 5 & 9 \end{bmatrix}$$

1. $A + B$
2. $2A - 3B$
3. $3A - A$
4. $4B - 2A$
5. $3(A - B) + B$
6. $2(A - B) - (A - 3B)$

Matrices A and B are given below. In Exercises 7 – 10, simplify the given expression.

$$A = \begin{bmatrix} 3 \\ 5 \end{bmatrix} \quad B = \begin{bmatrix} -2 \\ 4 \end{bmatrix}$$

7. $4B - 2A$
8. $-2A + 3A$

9. $-2A - 3A$

10. $-B + 3B - 2B$

Matrices A and B are given below. In Exercises 11 – 14, find X that satisfies the equation.

$$A = \begin{bmatrix} 3 & -1 \\ 2 & 5 \end{bmatrix} \quad B = \begin{bmatrix} 1 & 7 \\ 3 & -4 \end{bmatrix}$$

11. $2A + X = B$

12. $A - X = 3B$

13. $3A + 2X = -1B$

14. $A - \frac{1}{2}X = -B$

In Exercises 15 – 21, find values for the scalars a and b that satisfy the given equation.

15. $a\begin{bmatrix} 1 \\ 2 \end{bmatrix} + b\begin{bmatrix} -1 \\ 5 \end{bmatrix} = \begin{bmatrix} 1 \\ 9 \end{bmatrix}$

16. $a\begin{bmatrix} -3 \\ 1 \end{bmatrix} + b\begin{bmatrix} 8 \\ 4 \end{bmatrix} = \begin{bmatrix} 7 \\ 1 \end{bmatrix}$

17. $a\begin{bmatrix} 4 \\ -2 \end{bmatrix} + b\begin{bmatrix} -6 \\ 3 \end{bmatrix} = \begin{bmatrix} 10 \\ -5 \end{bmatrix}$

18. $a\begin{bmatrix} 1 \\ 1 \end{bmatrix} + b\begin{bmatrix} -1 \\ 3 \end{bmatrix} = \begin{bmatrix} 5 \\ 5 \end{bmatrix}$

19. $a\begin{bmatrix} 1 \\ 3 \end{bmatrix} + b\begin{bmatrix} -3 \\ -9 \end{bmatrix} = \begin{bmatrix} 4 \\ -12 \end{bmatrix}$

20. $a\begin{bmatrix} 1 \\ 2 \\ 3 \end{bmatrix} + b\begin{bmatrix} 1 \\ 1 \\ 2 \end{bmatrix} = \begin{bmatrix} 0 \\ -1 \\ -1 \end{bmatrix}$

21. $a\begin{bmatrix} 1 \\ 0 \\ 1 \end{bmatrix} + b\begin{bmatrix} 5 \\ 1 \\ 2 \end{bmatrix} = \begin{bmatrix} 3 \\ 4 \\ 7 \end{bmatrix}$

2.2 Matrix Multiplication

AS YOU READ . . .

1. T/F: Column vectors are used more in this text than row vectors, although some other texts do the opposite.

2. T/F: To multiply $A \times B$, the number of rows of A and B need to be the same.

3. T/F: The entry in the 2nd row and 3rd column of the product AB comes from multipling the 2nd row of A with the 3rd column of B.

4. Name two properties of matrix multiplication that also hold for "regular multiplication" of numbers.

5. Name a property of "regular multiplication" of numbers that does not hold for matrix multiplication.

6. T/F: $A^3 = A \cdot A \cdot A$

In the previous section we found that the definition of matrix addition was very intuitive, and we ended that section discussing the fact that eventually we'd like to know what it means to multiply matrices together.

In the spirit of the last section, take another wild stab: what do you think

$$\begin{bmatrix} 1 & 2 \\ 3 & 4 \end{bmatrix} \times \begin{bmatrix} 1 & -1 \\ 2 & 2 \end{bmatrix}$$

means?

You are likely to have guessed

$$\begin{bmatrix} 1 & -2 \\ 6 & 8 \end{bmatrix}$$

but this is, in fact, *not* right.[2] The actual answer is

$$\begin{bmatrix} 5 & 3 \\ 11 & 5 \end{bmatrix}.$$

If you can look at this one example and suddenly understand exactly how matrix multiplication works, then you are probably smarter than the author. While matrix multiplication isn't hard, it isn't nearly as intuitive as matrix addition is.

To further muddy the waters (before we clear them), consider

$$\begin{bmatrix} 1 & 2 \\ 3 & 4 \end{bmatrix} \times \begin{bmatrix} 1 & -1 & 0 \\ 2 & 2 & -1 \end{bmatrix}.$$

Our experience from the last section would lend us to believe that this is not defined, but our confidence is probably a bit shaken by now. In fact, this multiplication *is* defined, and it is

$$\begin{bmatrix} 5 & 3 & -2 \\ 11 & 5 & -4 \end{bmatrix}.$$

You may see some similarity in this answer to what we got before, but again, probably not enough to really figure things out.

So let's take a step back and progress slowly. The first thing we'd like to do is define a special type of matrix called a vector.

Definition 12

Column and Row Vectors

A $m \times 1$ matrix is called a *column vector*.

A $1 \times n$ matrix is called a *row vector*.

While it isn't obvious right now, column vectors are going to become far more useful to us than row vectors. Therefore, we often omit the word "column" when refering to column vectors, and we just call them "vectors."[3]

[2] I guess you *could* define multiplication this way. If you'd prefer this type of multiplication, write your own book.

[3] In this text, row vectors are only used in this section when we discuss matrix multiplication, whereas we'll make extensive use of column vectors. Other texts make great use of row vectors, but little use of column vectors. It is a matter of preference and tradition: "most" texts use column vectors more.

We have been using upper case letters to denote matrices; we use lower case letters with an arrow overtop to denote row and column vectors. An example of a row vector is

$$\vec{u} = \begin{bmatrix} 1 & 2 & -1 & 0 \end{bmatrix}$$

and an example of a column vector is

$$\vec{v} = \begin{bmatrix} 1 \\ 7 \\ 8 \end{bmatrix}.$$

Before we learn how to multiply matrices in general, we will learn what it means to multiply a row vector by a column vector.

Definition 13 **Multiplying a row vector by a column vector**

Let $\vec{u}$ be an $1 \times n$ row vector with entries $u_1, u_2, \cdots, u_n$ and let $\vec{v}$ be an $n \times 1$ column vector with entries $v_1, v_2, \cdots, v_n$. The *product of $\vec{u}$ and $\vec{v}$*, denoted $\vec{u} \cdot \vec{v}$ or $\vec{u}\vec{v}$, is

$$\sum_{i=1}^{n} u_i v_i = u_1 v_1 + u_2 v_2 + \cdots + u_n v_n.$$

Don't worry if this definition doesn't make immediate sense. It is really an easy concept; an example will make things more clear.

Example 24 Let

$$\vec{u} = \begin{bmatrix} 1 & 2 & 3 \end{bmatrix}, \vec{v} = \begin{bmatrix} 2 & 0 & 1 & -1 \end{bmatrix}, \vec{x} = \begin{bmatrix} -2 \\ 4 \\ 3 \end{bmatrix}, \vec{y} = \begin{bmatrix} 1 \\ 2 \\ 5 \\ 0 \end{bmatrix}.$$

Find the following products.

1. $\vec{u}\vec{x}$
2. $\vec{v}\vec{y}$
3. $\vec{u}\vec{y}$
4. $\vec{u}\vec{v}$
5. $\vec{x}\vec{u}$

Solution

1. $\vec{u}\vec{x} = \begin{bmatrix} 1 & 2 & 3 \end{bmatrix} \begin{bmatrix} -2 \\ 4 \\ 3 \end{bmatrix} = 1(-2) + 2(4) + 3(3) = 15$

2. $\vec{v}\vec{y} = \begin{bmatrix} 2 & 0 & 1 & -1 \end{bmatrix} \begin{bmatrix} 1 \\ 2 \\ 5 \\ 0 \end{bmatrix} = 2(1) + 0(2) + 1(5) - 1(0) = 7$

3. $\vec{u}\vec{y}$ is not defined; Definition 13 specifies that in order to multiply a row vector and column vector, they must have the same number of entries.

4. $\vec{u}\vec{v}$ is not defined; we only know how to multipy row vectors by column vectors. We haven't defined how to multiply two row vectors (in general, it can't be done).

5. The product $\vec{x}\vec{u}$ *is* defined, but we don't know how to do it yet. Right now, we only know how to multiply a row vector times a column vector; we don't know how to multiply a column vector times a row vector. (That's right: $\vec{u}\vec{x} \neq \vec{x}\vec{u}$!)

Now that we understand how to multiply a row vector by a column vector, we are ready to define matrix multiplication.

Definition 14 **Matrix Multiplication**

Let A be an $m \times r$ matrix, and let B be an $r \times n$ matrix. The *matrix product of A and B*, denoted $A \cdot B$, or simply AB, is the $m \times n$ matrix M whose entry in the ith row and j^{th} column is the product of the ith row of A and the jth column of B.

It may help to illustrate it in this way. Let matrix A have rows $\vec{a_1}, \vec{a_2}, \cdots, \vec{a_m}$ and let B have columns $\vec{b_1}, \vec{b_2}, \cdots, \vec{b_n}$. Thus A looks like

$$\begin{bmatrix} - & \vec{a_1} & - \\ - & \vec{a_2} & - \\ & \vdots & \\ - & \vec{a_m} & - \end{bmatrix}$$

where the "$-$" symbols just serve as reminders that the $\vec{a_i}$ represent rows, and B looks like

$$\begin{bmatrix} | & | & & | \\ \vec{b_1} & \vec{b_2} & \cdots & \vec{b_n} \\ | & | & & | \end{bmatrix}$$

where again, the "$|$" symbols just remind us that the $\vec{b_i}$ represent column vectors. Then

$$AB = \begin{bmatrix} \vec{a_1}\vec{b_1} & \vec{a_1}\vec{b_2} & \cdots & \vec{a_1}\vec{b_n} \\ \vec{a_2}\vec{b_1} & \vec{a_2}\vec{b_2} & \cdots & \vec{a_2}\vec{b_n} \\ \vdots & \vdots & \ddots & \vdots \\ \vec{a_m}\vec{b_1} & \vec{a_m}\vec{b_2} & \cdots & \vec{a_m}\vec{b_n} \end{bmatrix}.$$

Two quick notes about this definition. First, notice that in order to multiply A and B, the number of *columns* of A must be the same as the number of *rows* of B (we refer to these as the "inner dimensions"). Secondly, the resulting matrix has the same number of *rows* as A and the same number of *columns* as B (we refer to these as the "outer dimensions").

$$\overbrace{(m \times \underbrace{r) \times (r}_{\substack{\text{these inner dimensions} \\ \text{must match}}} \times n)}^{\substack{\text{final dimensions are the outer} \\ \text{dimensions}}}$$

Of course, this will make much more sense when we see an example.

Example 25 Revisit the matrix product we saw at the beginning of this section; multiply

$$\begin{bmatrix} 1 & 2 \\ 3 & 4 \end{bmatrix} \begin{bmatrix} 1 & -1 & 0 \\ 2 & 2 & -1 \end{bmatrix}.$$

SOLUTION Let's call our first matrix A and the second B. We should first check to see that we can actually perform this multiplication. Matrix A is 2×2 and B is 2×3. The "inner" dimensions match up, so we can compute the product; the "outer" dimensions tell us that the product will be 2×3. Let

$$AB = \begin{bmatrix} m_{11} & m_{12} & m_{13} \\ m_{21} & m_{22} & m_{23} \end{bmatrix}.$$

Let's find the value of each of the entries.

The entry m_{11} is in the first row and first column; therefore to find its value, we need to multiply the first row of A by the first column of B. Thus

$$m_{11} = \begin{bmatrix} 1 & 2 \end{bmatrix} \begin{bmatrix} 1 \\ 2 \end{bmatrix} = 1(1) + 2(2) = 5.$$

So now we know that

$$AB = \begin{bmatrix} 5 & m_{12} & m_{13} \\ m_{21} & m_{22} & m_{23} \end{bmatrix}.$$

Finishing out the first row, we have

$$m_{12} = \begin{bmatrix} 1 & 2 \end{bmatrix} \begin{bmatrix} -1 \\ 2 \end{bmatrix} = 1(-1) + 2(2) = 3$$

using the first row of A and the second column of B, and

$$m_{13} = \begin{bmatrix} 1 & 2 \end{bmatrix} \begin{bmatrix} 0 \\ -1 \end{bmatrix} = 1(0) + 2(-1) = -2$$

using the first row of A and the third column of B. Thus we have

$$AB = \begin{bmatrix} 5 & 3 & -2 \\ m_{21} & m_{22} & m_{23} \end{bmatrix}.$$

To compute the second row of AB, we multiply with the second row of A. We find

$$m_{21} = \begin{bmatrix} 3 & 4 \end{bmatrix} \begin{bmatrix} 1 \\ 2 \end{bmatrix} = 11,$$

$$m_{22} = \begin{bmatrix} 3 & 4 \end{bmatrix} \begin{bmatrix} -1 \\ 2 \end{bmatrix} = 5,$$

and

$$m_{23} = \begin{bmatrix} 3 & 4 \end{bmatrix} \begin{bmatrix} 0 \\ -1 \end{bmatrix} = -4.$$

Thus

$$AB = \begin{bmatrix} 1 & 2 \\ 3 & 4 \end{bmatrix} \begin{bmatrix} 1 & -1 & 0 \\ 2 & 2 & -1 \end{bmatrix} = \begin{bmatrix} 5 & 3 & -2 \\ 11 & 5 & -4 \end{bmatrix}.$$

Example 26 Multiply

$$\begin{bmatrix} 1 & -1 \\ 5 & 2 \\ -2 & 3 \end{bmatrix} \begin{bmatrix} 1 & 1 & 1 & 1 \\ 2 & 6 & 7 & 9 \end{bmatrix}.$$

Solution Let's first check to make sure this product is defined. Again calling the first matrix A and the second B, we see that A is a 3×2 matrix and B is a 2×4 matrix; the inner dimensions match so the product is defined, and the product will be a 3×4 matrix,

$$AB = \begin{bmatrix} m_{11} & m_{12} & m_{13} & m_{14} \\ m_{21} & m_{22} & m_{23} & m_{24} \\ m_{31} & m_{32} & m_{33} & m_{34} \end{bmatrix}.$$

We will demonstrate how to compute some of the entries, then give the final answer. The reader can fill in the details of how each entry was computed.

$$m_{11} = \begin{bmatrix} 1 & -1 \end{bmatrix} \begin{bmatrix} 1 \\ 2 \end{bmatrix} = -1.$$

$$m_{13} = \begin{bmatrix} 1 & -1 \end{bmatrix} \begin{bmatrix} 1 \\ 7 \end{bmatrix} = -6.$$

$$m_{23} = \begin{bmatrix} 5 & 2 \end{bmatrix} \begin{bmatrix} 1 \\ 7 \end{bmatrix} = 19.$$

$$m_{24} = \begin{bmatrix} 5 & 2 \end{bmatrix} \begin{bmatrix} 1 \\ 9 \end{bmatrix} = 23.$$

$$m_{32} = \begin{bmatrix} -2 & 3 \end{bmatrix} \begin{bmatrix} 1 \\ 6 \end{bmatrix} = 16.$$

$$m_{34} = \begin{bmatrix} -2 & 3 \end{bmatrix} \begin{bmatrix} 1 \\ 9 \end{bmatrix} = 25.$$

So far, we've computed this much of AB:

$$AB = \begin{bmatrix} -1 & m_{12} & -6 & m_{14} \\ m_{21} & m_{22} & 19 & 23 \\ m_{31} & 16 & m_{33} & 25 \end{bmatrix}.$$

The final product is

$$AB = \begin{bmatrix} -1 & -5 & -6 & -8 \\ 9 & 17 & 19 & 23 \\ 4 & 16 & 19 & 25 \end{bmatrix}.$$

Example 27 Multiply, if possible,

$$\begin{bmatrix} 2 & 3 & 4 \\ 9 & 8 & 7 \end{bmatrix} \begin{bmatrix} 3 & 6 \\ 5 & -1 \end{bmatrix}.$$

Solution Again, we'll call the first matrix A and the second B. Checking the dimensions of each matrix, we see that A is a 2×3 matrix, whereas B is a 2×2 matrix. The inner dimensions do not match, therefore this multiplication is not defined.

Example 28 In Example 24, we were told that the product $\vec{x}\vec{u}$ was defined, where

$$\vec{x} = \begin{bmatrix} -2 \\ 4 \\ 3 \end{bmatrix} \quad \text{and} \quad \vec{u} = \begin{bmatrix} 1 & 2 & 3 \end{bmatrix},$$

although we were not shown what that product was. Find $\vec{x}\vec{u}$.

Solution Again, we need to check to make sure the dimensions work correctly (remember that even though we are referring to $\vec{u}$ and $\vec{x}$ as vectors, they are, in fact, just matrices).

The column vector $\vec{x}$ has dimensions 3×1, whereas the row vector $\vec{u}$ has dimensions 1×3. Since the inner dimensions do match, the matrix product is defined; the outer dimensions tell us that the product will be a 3×3 matrix, as shown below:

$$\vec{x}\vec{u} = \begin{bmatrix} m_{11} & m_{12} & m_{13} \\ m_{21} & m_{22} & m_{23} \\ m_{31} & m_{32} & m_{33} \end{bmatrix}.$$

To compute the entry m_{11}, we multiply the first row of $\vec{x}$ by the first column of $\vec{u}$. What is the first row of $\vec{x}$? Simply the number -2. What is the first column of $\vec{u}$? Just the number 1. Thus $m_{11} = -2$. (This does seem odd, but through checking, you can see that we are indeed following the rules.)

What about the entry m_{12}? Again, we multiply the first row of $\vec{x}$ by the first column of $\vec{u}$; that is, we multiply $-2(2)$. So $m_{12} = -4$.

What about m_{23}? Multiply the second row of $\vec{x}$ by the third column of $\vec{u}$; multiply $4(3)$, so $m_{23} = 12$.

One final example: m_{31} comes from multiplying the third row of $\vec{x}$, which is 3, by the first column of $\vec{u}$, which is 1. Therefore $m_{31} = 3$.

So far we have computed

$$\vec{x}\vec{u} = \begin{bmatrix} -2 & -4 & m_{13} \\ m_{21} & m_{22} & 12 \\ 3 & m_{32} & m_{33} \end{bmatrix}.$$

After performing all 9 multiplications, we find

$$\vec{x}\vec{u} = \begin{bmatrix} -2 & -4 & -6 \\ 4 & 8 & 12 \\ 3 & 6 & 9 \end{bmatrix}.$$

In this last example, we saw a "nonstandard" multiplication (at least, it felt nonstandard). Studying the entries of this matrix, it seems that there are several different patterns that can be seen amongst the entries. (Remember that mathematicians like to look for patterns. Also remember that we often guess wrong at first; don't be scared and try to identify some patterns.)

In Section 2.1, we identified the zero matrix **0** that had a nice property in relation to matrix addition (i.e., $A + \mathbf{0} = A$ for any matrix A). In the following example we'll identify a matrix that works well with multiplication as well as some multiplicative properties. For instance, we've learned how $1 \cdot A = A$; is there a *matrix* that acts like the number 1? That is, can we find a matrix X where $X \cdot A = A$?[4]

Example 29 Let

$$A = \begin{bmatrix} 1 & 2 & 3 \\ 2 & -7 & 5 \\ -2 & -8 & 3 \end{bmatrix}, \quad B = \begin{bmatrix} 1 & 1 & 1 \\ 1 & 1 & 1 \\ 1 & 1 & 1 \end{bmatrix}$$

$$C = \begin{bmatrix} 1 & 0 & 2 \\ 2 & 1 & 0 \\ 0 & 2 & 1 \end{bmatrix}, \quad I = \begin{bmatrix} 1 & 0 & 0 \\ 0 & 1 & 0 \\ 0 & 0 & 1 \end{bmatrix}.$$

Find the following products.

1. AB
2. BA
3. $A\mathbf{0}_{3\times 4}$
4. AI
5. IA
6. I^2
7. BC
8. B^2

[4]We made a guess in Section 2.1 that maybe a matrix of all 1s would work.

Solution We will find each product, but we leave the details of each computation to the reader.

1. $AB = \begin{bmatrix} 1 & 2 & 3 \\ 2 & -7 & 5 \\ -2 & -8 & 3 \end{bmatrix} \begin{bmatrix} 1 & 1 & 1 \\ 1 & 1 & 1 \\ 1 & 1 & 1 \end{bmatrix} = \begin{bmatrix} 6 & 6 & 6 \\ 0 & 0 & 0 \\ -7 & -7 & -7 \end{bmatrix}$

2. $BA = \begin{bmatrix} 1 & 1 & 1 \\ 1 & 1 & 1 \\ 1 & 1 & 1 \end{bmatrix} \begin{bmatrix} 1 & 2 & 3 \\ 2 & -7 & 5 \\ -2 & -8 & 3 \end{bmatrix} = \begin{bmatrix} 1 & -13 & 11 \\ 1 & -13 & 11 \\ 1 & -13 & 11 \end{bmatrix}$

3. $A\mathbf{0}_{3\times 4} = \mathbf{0}_{3\times 4}$.

4. $AI = \begin{bmatrix} 1 & 2 & 3 \\ 2 & -7 & 5 \\ -2 & -8 & 3 \end{bmatrix} \begin{bmatrix} 1 & 0 & 0 \\ 0 & 1 & 0 \\ 0 & 0 & 1 \end{bmatrix} = \begin{bmatrix} 1 & 2 & 3 \\ 2 & -7 & 5 \\ -2 & -8 & 3 \end{bmatrix}$

5. $IA = \begin{bmatrix} 1 & 0 & 0 \\ 0 & 1 & 0 \\ 0 & 0 & 1 \end{bmatrix} \begin{bmatrix} 1 & 2 & 3 \\ 2 & -7 & 5 \\ -2 & -8 & 3 \end{bmatrix} = \begin{bmatrix} 1 & 2 & 3 \\ 2 & -7 & 5 \\ -2 & -8 & 3 \end{bmatrix}$

6. We haven't formally defined what I^2 means, but we could probably make the reasonable guess that $I^2 = I \cdot I$. Thus

$$I^2 = \begin{bmatrix} 1 & 0 & 0 \\ 0 & 1 & 0 \\ 0 & 0 & 1 \end{bmatrix} \begin{bmatrix} 1 & 0 & 0 \\ 0 & 1 & 0 \\ 0 & 0 & 1 \end{bmatrix} = \begin{bmatrix} 1 & 0 & 0 \\ 0 & 1 & 0 \\ 0 & 0 & 1 \end{bmatrix}$$

7. $BC = \begin{bmatrix} 1 & 1 & 1 \\ 1 & 1 & 1 \\ 1 & 1 & 1 \end{bmatrix} \begin{bmatrix} 1 & 0 & 2 \\ 2 & 1 & 0 \\ 0 & 2 & 1 \end{bmatrix} = \begin{bmatrix} 3 & 3 & 3 \\ 3 & 3 & 3 \\ 3 & 3 & 3 \end{bmatrix}$

8. $B^2 = BB = \begin{bmatrix} 1 & 1 & 1 \\ 1 & 1 & 1 \\ 1 & 1 & 1 \end{bmatrix} \begin{bmatrix} 1 & 1 & 1 \\ 1 & 1 & 1 \\ 1 & 1 & 1 \end{bmatrix} = \begin{bmatrix} 3 & 3 & 3 \\ 3 & 3 & 3 \\ 3 & 3 & 3 \end{bmatrix}$

This example is simply chock full of interesting ideas; it is almost hard to think about where to start.

Interesting Idea #1: Notice that in our example, $AB \neq BA$! When dealing with numbers, we were used to the idea that $ab = ba$. With matrices, multiplication is *not* commutative. (Of course, we can find special situations where it does work. In general, though, it doesn't.)

Interesting Idea #2: Right before this example we wondered if there was a matrix that "acted like the number 1," and guessed it may be a matrix of all 1s. However, we found out that such a matrix does not work in that way; in our example, $AB \neq A$. We did find that $AI = IA = A$. There is a Multiplicative Identity; it just isn't what we thought it would be. And just as $1^2 = 1$, $I^2 = I$.

Interesting Idea #3: When dealing with numbers, we are very familiar with the notion that "If $ax = bx$, then $a = b$." (As long as $x \neq 0$.) Notice that, in our example,

$BB = BC$, yet $B \neq C$. In general, just because $AX = BX$, we *cannot* conclude that $A = B$.

Matrix multiplication is turning out to be a very strange operation. We are very used to multiplying numbers, and we know a bunch of properties that hold when using this type of multiplication. When multiplying matrices, though, we probably find ourselves asking two questions, "What *does* work?" and "What *doesn't* work?" We'll answer these questions; first we'll do an example that demonstrates some of the things that do work.

Example 30 Let

$$A = \begin{bmatrix} 1 & 2 \\ 3 & 4 \end{bmatrix}, \quad B = \begin{bmatrix} 1 & 1 \\ 1 & -1 \end{bmatrix} \quad \text{and} \quad C = \begin{bmatrix} 2 & 1 \\ 1 & 2 \end{bmatrix}.$$

Find the following:

1. $A(B + C)$
2. $AB + AC$
3. $A(BC)$
4. $(AB)C$

Solution We'll compute each of these without showing all the intermediate steps. Keep in mind order of operations: things that appear inside of parentheses are computed first.

1.

$$\begin{aligned} A(B + C) &= \begin{bmatrix} 1 & 2 \\ 3 & 4 \end{bmatrix} \left(\begin{bmatrix} 1 & 1 \\ 1 & -1 \end{bmatrix} + \begin{bmatrix} 2 & 1 \\ 1 & 2 \end{bmatrix} \right) \\ &= \begin{bmatrix} 1 & 2 \\ 3 & 4 \end{bmatrix} \begin{bmatrix} 3 & 2 \\ 2 & 1 \end{bmatrix} \\ &= \begin{bmatrix} 7 & 4 \\ 17 & 10 \end{bmatrix} \end{aligned}$$

2.

$$\begin{aligned} AB + AC &= \begin{bmatrix} 1 & 2 \\ 3 & 4 \end{bmatrix} \begin{bmatrix} 1 & 1 \\ 1 & -1 \end{bmatrix} + \begin{bmatrix} 1 & 2 \\ 3 & 4 \end{bmatrix} \begin{bmatrix} 2 & 1 \\ 1 & 2 \end{bmatrix} \\ &= \begin{bmatrix} 3 & -1 \\ 7 & -1 \end{bmatrix} + \begin{bmatrix} 4 & 5 \\ 10 & 11 \end{bmatrix} \\ &= \begin{bmatrix} 7 & 4 \\ 17 & 10 \end{bmatrix} \end{aligned}$$

3.

$$A(BC) = \begin{bmatrix} 1 & 2 \\ 3 & 4 \end{bmatrix} \left(\begin{bmatrix} 1 & 1 \\ 1 & -1 \end{bmatrix} \begin{bmatrix} 2 & 1 \\ 1 & 2 \end{bmatrix} \right)$$
$$= \begin{bmatrix} 1 & 2 \\ 3 & 4 \end{bmatrix} \begin{bmatrix} 3 & 3 \\ 1 & -1 \end{bmatrix}$$
$$= \begin{bmatrix} 5 & 1 \\ 13 & 5 \end{bmatrix}$$

4.

$$(AB)\,C = \left(\begin{bmatrix} 1 & 2 \\ 3 & 4 \end{bmatrix} \begin{bmatrix} 1 & 1 \\ 1 & -1 \end{bmatrix} \right) \begin{bmatrix} 2 & 1 \\ 1 & 2 \end{bmatrix}$$
$$= \begin{bmatrix} 3 & -1 \\ 7 & -1 \end{bmatrix} \begin{bmatrix} 2 & 1 \\ 1 & 2 \end{bmatrix}$$
$$= \begin{bmatrix} 5 & 1 \\ 13 & 5 \end{bmatrix}$$

In looking at our example, we should notice two things. First, it looks like the "distributive property" holds; that is, $A(B + C) = AB + AC$. This is nice as many algebraic techniques we have learned about in the past (when doing "ordinary algebra") will still work. Secondly, it looks like the "associative property" holds; that is, $A(BC) = (AB)C$. This is nice, for it tells us that when we are multiplying several matrices together, we don't have to be particularly careful in what order we multiply certain pairs of matrices together.[5]

In leading to an important theorem, let's define a matrix we saw in an earlier example.[6]

Definition 15

Identity Matrix

The $n \times n$ matrix with 1's on the diagonal and zeros elsewhere is the $n \times n$ *identity matrix*, denoted I_n. When the context makes the dimension of the identity clear, the subscript is generally omitted.

Note that while the zero matrix can come in all different shapes and sizes, the

[5]Be careful: in computing ABC together, we can first multiply AB or BC, but we cannot change the *order* in which these matrices appear. We cannot multiply BA or AC, for instance.

[6]The following definition uses a term we won't define until Definition 20 on page 123: *diagonal*. In short, a "diagonal matrix" is one in which the only nonzero entries are the "diagonal entries." The examples given here and in the exercises should suffice until we meet the full definition later.

identity matrix is always a square matrix. We show a few identity matrices below.

$$I_2 = \begin{bmatrix} 1 & 0 \\ 0 & 1 \end{bmatrix}, \quad I_3 = \begin{bmatrix} 1 & 0 & 0 \\ 0 & 1 & 0 \\ 0 & 0 & 1 \end{bmatrix}, \quad I_4 = \begin{bmatrix} 1 & 0 & 0 & 0 \\ 0 & 1 & 0 & 0 \\ 0 & 0 & 1 & 0 \\ 0 & 0 & 0 & 1 \end{bmatrix}$$

In our examples above, we have seen examples of things that do and do not work. We should be careful about what examples *prove*, though. If someone were to claim that $AB = BA$ is always true, one would only need to show them one example where they were false, and we would know the person was wrong. However, if someone claims that $A(B + C) = AB + AC$ is always true, we can't prove this with just one example. We need something more powerful; we need a true proof.

In this text, we forgo most proofs. The reader should know, though, that when we state something in a theorem, there is a proof that backs up what we state. Our justification comes from something stronger than just examples.

Now we give the good news of what does work when dealing with matrix multiplication.

Theorem 3

Properties of Matrix Multiplication

Let A, B and C be matrices with dimensions so that the following operations make sense, and let k be a scalar. The following equalities hold:

1. $A(BC) = (AB)C$ (Associative Property)
2. $A(B + C) = AB + AB$ and
   $(B + C)A = BA + CA$ (Distributive Property)
3. $k(AB) = (kA)B = A(kB)$
4. $AI = IA = A$

The above box contains some very good news, and probably some very surprising news. Matrix multiplication probably seems to us like a very odd operation, so we probably wouldn't have been surprised if we were told that $A(BC) \neq (AB)C$. It is a very nice thing that the Associative Property does hold.

As we near the end of this section, we raise one more issue of notation. We define $A^0 = I$. If n is a positive integer, we define

$$A^n = \underbrace{A \cdot A \cdot \cdots \cdot A}_{n \text{ times}}.$$

With numbers, we are used to $a^{-n} = \frac{1}{a^n}$. Do negative exponents work with matrices, too? The answer is yes, sort of. We'll have to be careful, and we'll cover the topic in detail once we define the inverse of a matrix. For now, though, we recognize the fact that $A^{-1} \neq \frac{1}{A}$, for $\frac{1}{A}$ makes no sense; we don't know how to "divide" by a matrix.

We end this section with a reminder of some of the things that do not work with matrix multiplication. The good news is that there are really only two things on this list.

1. Matrix multiplication is not commutative; that is, $AB \neq BA$.
2. In general, just because $AX = BX$, we cannot conclude that $A = B$.

The bad news is that these ideas pop up in many places where we don't expect them. For instance, we are used to

$$(a+b)^2 = a^2 + 2ab + b^2.$$

What about $(A+B)^2$? All we'll say here is that

$$(A+B)^2 \neq A^2 + 2AB + B^2;$$

we leave it to the reader to figure out why.

The next section is devoted to visualizing column vectors and "seeing" how some of these arithmetic properties work together.

Exercises 2.2

In Exercises 1 – 12, row and column vectors $\vec{u}$ and $\vec{v}$ are defined. Find the product $\vec{u}\vec{v}$, where possible.

1. $\vec{u} = \begin{bmatrix} 1 & -4 \end{bmatrix}$ $\vec{v} = \begin{bmatrix} -2 \\ 5 \end{bmatrix}$
2. $\vec{u} = \begin{bmatrix} 2 & 3 \end{bmatrix}$ $\vec{v} = \begin{bmatrix} 7 \\ -4 \end{bmatrix}$
3. $\vec{u} = \begin{bmatrix} 1 & -1 \end{bmatrix}$ $\vec{v} = \begin{bmatrix} 3 \\ 3 \end{bmatrix}$
4. $\vec{u} = \begin{bmatrix} 0.6 & 0.8 \end{bmatrix}$ $\vec{v} = \begin{bmatrix} 0.6 \\ 0.8 \end{bmatrix}$
5. $\vec{u} = \begin{bmatrix} 1 & 2 & -1 \end{bmatrix}$ $\vec{v} = \begin{bmatrix} 2 \\ 1 \\ -1 \end{bmatrix}$
6. $\vec{u} = \begin{bmatrix} 3 & 2 & -2 \end{bmatrix}$ $\vec{v} = \begin{bmatrix} -1 \\ 0 \\ 9 \end{bmatrix}$
7. $\vec{u} = \begin{bmatrix} 8 & -4 & 3 \end{bmatrix}$ $\vec{v} = \begin{bmatrix} 2 \\ 4 \\ 5 \end{bmatrix}$
8. $\vec{u} = \begin{bmatrix} -3 & 6 & 1 \end{bmatrix}$ $\vec{v} = \begin{bmatrix} 1 \\ -1 \\ 1 \end{bmatrix}$
9. $\vec{u} = \begin{bmatrix} 1 & 2 & 3 & 4 \end{bmatrix}$ $\vec{v} = \begin{bmatrix} 1 \\ -1 \\ 1 \\ -1 \end{bmatrix}$
10. $\vec{u} = \begin{bmatrix} 6 & 2 & -1 & 2 \end{bmatrix}$ $\vec{v} = \begin{bmatrix} 3 \\ 2 \\ 9 \\ 5 \end{bmatrix}$
11. $\vec{u} = \begin{bmatrix} 1 & 2 & 3 \end{bmatrix}$ $\vec{v} = \begin{bmatrix} 3 \\ 2 \end{bmatrix}$
12. $\vec{u} = \begin{bmatrix} 2 & -5 \end{bmatrix}$ $\vec{v} = \begin{bmatrix} 1 \\ 1 \\ 1 \end{bmatrix}$

In Exercises 13 – 27, matrices A and B are defined.

(a) Give the dimensions of A and B. If the dimensions properly match, give the dimensions of AB and BA.

(b) Find the products AB and BA, if possible.

13. $A = \begin{bmatrix} 1 & 2 \\ -1 & 4 \end{bmatrix}$ $B = \begin{bmatrix} 2 & 5 \\ 3 & -1 \end{bmatrix}$

14. $A = \begin{bmatrix} 3 & 7 \\ 2 & 5 \end{bmatrix}$ $B = \begin{bmatrix} 1 & -1 \\ 3 & -3 \end{bmatrix}$

15. $A = \begin{bmatrix} 3 & -1 \\ 2 & 2 \end{bmatrix}$

$B = \begin{bmatrix} 1 & 0 & 7 \\ 4 & 2 & 9 \end{bmatrix}$

16. $A = \begin{bmatrix} 0 & 1 \\ 1 & -1 \\ -2 & -4 \end{bmatrix}$

$B = \begin{bmatrix} -2 & 0 \\ 3 & 8 \end{bmatrix}$

17. $A = \begin{bmatrix} 9 & 4 & 3 \\ 9 & -5 & 9 \end{bmatrix}$

$B = \begin{bmatrix} -2 & 5 \\ -2 & -1 \end{bmatrix}$

18. $A = \begin{bmatrix} -2 & -1 \\ 9 & -5 \\ 3 & -1 \end{bmatrix}$

$B = \begin{bmatrix} -5 & 6 & -4 \\ 0 & 6 & -3 \end{bmatrix}$

19. $A = \begin{bmatrix} 2 & 6 \\ 6 & 2 \\ 5 & -1 \end{bmatrix}$

$B = \begin{bmatrix} -4 & 5 & 0 \\ -4 & 4 & -4 \end{bmatrix}$

20. $A = \begin{bmatrix} -5 & 2 \\ -5 & -2 \\ -5 & -4 \end{bmatrix}$

$B = \begin{bmatrix} 0 & -5 & 6 \\ -5 & -3 & -1 \end{bmatrix}$

21. $A = \begin{bmatrix} 8 & -2 \\ 4 & 5 \\ 2 & -5 \end{bmatrix}$

$B = \begin{bmatrix} -5 & 1 & -5 \\ 8 & 3 & -2 \end{bmatrix}$

22. $A = \begin{bmatrix} 1 & 4 \\ 7 & 6 \end{bmatrix}$

$B = \begin{bmatrix} 1 & -1 & -5 & 5 \\ -2 & 1 & 3 & -5 \end{bmatrix}$

23. $A = \begin{bmatrix} -1 & 5 \\ 6 & 7 \end{bmatrix}$

$B = \begin{bmatrix} 5 & -3 & -4 & -4 \\ -2 & -5 & -5 & -1 \end{bmatrix}$

24. $A = \begin{bmatrix} -1 & 2 & 1 \\ -1 & 2 & -1 \\ 0 & 0 & -2 \end{bmatrix}$

$B = \begin{bmatrix} 0 & 0 & -2 \\ 1 & 2 & -1 \\ 1 & 0 & 0 \end{bmatrix}$

25. $A = \begin{bmatrix} -1 & 1 & 1 \\ -1 & -1 & -2 \\ 1 & 1 & -2 \end{bmatrix}$

$B = \begin{bmatrix} -2 & -2 & -2 \\ 0 & -2 & 0 \\ -2 & 0 & 2 \end{bmatrix}$

26. $A = \begin{bmatrix} -4 & 3 & 3 \\ -5 & -1 & -5 \\ -5 & 0 & -1 \end{bmatrix}$

$B = \begin{bmatrix} 0 & 5 & 0 \\ -5 & -4 & 3 \\ 5 & -4 & 3 \end{bmatrix}$

27. $A = \begin{bmatrix} -4 & -1 & 3 \\ 2 & -3 & 5 \\ 1 & 5 & 3 \end{bmatrix}$

$B = \begin{bmatrix} -2 & 4 & 3 \\ -1 & 1 & -1 \\ 4 & 0 & 2 \end{bmatrix}$

In Exercises 28 – 33, a *diagonal* matrix D and a matrix A are given. Find the products DA and AD, where possible.

28. $D = \begin{bmatrix} 3 & 0 \\ 0 & -1 \end{bmatrix}$

$A = \begin{bmatrix} 2 & 4 \\ 6 & 8 \end{bmatrix}$

29. $D = \begin{bmatrix} 4 & 0 \\ 0 & -3 \end{bmatrix}$

$A = \begin{bmatrix} 1 & 2 \\ 1 & 2 \end{bmatrix}$

30. $D = \begin{bmatrix} -1 & 0 & 0 \\ 0 & 2 & 0 \\ 0 & 0 & 3 \end{bmatrix}$

$A = \begin{bmatrix} 1 & 2 & 3 \\ 4 & 5 & 6 \\ 7 & 8 & 9 \end{bmatrix}$

31. $D = \begin{bmatrix} 1 & 1 & 1 \\ 2 & 2 & 2 \\ -3 & -3 & -3 \end{bmatrix}$

$A = \begin{bmatrix} 2 & 0 & 0 \\ 0 & -3 & 0 \\ 0 & 0 & 5 \end{bmatrix}$

32. $D = \begin{bmatrix} d_1 & 0 \\ 0 & d_2 \end{bmatrix}$

$A = \begin{bmatrix} a & b \\ c & d \end{bmatrix}$

33. $D = \begin{bmatrix} d_1 & 0 & 0 \\ 0 & d_2 & 0 \\ 0 & 0 & d_3 \end{bmatrix}$

$A = \begin{bmatrix} a & b & c \\ d & e & f \\ g & h & i \end{bmatrix}$

In Exercises 34 – 39, a matrix A and a vector $\vec{x}$ are given. Find the product $A\vec{x}$.

34. $A = \begin{bmatrix} 2 & 3 \\ 1 & -1 \end{bmatrix}, \quad \vec{x} = \begin{bmatrix} 4 \\ 9 \end{bmatrix}$

35. $A = \begin{bmatrix} -1 & 4 \\ 7 & 3 \end{bmatrix}, \quad \vec{x} = \begin{bmatrix} 2 \\ -1 \end{bmatrix}$

36. $A = \begin{bmatrix} 2 & 0 & 3 \\ 1 & 1 & 1 \\ 3 & -1 & 2 \end{bmatrix}, \quad \vec{x} = \begin{bmatrix} 1 \\ 4 \\ 2 \end{bmatrix}$

37. $A = \begin{bmatrix} -2 & 0 & 3 \\ 1 & 1 & -2 \\ 4 & 2 & -1 \end{bmatrix}, \quad \vec{x} = \begin{bmatrix} 4 \\ 3 \\ 1 \end{bmatrix}$

38. $A = \begin{bmatrix} 2 & -1 \\ 4 & 3 \end{bmatrix}, \quad \vec{x} = \begin{bmatrix} x_1 \\ x_2 \end{bmatrix}$

39. $A = \begin{bmatrix} 1 & 2 & 3 \\ 1 & 0 & 2 \\ 2 & 3 & 1 \end{bmatrix}, \quad \vec{x} = \begin{bmatrix} x_1 \\ x_2 \\ x_3 \end{bmatrix}$

40. Let $A = \begin{bmatrix} 0 & 1 \\ 1 & 0 \end{bmatrix}$. Find A^2 and A^3.

41. Let $A = \begin{bmatrix} 2 & 0 \\ 0 & 3 \end{bmatrix}$. Find A^2 and A^3.

42. Let $A = \begin{bmatrix} -1 & 0 & 0 \\ 0 & 3 & 0 \\ 0 & 0 & 5 \end{bmatrix}$. Find A^2 and A^3.

43. Let $A = \begin{bmatrix} 0 & 1 & 0 \\ 0 & 0 & 1 \\ 1 & 0 & 0 \end{bmatrix}$. Find A^2 and A^3.

44. Let $A = \begin{bmatrix} 0 & 0 & 1 \\ 0 & 0 & 0 \\ 0 & 1 & 0 \end{bmatrix}$. Find A^2 and A^3.

45. In the text we state that $(A + B)^2 \neq A^2 + 2AB + B^2$. We investigate that claim here.

(a) Let $A = \begin{bmatrix} 5 & 3 \\ -3 & -2 \end{bmatrix}$ and let $B = \begin{bmatrix} -5 & -5 \\ -2 & 1 \end{bmatrix}$. Compute $A + B$.

(b) Find $(A + B)^2$ by using your answer from (a).

(c) Compute $A^2 + 2AB + B^2$.

(d) Are the results from (a) and (b) the same?

(e) Carefully expand the expression $(A + B)^2 = (A + B)(A + B)$ and show why this is not equal to $A^2 + 2AB + B^2$.

2.3 Visualizing Matrix Arithmetic in 2D

1. T/F: Two vectors with the same length and direction are equal even if they start from different places.
2. One can visualize vector addition using what law?
3. T/F: Multiplying a vector by 2 doubles its length.
4. What do mathematicians do?
5. T/F: Multiplying a vector by a matrix always changes its length and direction.

When we first learned about adding numbers together, it was useful to picture a number line: $2+3=5$ could be pictured by starting at 0, going out 2 tick marks, then another 3, and then realizing that we moved 5 tick marks from 0. Similar visualizations helped us understand what $2-3$ meant and what 2×3 meant.

We now investigate a way to picture matrix arithmetic – in particular, operations involving column vectors. This not only will help us better understand the arithmetic operations, it will open the door to a great wealth of interesting study. Visualizing matrix arithmetic has a wide variety of applications, the most common being computer graphics. While we often think of these graphics in terms of video games, there are numerous other important applications. For example, chemists and biologists often use computer models to "visualize" complex molecules to "see" how they interact with other molecules.

We will start with vectors in two dimensions (2D) – that is, vectors with only two entries. We assume the reader is familiar with the Cartesian plane, that is, plotting points and graphing functions on "the x–y plane." We graph vectors in a manner very similar to plotting points. Given the vector

$$\vec{x}=\begin{bmatrix}1\\2\end{bmatrix},$$

we draw $\vec{x}$ by drawing an arrow whose tip is 1 unit to the right and 2 units up from its origin.[7]

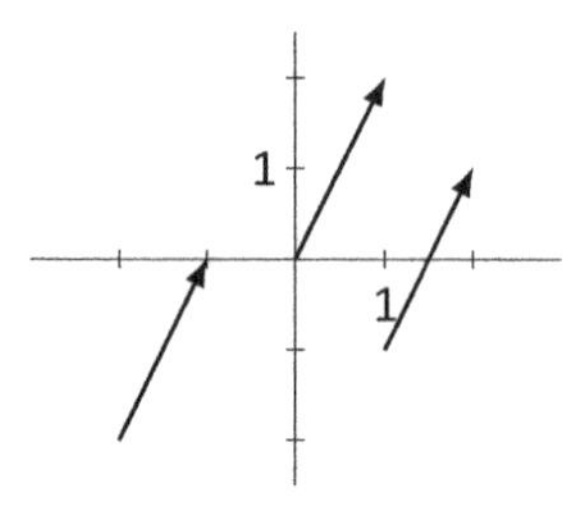

Figure 2.1: Various drawings of $\vec{x}$

[7] To help reduce clutter, in all figures each tick mark represents one unit.

When drawing vectors, we do not specify where you start drawing; all we specify is where the tip lies based on where we started. Figure 2.1 shows vector $\vec{x}$ drawn 3 ways. In some ways, the "most common" way to draw a vector has the arrow start at the origin,but this is by no means the only way of drawing the vector.

Let's practice this concept by drawing various vectors from given starting points.

Example 31 Let

$$\vec{x} = \begin{bmatrix} 1 \\ -1 \end{bmatrix} \quad \vec{y} = \begin{bmatrix} 2 \\ 3 \end{bmatrix} \quad \text{and} \quad \vec{z} = \begin{bmatrix} -3 \\ 2 \end{bmatrix}.$$

Draw $\vec{x}$ starting from the point $(0, -1)$; draw $\vec{y}$ starting from the point $(-1, -1)$, and draw $\vec{z}$ starting from the point $(2, -1)$.

SOLUTION To draw $\vec{x}$, start at the point $(0, -1)$ as directed, then move to the right one unit and down one unit and draw the tip. Thus the arrow "points" from $(0, -1)$ to $(1, -2)$.

To draw $\vec{y}$, we are told to start and the point $(-1, -1)$. We draw the tip by moving to the right 2 units and up 3 units; hence $\vec{y}$ points from $(-1, -1)$ to (1,2).

To draw $\vec{z}$, we start at $(2, -1)$ and draw the tip 3 units to the left and 2 units up; $\vec{z}$ points from $(2, -1)$ to $(-1, 1)$.

Each vector is drawn as shown in Figure 2.2.

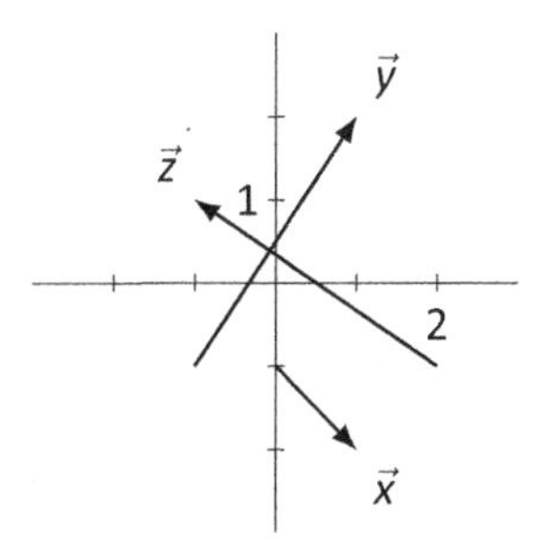

Figure 2.2: Drawing vectors $\vec{x}$, $\vec{y}$ and $\vec{z}$ in Example 31

How does one draw the zero vector, $\vec{0} = \begin{bmatrix} 0 \\ 0 \end{bmatrix}$?[8] Following our basic procedure, we start by going 0 units in the x direction, followed by 0 units in the y direction. In other words, we don't go anywhere. In general, we don't actually draw $\vec{0}$. At best, one can draw a dark circle at the origin to convey the idea that $\vec{0}$, when starting at the origin, points to the origin.

In section 2.1 we learned about matrix arithmetic operations: matrix addition and scalar multiplication. Let's investigate how we can "draw" these operations.

[8]Vectors are just special types of matrices. The zero vector, $\vec{0}$, is a special type of zero matrix, **0**. It helps to distinguish the two by using different notation.

Vector Addition

Given two vectors $\vec{x}$ and $\vec{y}$, how do we draw the vector $\vec{x}+\vec{y}$? Let's look at this in the context of an example, then study the result.

Example 32 Let

$$\vec{x}=\begin{bmatrix}1\\1\end{bmatrix} \quad \text{and} \quad \vec{y}=\begin{bmatrix}3\\1\end{bmatrix}.$$

Sketch $\vec{x}$, $\vec{y}$ and $\vec{x}+\vec{y}$.

Solution A starting point for drawing each vector was not given; by default, we'll start at the origin. (This is in many ways nice; this means that the *vector* $\begin{bmatrix}3\\1\end{bmatrix}$ "points" to the *point* (3,1).) We first compute $\vec{x}+\vec{y}$:

$$\vec{x}+\vec{y}=\begin{bmatrix}1\\1\end{bmatrix}+\begin{bmatrix}3\\1\end{bmatrix}=\begin{bmatrix}4\\2\end{bmatrix}$$

Sketching each gives the picture in Figure 2.3.

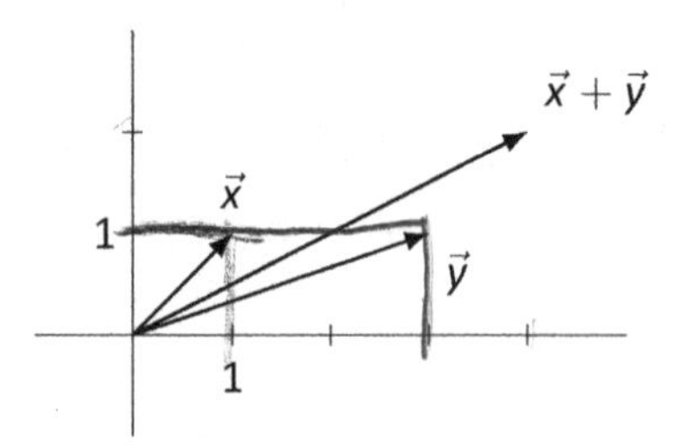

Figure 2.3: Adding vectors $\vec{x}$ and $\vec{y}$ in Example 32

This example is pretty basic; we were given two vectors, told to add them together, then sketch all three vectors. Our job now is to go back and try to see a relationship between the drawings of $\vec{x}$, $\vec{y}$ and $\vec{x}+\vec{y}$. Do you see any?

Here is one way of interpreting the adding of $\vec{x}$ to $\vec{y}$. Regardless of where we start, we draw $\vec{x}$. Now, from the tip of $\vec{x}$, draw $\vec{y}$. The vector $\vec{x}+\vec{y}$ is the vector found by drawing an arrow from the *origin* of $\vec{x}$ to the *tip* of $\vec{y}$. Likewise, we could start by drawing $\vec{y}$. Then, starting from the tip of $\vec{y}$, we can draw $\vec{x}$. Finally, draw $\vec{x}+\vec{y}$ by drawing the vector that starts at the origin of $\vec{y}$ and ends at the tip of $\vec{x}$.

The picture in Figure 2.4 illustrates this. The gray vectors demonstrate drawing the second vector from the tip of the first; we draw the vector $\vec{x}+\vec{y}$ dashed to set it apart from the rest. We also lightly filled the *parallelogram* whose opposing sides are the vectors $\vec{x}$ and $\vec{y}$. This highlights what is known as the *Parallelogram Law*.

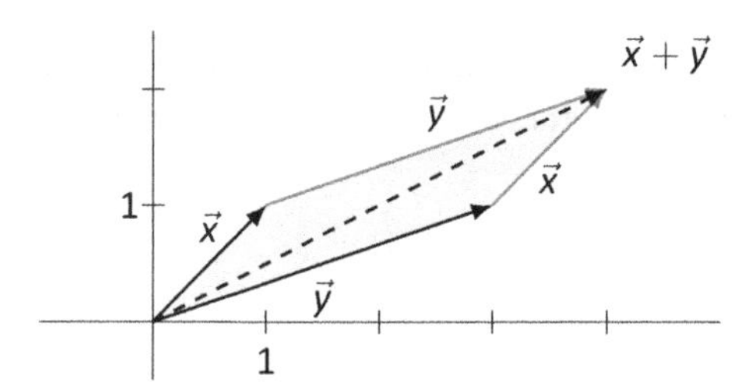

Figure 2.4: Adding vectors graphically using the Parallelogram Law

Key Idea 5

Parallelogram Law

To draw the vector $\vec{x}+\vec{y}$, one can draw the parallelogram with $\vec{x}$ and $\vec{y}$ as its sides. The vector that points from the vertex where $\vec{x}$ and $\vec{y}$ originate to the vertex where $\vec{x}$ and $\vec{y}$ meet is the vector $\vec{x}+\vec{y}$.

Knowing all of this allows us to draw the sum of two vectors without knowing specifically what the vectors are, as we demonstrate in the following example.

Example 33 Consider the vectors $\vec{x}$ and $\vec{y}$ as drawn in Figure 2.5. Sketch the vector $\vec{x}+\vec{y}$.

Solution

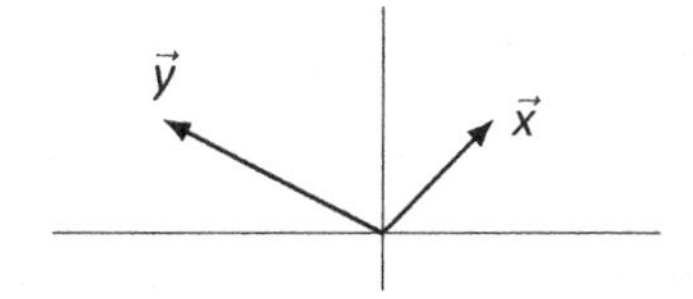

Figure 2.5: Vectors $\vec{x}$ and $\vec{y}$ in Example 33

We'll apply the Parallelogram Law, as given in Key Idea 5. As before, we draw $\vec{x}+\vec{y}$ dashed to set it apart. The result is given in Figure 2.6.

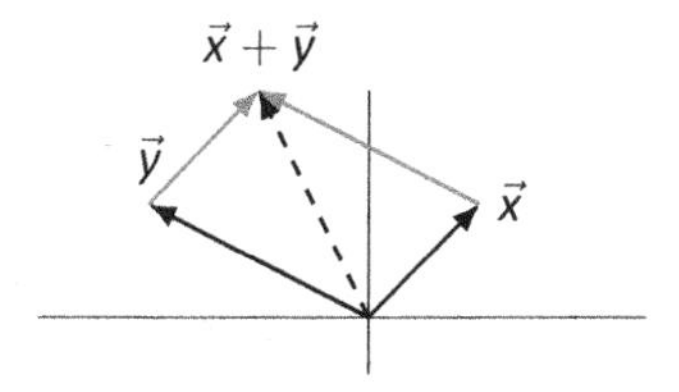

Figure 2.6: Vectors $\vec{x}$, $\vec{y}$ and $\vec{x}+\vec{y}$ in Example 33

Scalar Multiplication

After learning about matrix addition, we learned about scalar multiplication. We apply that concept now to vectors and see how this is represented graphically.

Example 34 Let

$$\vec{x} = \begin{bmatrix} 1 \\ 1 \end{bmatrix} \quad \text{and} \quad \vec{y} = \begin{bmatrix} -2 \\ 1 \end{bmatrix}.$$

Sketch $\vec{x}$, $\vec{y}$, $3\vec{x}$ and $-1\vec{y}$.

SOLUTION We begin by computing $3\vec{x}$ and $-\vec{y}$:

$$3\vec{x} = \begin{bmatrix} 3 \\ 3 \end{bmatrix} \quad \text{and} \quad -\vec{y} = \begin{bmatrix} 2 \\ -1 \end{bmatrix}.$$

All four vectors are sketched in Figure 2.7.

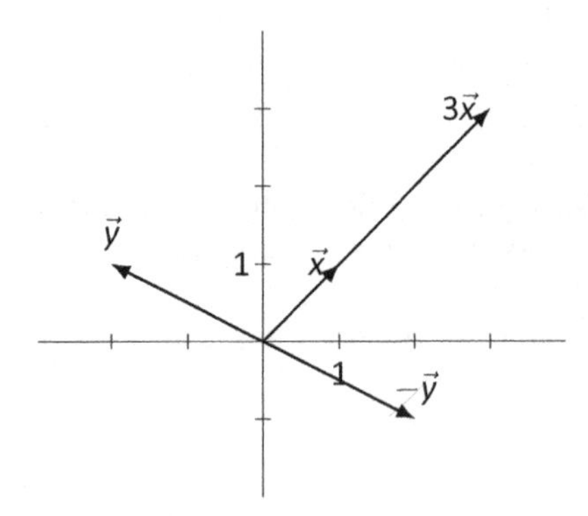

Figure 2.7: Vectors $\vec{x}$, $\vec{y}$, $3\vec{x}$ and $-\vec{y}$ in Example 34

As we often do, let us look at the previous example and see what we can learn from it. We can see that $\vec{x}$ and $3\vec{x}$ point in the same direction (they lie on the same line), but $3\vec{x}$ is just longer than $\vec{x}$. (In fact, it looks like $3\vec{x}$ is *3* times longer than $\vec{x}$. Is it? How do we measure length?)

We also see that $\vec{y}$ and $-\vec{y}$ seem to have the same length and lie on the same line, but point in the opposite direction.

A vector inherently conveys two pieces of information: length and direction. Multiplying a vector by a positive scalar c stretches the vectors by a factor of c; multiplying by a negative scalar c both stretches the vector and makes it point in the opposite direction.

Knowing this, we can sketch scalar multiples of vectors without knowing specifically what they are, as we do in the following example.

Example 35 Let vectors $\vec{x}$ and $\vec{y}$ be as in Figure 2.8. Draw $3\vec{x}$, $-2\vec{x}$, and $\frac{1}{2}\vec{y}$.

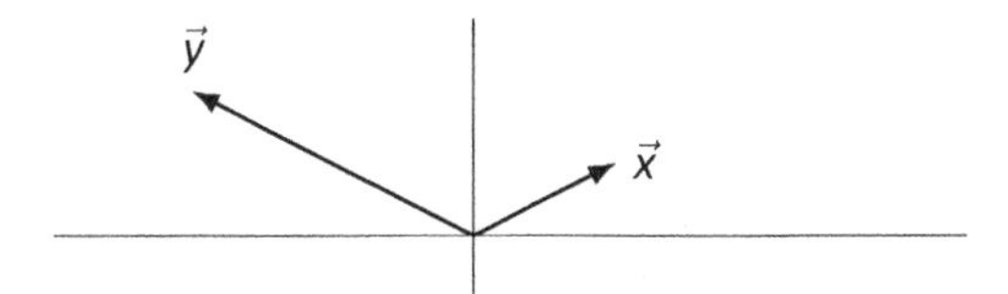

Figure 2.8: Vectors $\vec{x}$ and $\vec{y}$ in Example 35

SOLUTION To draw $3\vec{x}$, we draw a vector in the same direction as $\vec{x}$, but 3 times as long. To draw $-2\vec{x}$, we draw a vector twice as long as $\vec{x}$ in the opposite direction; to draw $\frac{1}{2}\vec{y}$, we draw a vector half the length of $\vec{y}$ in the same direction as $\vec{y}$. We again use the default of drawing all the vectors starting at the origin. All of this is shown in Figure 2.9.

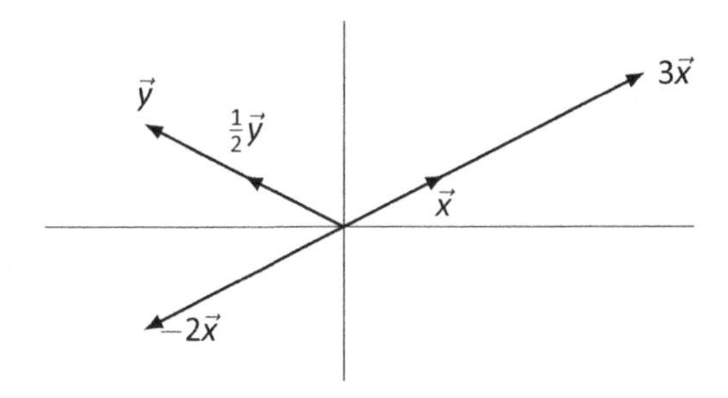

Figure 2.9: Vectors $\vec{x}$, $\vec{y}$, $3\vec{x}$, $-2x$ and $\frac{1}{2}\vec{x}$ in Example 35

Vector Subtraction

The final basic operation to consider between two vectors is that of vector subtraction: given vectors $\vec{x}$ and $\vec{y}$, how do we draw $\vec{x} - \vec{y}$?

If we know explicitly what $\vec{x}$ and $\vec{y}$ are, we can simply compute what $\vec{x} - \vec{y}$ is and then draw it. We can also think in terms of vector addition and scalar multiplication: we can *add* the vectors $\vec{x} + (-1)\vec{y}$. That is, we can draw $\vec{x}$ and draw $-\vec{y}$, then add them as we did in Example 33. This is especially useful we don't know explicitly what $\vec{x}$ and $\vec{y}$ are.

Example 36 Let vectors $\vec{x}$ and $\vec{y}$ be as in Figure 2.10. Draw $\vec{x} - \vec{y}$.

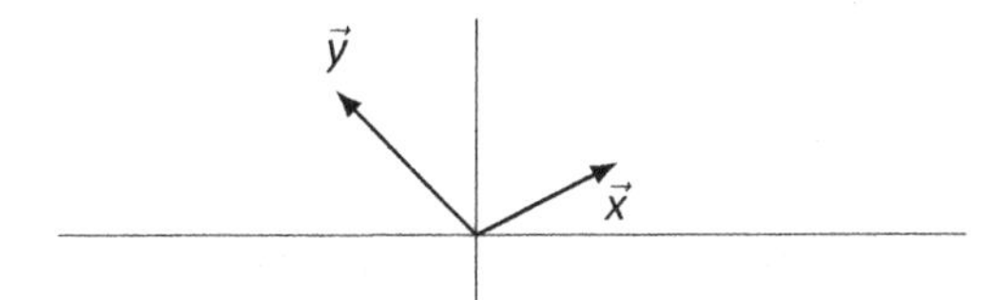

Figure 2.10: Vectors $\vec{x}$ and $\vec{y}$ in Example 36

Solution To draw $\vec{x} - \vec{y}$, we will first draw $-\vec{y}$ and then apply the Parallelogram Law to add $\vec{x}$ to $-\vec{y}$. See Figure 2.11.

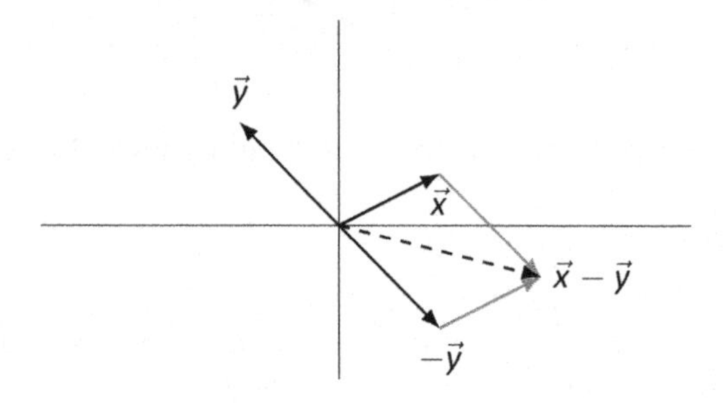

Figure 2.11: Vectors $\vec{x}$, $\vec{y}$ and $\vec{x} - \vec{y}$ in Example 36

In Figure 2.12, we redraw Figure 2.11 from Example 36 but remove the gray vectors that tend to add clutter, and we redraw the vector $\vec{x}-\vec{y}$ dotted so that it starts from the tip of $\vec{y}$.[9] Note that the dotted version of $\vec{x} - \vec{y}$ points from $\vec{y}$ to $\vec{x}$. This is a "shortcut" to drawing $\vec{x} - \vec{y}$; simply draw the vector that starts at the tip of $\vec{y}$ and ends at the tip of $\vec{x}$. This is important so we make it a Key Idea.

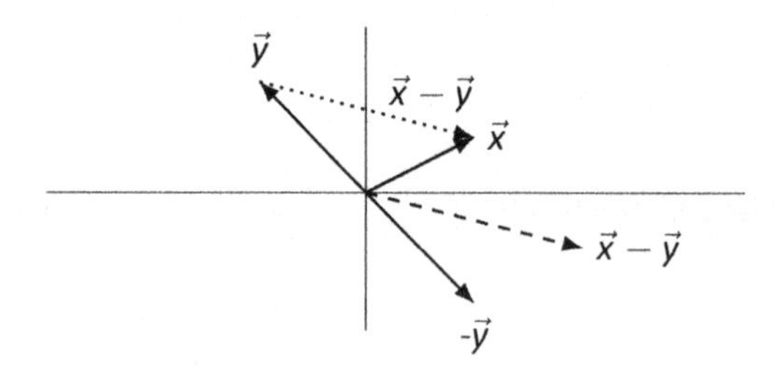

Figure 2.12: Redrawing vector $\vec{x} - \vec{y}$

Key Idea 6

Vector Subtraction

To draw the vector $\vec{x} - \vec{y}$, draw $\vec{x}$ and $\vec{y}$ so that they have the same origin. The vector $\vec{x} - \vec{y}$ is the vector that starts from the tip of $\vec{y}$ and points to the tip of $\vec{x}$.

Let's practice this once more with a quick example.

Example 37 Let $\vec{x}$ and $\vec{y}$ be as in Figure **??** (a). Draw $\vec{x} - \vec{y}$.

Solution We simply apply Key Idea 6: we draw an arrow from $\vec{y}$ to $\vec{x}$. We do so in Figure 2.13; $\vec{x} - \vec{y}$ is dashed.

[9]Remember that we can draw vectors starting from anywhere.

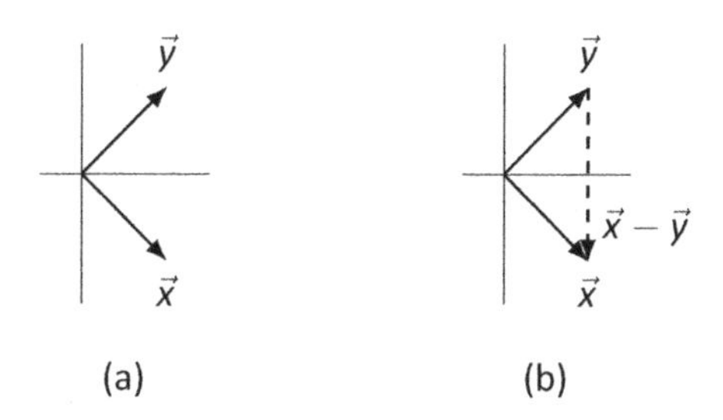

Figure 2.13: Vectors $\vec{x}$, $\vec{y}$ and $\vec{x} - \vec{y}$ in Example 37

Vector Length

When we discussed scalar multiplication, we made reference to a fundamental question: How do we measure the length of a vector? Basic geometry gives us an answer in the two dimensional case that we are dealing with right now, and later we can extend these ideas to higher dimensions.

Consider Figure 2.14. A vector $\vec{x}$ is drawn in black, and dashed and dotted lines have been drawn to make it the hypotenuse of a right triangle.

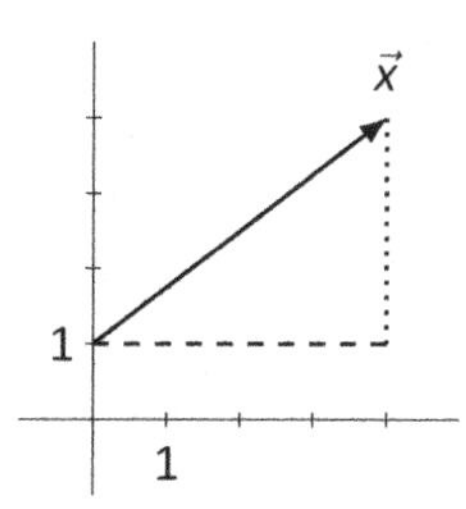

Figure 2.14: Measuring the length of a vector

It is easy to see that the dashed line has length 4 and the dotted line has length 3. We'll let c denote the length of $\vec{x}$; according to the Pythagorean Theorem, $4^2+3^2=c^2$. Thus $c^2=25$ and we quickly deduce that $c=5$.

Notice that in our figure, $\vec{x}$ goes to the right 4 units and then up 3 units. In other words, we can write

$$\vec{x}=\begin{bmatrix}4\\3\end{bmatrix}.$$

We learned above that the length of $\vec{x}$ is $\sqrt{4^2+3^2}$.[10] This hints at a basic calculation that works for all vectors $\vec{x}$, and we define the length of a vector according to this rule.

[10]Remember that $\sqrt{4^2+3^2}\neq 4+3$!

Definition 16 **Vector Length**

Let
$$\vec{x} = \begin{bmatrix} x_1 \\ x_2 \end{bmatrix}.$$
The *length* of $\vec{x}$, denoted $||\vec{x}||$, is
$$||\vec{x}|| = \sqrt{x_1^2 + x_2^2}.$$

Example 38 Find the length of each of the vectors given below.

$$\vec{x_1} = \begin{bmatrix} 1 \\ 1 \end{bmatrix} \quad \vec{x_2} = \begin{bmatrix} 2 \\ -3 \end{bmatrix} \quad \vec{x_3} = \begin{bmatrix} .6 \\ .8 \end{bmatrix} \quad \vec{x_4} = \begin{bmatrix} 3 \\ 0 \end{bmatrix}$$

Solution We apply Definition 16 to each vector.

$$||\vec{x_1}|| = \sqrt{1^2 + 1^2} = \sqrt{2}.$$
$$||\vec{x_2}|| = \sqrt{2^2 + (-3)^2} = \sqrt{13}.$$
$$||\vec{x_3}|| = \sqrt{.6^2 + .8^2} = \sqrt{.36 + .64} = 1.$$
$$||\vec{x_4}|| = \sqrt{3^2 + 0} = 3.$$

Now that we know how to compute the length of a vector, let's revisit a statement we made as we explored Examples 34 and 35: "Multiplying a vector by a positive scalar c stretches the vectors by a factor of c ..." At that time, we did not know how to measure the length of a vector, so our statement was unfounded. In the following example, we will confirm the truth of our previous statement.

Example 39 Let $\vec{x} = \begin{bmatrix} 2 \\ -1 \end{bmatrix}$. Compute $||\vec{x}||$, $||3\vec{x}||$, $||-2\vec{x}||$, and $||c\vec{x}||$, where c is a scalar.

Solution We apply Definition 16 to each of the vectors.

$||\vec{x}|| = \sqrt{4+1} = \sqrt{5}$.

Before computing the length of $||3\vec{x}||$, we note that $3\vec{x} = \begin{bmatrix} 6 \\ -3 \end{bmatrix}$.

$||3\vec{x}|| = \sqrt{36+9} = \sqrt{45} = 3\sqrt{5} = 3||\vec{x}||$.

Before computing the length of $||-2\vec{x}||$, we note that $-2\vec{x} = \begin{bmatrix} -4 \\ 2 \end{bmatrix}$.

$||-2\vec{x}|| = \sqrt{16+4} = \sqrt{20} = 2\sqrt{5} = 2||\vec{x}||$.

Finally, to compute $||c\vec{x}||$, we note that $c\vec{x} = \begin{bmatrix} 2c \\ -c \end{bmatrix}$. Thus:

$||c\vec{x}|| = \sqrt{(2c)^2 + (-c)^2} = \sqrt{4c^2 + c^2} = \sqrt{5c^2} = |c|\sqrt{5}$.

This last line is true because the square root of any number squared is the *absolute value* of that number (for example, $\sqrt{(-3)^2} = 3$).

The last computation of our example is the most important one. It shows that, in general, multiplying a vector $\vec{x}$ by a scalar c stretches $\vec{x}$ by a factor of $|c|$ (and the direction will change if c is negative). This is important so we'll make it a Theorem.

Theorem 4

Vector Length and Scalar Multiplication

Let $\vec{x}$ be a vector and let c be a scalar. Then the length of $c\vec{x}$ is

$$||c\vec{x}|| = |c| \cdot ||\vec{x}||.$$

Matrix – Vector Multiplication

The last arithmetic operation to consider visualizing is matrix multiplication. Specifically, we want to visualize the result of multiplying a vector by a matrix. In order to multiply a 2D vector by a matrix and get a 2D vector back, our matrix must be a square, 2×2 matrix.[11]

We'll start with an example. Given a matrix A and several vectors, we'll graph the vectors before and after they've been multiplied by A and see what we learn.

Example 40 Let A be a matrix, and $\vec{x}$, $\vec{y}$, and $\vec{z}$ be vectors as given below.

$$A = \begin{bmatrix} 1 & 4 \\ 2 & 3 \end{bmatrix}, \quad \vec{x} = \begin{bmatrix} 1 \\ 1 \end{bmatrix}, \quad \vec{y} = \begin{bmatrix} -1 \\ 1 \end{bmatrix}, \quad \vec{z} = \begin{bmatrix} 3 \\ -1 \end{bmatrix}$$

Graph $\vec{x}$, $\vec{y}$ and $\vec{z}$, as well as $A\vec{x}$, $A\vec{y}$ and $A\vec{z}$.

SOLUTION

[11] We can multiply a 3×2 matrix by a 2D vector and get a 3D vector back, and this gives very interesting results. See section 5.2.

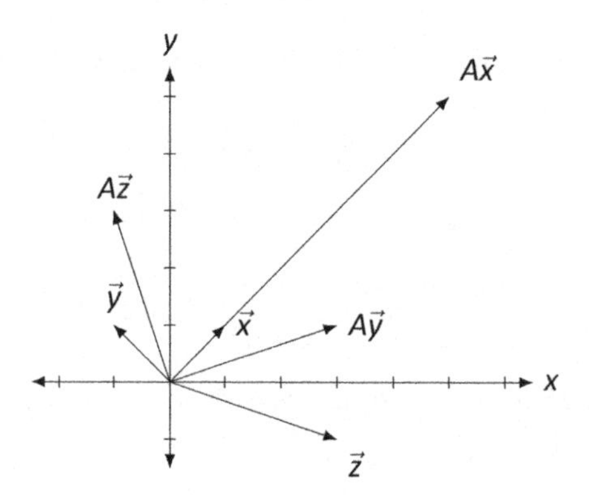

Figure 2.15: Multiplying vectors by a matrix in Example 40.

It is straightforward to compute:

$$A\vec{x} = \begin{bmatrix} 5 \\ 5 \end{bmatrix}, \quad A\vec{y} = \begin{bmatrix} 3 \\ 1 \end{bmatrix}, \quad \text{and} \quad A\vec{z} = \begin{bmatrix} -1 \\ 3 \end{bmatrix}.$$

The vectors are sketched in Figure 2.15

There are several things to notice. When each vector is multiplied by A, the result is a vector with a different length (in this example, always longer), and in two of the cases (for $\vec{y}$ and $\vec{z}$), the resulting vector points in a different direction.

This isn't surprising. In the previous section we learned about matrix multiplication, which is a strange and seemingly unpredictable operation. Would you expect to see some sort of immediately recognizable pattern appear from multiplying a matrix and a vector?[12] In fact, the surprising thing from the example is that $\vec{x}$ and $A\vec{x}$ point in the same direction! Why does the direction of $\vec{x}$ not change after multiplication by A? (We'll answer this in Section 4.1 when we learn about something called "eigenvectors.")

Different matrices act on vectors in different ways.[13] Some always increase the length of a vector through multiplication, others always decrease the length, others increase the length of some vectors and decrease the length of others, and others still don't change the length at all. A similar statement can be made about how matrices affect the direction of vectors through multiplication: some change every vector's direction, some change "most" vector's direction but leave some the same, and others still don't change the direction of any vector.

How do we set about studying how matrix multiplication affects vectors? We could just create lots of different matrices and lots of different vectors, multiply, then graph, but this would be a lot of work with very little useful result. It would be too hard to find a pattern of behavior in this.[14]

[12] This is a rhetorical question; the expected answer is "No."

[13] That's one reason we call them "different."

[14] Remember, that's what mathematicians do. We look for patterns.

Instead, we'll begin by using a technique we've employed often in the past. We have a "new" operation; let's explore how it behaves with "old" operations. Specifically, we know how to sketch vector addition. What happens when we throw matrix multiplication into the mix? Let's try an example.

Example 41 Let A be a matrix and $\vec{x}$ and $\vec{y}$ be vectors as given below.

$$A = \begin{bmatrix} 1 & 1 \\ 1 & 2 \end{bmatrix}, \quad \vec{x} = \begin{bmatrix} 2 \\ 1 \end{bmatrix}, \quad \vec{y} = \begin{bmatrix} -1 \\ 1 \end{bmatrix}$$

Sketch $\vec{x} + \vec{y}$, $A\vec{x}$, $A\vec{y}$, and $A(\vec{x} + \vec{y})$.

SOLUTION It is pretty straightforward to compute:

$$\vec{x} + \vec{y} = \begin{bmatrix} 1 \\ 2 \end{bmatrix}; \quad A\vec{x} = \begin{bmatrix} 3 \\ 4 \end{bmatrix}; \quad A\vec{y} = \begin{bmatrix} 0 \\ 1 \end{bmatrix}, \quad A(\vec{x} + \vec{y}) = \begin{bmatrix} 3 \\ 5 \end{bmatrix}.$$

In Figure 2.16, we have graphed the above vectors and have included dashed gray vectors to highlight the additive nature of $\vec{x} + \vec{y}$ and $A(\vec{x} + \vec{y})$. Does anything strike you as interesting?

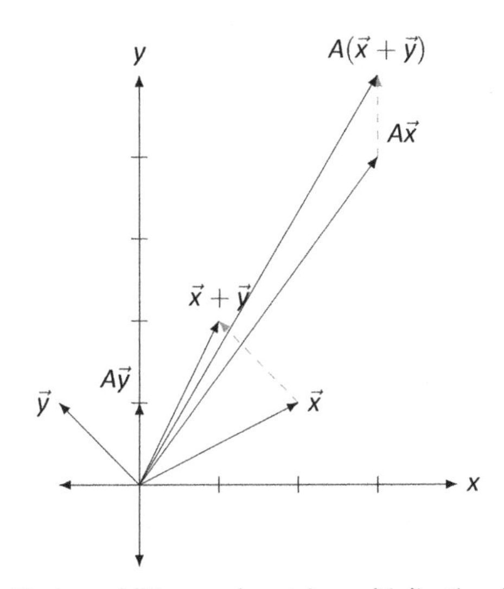

Figure 2.16: Vector addition and matrix multiplication in Example 41.

Let's not focus on things which don't matter right now: let's not focus on how long certain vectors became, nor necessarily how their direction changed. Rather, think about how matrix multiplication interacted with the vector addition.

In some sense, we started with three vectors, $\vec{x}$, $\vec{y}$, and $\vec{x} + \vec{y}$. This last vector is special; it is the sum of the previous two. Now, multiply all three by A. What happens? We get three new vectors, but the significant thing is this: the last vector is still the sum of the previous two! (We emphasize this by drawing dotted vectors to represent part of the Parallelogram Law.)

Of course, we knew this already: we already knew that $A\vec{x} + A\vec{y} = A(\vec{x} + \vec{y})$, for this is just the Distributive Property. However, now we get to see this graphically.

In Section 5.1 we'll study in greater depth how matrix multiplication affects vectors and the whole Cartesian plane. For now, we'll settle for simple practice: given a matrix and some vectors, we'll multiply and graph. Let's do one more example.

Example 42 Let A, $\vec{x}$, $\vec{y}$, and $\vec{z}$ be as given below.

$$A = \begin{bmatrix} 1 & -1 \\ 1 & -1 \end{bmatrix}, \quad \vec{x} = \begin{bmatrix} 1 \\ 1 \end{bmatrix}, \quad \vec{y} = \begin{bmatrix} -1 \\ 1 \end{bmatrix}, \quad \vec{z} = \begin{bmatrix} 4 \\ 1 \end{bmatrix}$$

Graph $\vec{x}$, $\vec{y}$ and $\vec{z}$, as well as $A\vec{x}$, $A\vec{y}$ and $A\vec{z}$.

SOLUTION

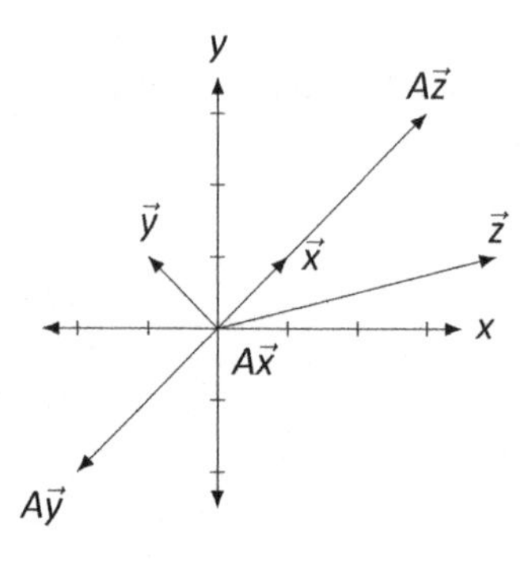

Figure 2.17: Multiplying vectors by a matrix in Example 42.

It is straightforward to compute:

$$A\vec{x} = \begin{bmatrix} 0 \\ 0 \end{bmatrix}, \quad A\vec{y} = \begin{bmatrix} -2 \\ -2 \end{bmatrix}, \quad \text{and} \quad A\vec{z} = \begin{bmatrix} 3 \\ 3 \end{bmatrix}.$$

The vectors are sketched in Figure 2.17.

These results are interesting. While we won't explore them in great detail here, notice how $\vec{x}$ got sent to the zero vector. Notice also that $A\vec{x}$, $A\vec{y}$ and $A\vec{z}$ are all in a line (as well as $\vec{x}$!). Why is that? Are $\vec{x}$, $\vec{y}$ and $\vec{z}$ just special vectors, or would any other vector get sent to the same line when multiplied by A?[15]

This section has focused on vectors in two dimensions. Later on in this book, we'll extend these ideas into three dimensions (3D).

In the next section we'll take a new idea (matrix multiplication) and apply it to an old idea (solving systems of linear equations). This will allow us to view an old idea in a new way – and we'll even get to "visualize" it.

[15] Don't just sit there, try it out!

Exercises 2.3

In Exercises 1 – 4, vectors $\vec{x}$ and $\vec{y}$ are given. Sketch $\vec{x}$, $\vec{y}$, $\vec{x}+\vec{y}$, and $\vec{x}-\vec{y}$ on the same Cartesian axes.

1. $\vec{x} = \begin{bmatrix} 1 \\ 1 \end{bmatrix}, \vec{y} = \begin{bmatrix} -2 \\ 3 \end{bmatrix}$

2. $\vec{x} = \begin{bmatrix} 3 \\ 1 \end{bmatrix}, \vec{y} = \begin{bmatrix} 1 \\ -2 \end{bmatrix}$

3. $\vec{x} = \begin{bmatrix} -1 \\ 1 \end{bmatrix}, \vec{y} = \begin{bmatrix} -2 \\ 2 \end{bmatrix}$

4. $\vec{x} = \begin{bmatrix} 2 \\ 0 \end{bmatrix}, \vec{y} = \begin{bmatrix} 1 \\ 3 \end{bmatrix}$

In Exercises 5 – 8, vectors $\vec{x}$ and $\vec{y}$ are drawn. Sketch $2\vec{x}$, $-\vec{y}$, $\vec{x}+\vec{y}$, and $\vec{x}-\vec{y}$ on the same Cartesian axes.

5.
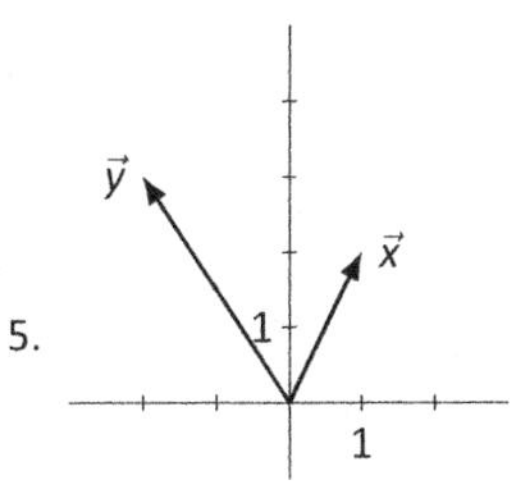

6.
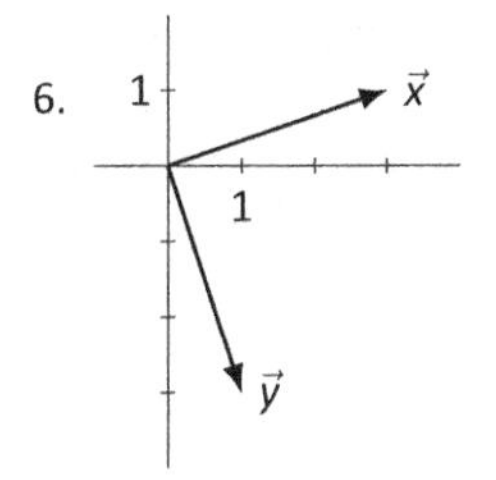

7.
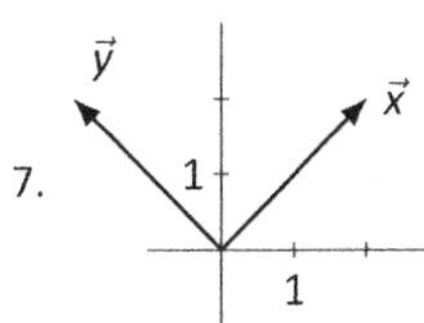

8.
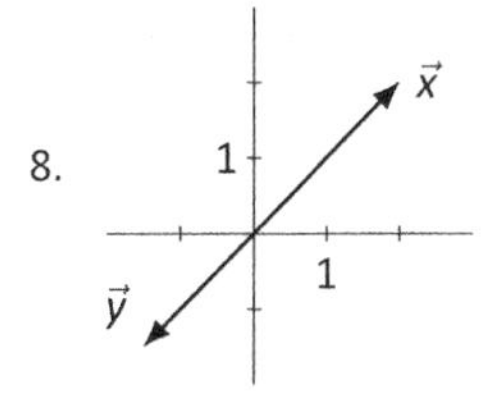

In Exercises 9 – 12, a vector $\vec{x}$ and a scalar a are given. Using Definition 16, compute the lengths of $\vec{x}$ and $a\vec{x}$, then compare these lengths.

9. $\vec{x} = \begin{bmatrix} 2 \\ 1 \end{bmatrix}, a = 3.$

10. $\vec{x} = \begin{bmatrix} 4 \\ 7 \end{bmatrix}, a = -2.$

11. $\vec{x} = \begin{bmatrix} -3 \\ 5 \end{bmatrix}, a = -1.$

12. $\vec{x} = \begin{bmatrix} 3 \\ -9 \end{bmatrix}, a = \frac{1}{3}.$

13. Four pairs of vectors $\vec{x}$ and $\vec{y}$ are given below. For each pair, compute $||\vec{x}||$, $||\vec{y}||$, and $||\vec{x}+\vec{y}||$. Use this information to answer: Is it always, sometimes, or never true that $||\vec{x}|| + ||\vec{y}|| = ||\vec{x}+\vec{y}||$? If it always or never true, explain why. If it is sometimes true, explain when it is true.

    (a) $\vec{x} = \begin{bmatrix} 1 \\ 1 \end{bmatrix}, \vec{y} = \begin{bmatrix} 2 \\ 3 \end{bmatrix}$

    (b) $\vec{x} = \begin{bmatrix} 1 \\ -2 \end{bmatrix}, \vec{y} = \begin{bmatrix} 3 \\ -6 \end{bmatrix}$

    (c) $\vec{x} = \begin{bmatrix} -1 \\ 3 \end{bmatrix}, \vec{y} = \begin{bmatrix} 2 \\ 5 \end{bmatrix}$

    (d) $\vec{x} = \begin{bmatrix} 2 \\ 1 \end{bmatrix}, \vec{y} = \begin{bmatrix} -4 \\ -2 \end{bmatrix}$

In Exercises 14 – 17, a matrix A is given. Sketch $\vec{x}$, $\vec{y}$, $A\vec{x}$ and $A\vec{y}$ on the same Cartesian axes, where

$$\vec{x} = \begin{bmatrix} 1 \\ 1 \end{bmatrix} \text{ and } \vec{y} = \begin{bmatrix} -1 \\ 2 \end{bmatrix}.$$

14. $A = \begin{bmatrix} 1 & -1 \\ 2 & 3 \end{bmatrix}$

15. $A = \begin{bmatrix} 2 & 0 \\ -1 & 3 \end{bmatrix}$

16. $A = \begin{bmatrix} 1 & 1 \\ 1 & 1 \end{bmatrix}$

17. $A = \begin{bmatrix} 1 & 2 \\ -1 & -2 \end{bmatrix}$

2.4 Vector Solutions to Linear Systems

AS YOU READ . . .

1. T/F: The equation $A\vec{x} = \vec{b}$ is just another way of writing a system of linear equations.

2. T/F: In solving $A\vec{x} = \vec{0}$, if there are 3 free variables, then the solution will be "pulled apart" into 3 vectors.

3. T/F: A homogeneous system of linear equations is one in which all of the coefficients are 0.

4. Whether or not the equation $A\vec{x} = \vec{b}$ has a solution depends on an intrinsic property of ________.

The first chapter of this text was spent finding solutions to systems of linear equations. We have spent the first two sections of this chapter learning operations that can be performed with matrices. One may have wondered "Are the ideas of the first chapter related to what we have been doing recently?" The answer is yes, these ideas are related. This section begins to show that relationship.

We have often hearkened back to previous algebra experience to help understand matrix algebra concepts. We do that again here. Consider the equation $ax = b$, where $a = 3$ and $b = 6$. If we asked one to "solve for x," what exactly would we be asking? We would want to find a number, which we call x, where a times x gives b; in this case, it is a number, when multiplied by 3, returns 6.

Now we consider matrix algebra expressions. We'll eventually consider solving equations like $AX = B$, where we know what the matrices A and B are and we want to find the matrix X. For now, we'll only consider equations of the type $A\vec{x} = \vec{b}$, where we know the matrix A and the vector $\vec{b}$. We will want to find what vector $\vec{x}$ satisfies this equation; we want to "solve for $\vec{x}$."

To help understand what this is asking, we'll consider an example. Let

$$A = \begin{bmatrix} 1 & 1 & 1 \\ 1 & -1 & 2 \\ 2 & 0 & 1 \end{bmatrix}, \quad \vec{b} = \begin{bmatrix} 2 \\ -3 \\ 1 \end{bmatrix} \quad \text{and} \quad \vec{x} = \begin{bmatrix} x_1 \\ x_2 \\ x_3 \end{bmatrix}.$$

(We don't know what $\vec{x}$ is, so we have to represent it's entries with the variables x_1, x_2 and x_3.) Let's "solve for $\vec{x}$," given the equation $A\vec{x} = \vec{b}$.

We can multiply out the left hand side of this equation. We find that

$$A\vec{x} = \begin{bmatrix} x_1 + x_2 + x_3 \\ x_1 - x_2 + 2x_3 \\ 2x_1 + x_3 \end{bmatrix}.$$

Be sure to note that the product is just a vector; it has just one column.

Since $A\vec{x}$ is equal to $\vec{b}$, we have

$$\begin{bmatrix} x_1 + x_2 + x_3 \\ x_1 - x_2 + 2x_3 \\ 2x_1 + x_3 \end{bmatrix} = \begin{bmatrix} 2 \\ -3 \\ 1 \end{bmatrix}.$$

Knowing that two vectors are equal only when their corresponding entries are equal, we know

$$\begin{aligned} x_1 + x_2 + x_3 &= 2 \\ x_1 - x_2 + 2x_3 &= -3 \\ 2x_1 + x_3 &= 1. \end{aligned}$$

This should look familiar; it is a system of linear equations! Given the matrix-vector equation $A\vec{x} = \vec{b}$, we can recognize A as the coefficient matrix from a linear system and $\vec{b}$ as the vector of the constants from the linear system. To solve a matrix–vector equation (and the corresponding linear system), we simply augment the matrix A with the vector $\vec{b}$, put this matrix into reduced row echelon form, and interpret the results.

We convert the above linear system into an augmented matrix and find the reduced row echelon form:

$$\begin{bmatrix} 1 & 1 & 1 & 2 \\ 1 & -1 & 2 & -3 \\ 2 & 0 & 1 & 1 \end{bmatrix} \quad \overrightarrow{\text{rref}} \quad \begin{bmatrix} 1 & 0 & 0 & 1 \\ 0 & 1 & 0 & 2 \\ 0 & 0 & 1 & -1 \end{bmatrix}.$$

This tells us that $x_1 = 1$, $x_2 = 2$ and $x_3 = -1$, so

$$\vec{x} = \begin{bmatrix} 1 \\ 2 \\ -1 \end{bmatrix}.$$

We should check our work; multiply out $A\vec{x}$ and verify that we indeed get $\vec{b}$:

$$\begin{bmatrix} 1 & 1 & 1 \\ 1 & -1 & 2 \\ 2 & 0 & 1 \end{bmatrix} \begin{bmatrix} 1 \\ 2 \\ -1 \end{bmatrix} \quad \text{does equal} \quad \begin{bmatrix} 2 \\ -3 \\ 1 \end{bmatrix}.$$

We should practice.

Example 43 Solve the equation $A\vec{x} = \vec{b}$ for $\vec{x}$ where

$$A = \begin{bmatrix} 1 & 2 & 3 \\ -1 & 2 & 1 \\ 1 & 1 & 0 \end{bmatrix} \quad \text{and} \quad \begin{bmatrix} 5 \\ -1 \\ 2 \end{bmatrix}.$$

Solution The solution is rather straightforward, even though we did a lot of work before to find the answer. Form the augmented matrix $\begin{bmatrix} A & \vec{b} \end{bmatrix}$ and interpret its reduced row echelon form.

$$\begin{bmatrix} 1 & 2 & 3 & 5 \\ -1 & 2 & 1 & -1 \\ 1 & 1 & 0 & 2 \end{bmatrix} \quad \overrightarrow{\text{rref}} \quad \begin{bmatrix} 1 & 0 & 0 & 2 \\ 0 & 1 & 0 & 0 \\ 0 & 0 & 1 & 1 \end{bmatrix}$$

In previous sections we were fine stating that the result as

$$x_1 = 2, \quad x_2 = 0, \quad x_3 = 1,$$

but we were asked to find $\vec{x}$; therefore, we state the solution as

$$\vec{x} = \begin{bmatrix} 2 \\ 0 \\ 1 \end{bmatrix}.$$

This probably seems all well and good. While asking one to solve the equation $A\vec{x} = \vec{b}$ for $\vec{x}$ seems like a new problem, in reality it is just asking that we solve a system of linear equations. Our variables x_1, etc., appear not individually but as the entries of our vector $\vec{x}$. We are simply writing an old problem in a new way.

In line with this new way of writing the problem, we have a new way of writing the solution. Instead of listing, individually, the values of the unknowns, we simply list them as the elements of our vector $\vec{x}$.

These are important ideas, so we state the basic principle once more: solving the equation $A\vec{x} = \vec{b}$ for $\vec{x}$ is the same thing as solving a linear system of equations. Equivalently, any system of linear equations can be written in the form $A\vec{x} = \vec{b}$ for some matrix A and vector $\vec{b}$.

Since these ideas are equivalent, we'll refer to $A\vec{x} = \vec{b}$ both as a matrix–vector equation and as a system of linear equations: they are the same thing.

We've seen two examples illustrating this idea so far, and in both cases the linear system had exactly one solution. We know from Theorem 1 that any linear system has either one solution, infinite solutions, or no solution. So how does our new method of writing a solution work with infinite solutions and no solutions?

Certainly, if $A\vec{x} = \vec{b}$ has no solution, we simply say that the linear system has no solution. There isn't anything special to write. So the only other option to consider is the case where we have infinite solutions. We'll learn how to handle these situations through examples.

Example 44 Solve the linear system $A\vec{x} = \vec{0}$ for $\vec{x}$ and write the solution in vector form, where

$$A = \begin{bmatrix} 1 & 2 \\ 2 & 4 \end{bmatrix} \quad \text{and} \quad \vec{0} = \begin{bmatrix} 0 \\ 0 \end{bmatrix}.$$

SOLUTION (Note: we didn't really need to specify that

$$\vec{0} = \begin{bmatrix} 0 \\ 0 \end{bmatrix},$$

but we did just to eliminate any uncertainty.)

To solve this system, put the augmented matrix into reduced row echelon form, which we do below.

$$\begin{bmatrix} 1 & 2 & 0 \\ 2 & 4 & 0 \end{bmatrix} \xrightarrow{\text{rref}} \begin{bmatrix} 1 & 2 & 0 \\ 0 & 0 & 0 \end{bmatrix}$$

We interpret the reduced row echelon form of this matrix to write the solution as

$$x_1 = -2x_2$$
$$x_2 \text{ is free.}$$

We are not done; we need to write the solution in vector form, for our solution is the vector $\vec{x}$. Recall that

$$\vec{x} = \begin{bmatrix} x_1 \\ x_2 \end{bmatrix}.$$

From above we know that $x_1 = -2x_2$, so we replace the x_1 in $\vec{x}$ with $-2x_2$. This gives our solution as

$$\vec{x} = \begin{bmatrix} -2x_2 \\ x_2 \end{bmatrix}.$$

Now we pull the x_2 out of the vector (it is just a scalar) and write $\vec{x}$ as

$$\vec{x} = x_2 \begin{bmatrix} -2 \\ 1 \end{bmatrix}.$$

For reasons that will become more clear later, set

$$\vec{v} = \begin{bmatrix} -2 \\ 1 \end{bmatrix}.$$

Thus our solution can be written as

$$\vec{x} = x_2\vec{v}.$$

Recall that since our system was consistent and had a free variable, we have infinite solutions. This form of the solution highlights this fact; pick any value for x_2 and we get a different solution.

For instance, by setting $x_2 = -1$, 0, and 5, we get the solutions

$$\vec{x} = \begin{bmatrix} 2 \\ -1 \end{bmatrix}, \quad \begin{bmatrix} 0 \\ 0 \end{bmatrix}, \quad \text{and} \quad \begin{bmatrix} -10 \\ 5 \end{bmatrix},$$

respectively.

We should check our work; multiply each of the above vectors by A to see if we indeed get $\vec{0}$.

We have officially solved this problem; we have found the solution to $A\vec{x} = \vec{0}$ and written it properly. One final thing we will do here is *graph* the solution, using our skills learned in the previous section.

Our solution is

$$\vec{x} = x_2 \begin{bmatrix} -2 \\ 1 \end{bmatrix}.$$

This means that any scalar multiply of the vector $\vec{v} = \begin{bmatrix} -2 \\ 1 \end{bmatrix}$ is a solution; we know how to sketch the scalar multiples of $\vec{v}$. This is done in Figure 2.18.

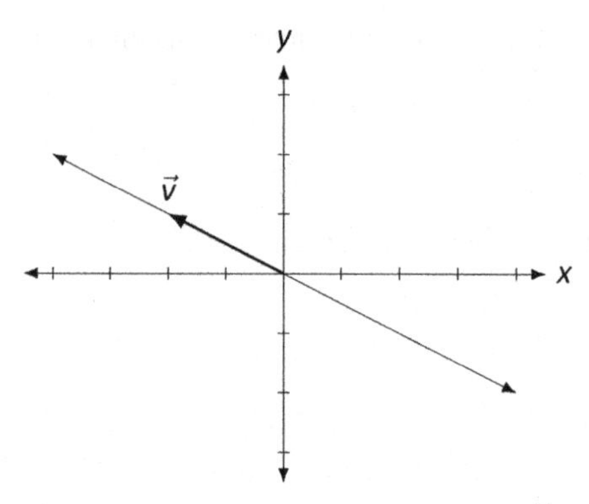

Figure 2.18: The solution, as a line, to $A\vec{x} = \vec{0}$ in Example 44.

Here vector $\vec{v}$ is drawn as well as the line that goes through the origin in the direction of $\vec{v}$. Any vector along this line is a solution. So in some sense, we can say that the solution to $A\vec{x} = \vec{0}$ is *a line*.

Let's practice this again.

Example 45 Solve the linear system $A\vec{x} = \vec{0}$ and write the solution in vector form, where

$$A = \begin{bmatrix} 2 & -3 \\ -2 & 3 \end{bmatrix}.$$

Solution Again, to solve this problem, we form the proper augmented matrix and we put it into reduced row echelon form, which we do below.

$$\begin{bmatrix} 2 & -3 & 0 \\ -2 & 3 & 0 \end{bmatrix} \quad \overrightarrow{\text{rref}} \quad \begin{bmatrix} 1 & -3/2 & 0 \\ 0 & 0 & 0 \end{bmatrix}$$

We interpret the reduced row echelon form of this matrix to find that

$$\begin{aligned} x_1 &= 3/2x_2 \\ x_2 &\text{ is free.} \end{aligned}$$

As before,

$$\vec{x} = \begin{bmatrix} x_1 \\ x_2 \end{bmatrix}.$$

Since $x_1 = 3/2x_2$, we replace x_1 in $\vec{x}$ with $3/2x_2$:

$$\vec{x} = \begin{bmatrix} 3/2x_2 \\ x_2 \end{bmatrix}.$$

Now we pull out the x_2 and write the solution as

$$\vec{x} = x_2 \begin{bmatrix} 3/2 \\ 1 \end{bmatrix}.$$

As before, let's set

$$\vec{v} = \begin{bmatrix} 3/2 \\ 1 \end{bmatrix}$$

so we can write our solution as

$$\vec{x} = x_2\vec{v}.$$

Again, we have infinite solutions; any choice of x_2 gives us one of these solutions. For instance, picking $x_2 = 2$ gives the solution

$$\vec{x} = \begin{bmatrix} 3 \\ 2 \end{bmatrix}.$$

(This is a particularly nice solution, since there are no fractions. . .)

As in the previous example, our solutions are multiples of a vector, and hence we can graph this, as done in Figure 2.19.

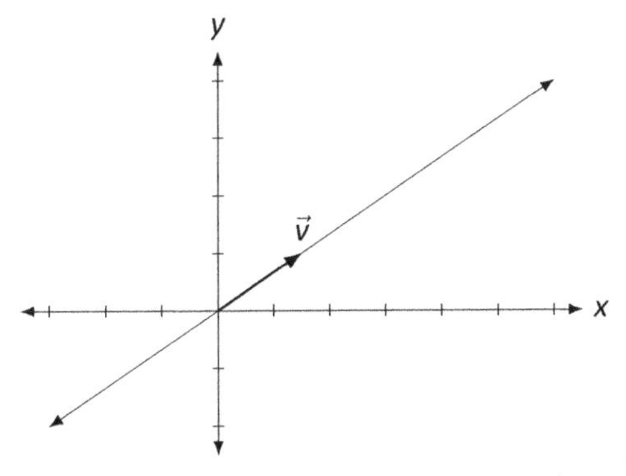

Figure 2.19: The solution, as a line, to $A\vec{x} = \vec{0}$ in Example 45.

Let's practice some more; this time, we won't solve a system of the form $A\vec{x} = \vec{0}$, but instead $A\vec{x} = \vec{b}$, for some vector $\vec{b}$.

Example 46 Solve the linear system $A\vec{x} = \vec{b}$, where

$$A = \begin{bmatrix} 1 & 2 \\ 2 & 4 \end{bmatrix} \quad \text{and} \quad \vec{b} = \begin{bmatrix} 3 \\ 6 \end{bmatrix}.$$

Solution (Note that this is the same matrix A that we used in Example 44. This will be important later.)

Our methodology is the same as before; we form the augmented matrix and put it into reduced row echelon form.

$$\begin{bmatrix} 1 & 2 & 3 \\ 2 & 4 & 6 \end{bmatrix} \quad \overrightarrow{\text{rref}} \quad \begin{bmatrix} 1 & 2 & 3 \\ 0 & 0 & 0 \end{bmatrix}$$

Interpreting this reduced row echelon form, we find that

$$x_1 = 3 - 2x_2$$
$$x_2 \text{ is free.}$$

Again,

$$\vec{x} = \begin{bmatrix} x_1 \\ x_2 \end{bmatrix},$$

and we replace x_1 with $3 - 2x_2$, giving

$$\vec{x} = \begin{bmatrix} 3 - 2x_2 \\ x_2 \end{bmatrix}.$$

This solution is different than what we've seen in the past two examples; we can't simply pull out a x_2 since there is a 3 in the first entry. Using the properties of matrix addition, we can "pull apart" this vector and write it as the sum of two vectors: one which contains only constants, and one that contains only "x_2 stuff." We do this below.

$$\begin{aligned} \vec{x} &= \begin{bmatrix} 3 - 2x_2 \\ x_2 \end{bmatrix} \\ &= \begin{bmatrix} 3 \\ 0 \end{bmatrix} + \begin{bmatrix} -2x_2 \\ x_2 \end{bmatrix} \\ &= \begin{bmatrix} 3 \\ 0 \end{bmatrix} + x_2 \begin{bmatrix} -2 \\ 1 \end{bmatrix}. \end{aligned}$$

Once again, let's give names to the different component vectors of this solution (we are getting near the explanation of why we are doing this). Let

$$\vec{x_p} = \begin{bmatrix} 3 \\ 0 \end{bmatrix} \quad \text{and} \quad \vec{v} = \begin{bmatrix} -2 \\ 1 \end{bmatrix}.$$

We can then write our solution in the form

$$\vec{x} = \vec{x_p} + x_2\vec{v}.$$

We still have infinite solutions; by picking a value for x_2 we get one of these solutions. For instance, by letting $x_2 = -1$, 0, or 2, we get the solutions

$$\begin{bmatrix} 5 \\ -1 \end{bmatrix}, \quad \begin{bmatrix} 3 \\ 0 \end{bmatrix} \quad \text{and} \quad \begin{bmatrix} -1 \\ 2 \end{bmatrix}.$$

We have officially solved the problem; we have solved the equation $A\vec{x} = \vec{b}$ for $\vec{x}$ and have written the solution in vector form. As an additional visual aid, we will graph this solution.

Each vector in the solution can be written as the sum of two vectors: $\vec{x_p}$ and a multiple of $\vec{v}$. In Figure 2.20, $\vec{x_p}$ is graphed and $\vec{v}$ is graphed with its origin starting at the tip of $\vec{x_p}$. Finally, a line is drawn in the direction of $\vec{v}$ from the tip of $\vec{x_p}$; any vector pointing to any point on this line is a solution to $A\vec{x} = \vec{b}$.

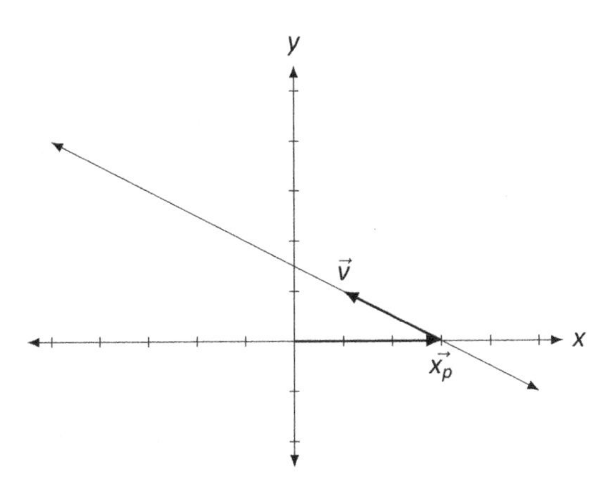

Figure 2.20: The solution, as a line, to $A\vec{x} = \vec{b}$ in Example 46.

The previous examples illustrate some important concepts. One is that we can "see" the solution to a system of linear equations in a new way. Before, when we had infinite solutions, we knew we could arbitrarily pick values for our free variables and get different solutions. We knew this to be true, and we even practiced it, but the result was not very "tangible." Now, we can view our solution as a vector; by picking different values for our free variables, we see this as multiplying certain important vectors by a scalar which gives a different solution.

Another important concept that these examples demonstrate comes from the fact that Examples 44 and 46 were only "slightly different" and hence had only "slightly different" answers. Both solutions had

$$x_2 \begin{bmatrix} -2 \\ 1 \end{bmatrix}$$

in them; in Example 46 the solution also had another vector added to this. Was this coincidence, or is there a definite pattern here?

Of course there is a pattern! Now . . . what exactly is it? First, we define a term.

Definition 17 **Homogeneous Linear System of Equations**

A system of linear equations is *homogeneous* if the constants in each equation are zero.

Note: a homogeneous system of equations can be written in vector form as $A\vec{x} = \vec{0}$.

The term *homogeneous* comes from two Greek words; *homo* meaning "same" and *genus* meaning "type." A homogeneous system of equations is a system in which each equation is of the same type – all constants are 0. Notice that the system of equations in Examples 44 and 46 are homogeneous.

Note that $A\vec{0} = \vec{0}$; that is, if we set $\vec{x} = \vec{0}$, we have a solution to a homogeneous set of equations. This fact is important; the zero vector is *always* a solution to a homogeneous linear system. Therefore a homogeneous system is always consistent; we need only to determine whether we have exactly one solution (just $\vec{0}$) or infinite solutions. This idea is important so we give it it's own box.

Key Idea 7

Homogeneous Systems and Consistency

All homogeneous linear systems are consistent.

How do we determine if we have exactly one or infinite solutions? Recall Key Idea 2: if the solution has any free variables, then it will have infinite solutions. How can we tell if the system has free variables? Form the augmented matrix $\begin{bmatrix} A & \vec{0} \end{bmatrix}$, put it into reduced row echelon form, and interpret the result.

It may seem that we've brought up a new question, "When does $A\vec{x} = \vec{0}$ have exactly one or infinite solutions?" only to answer with "Look at the reduced row echelon form of A and interpret the results, just as always." Why bring up a new question if the answer is an old one?

While the new question has an old solution, it does lead to a great idea. Let's refresh our memory; earlier we solved two linear systems,

$$A\vec{x} = \vec{0} \quad \text{and} \quad A\vec{x} = \vec{b}$$

where

$$A = \begin{bmatrix} 1 & 2 \\ 2 & 4 \end{bmatrix} \quad \text{and} \quad \vec{b} = \begin{bmatrix} 3 \\ 6 \end{bmatrix}.$$

The solution to the first system of equations, $A\vec{x} = \vec{0}$, is

$$\vec{x} = x_2 \begin{bmatrix} -2 \\ 1 \end{bmatrix}$$

and the solution to the second set of equations, $A\vec{x} = \vec{b}$, is

$$\vec{x} = \begin{bmatrix} 3 \\ 0 \end{bmatrix} + x_2 \begin{bmatrix} -2 \\ 1 \end{bmatrix},$$

for all values of x_2.

Recalling our notation used earlier, set

$$\vec{x_p} = \begin{bmatrix} 3 \\ 0 \end{bmatrix} \quad \text{and let} \quad \vec{v} = \begin{bmatrix} -2 \\ 1 \end{bmatrix}.$$

Thus our solution to the linear system $A\vec{x} = \vec{b}$ is

$$\vec{x} = \vec{x_p} + x_2\vec{v}.$$

Let us see how exactly this solution works; let's see why $A\vec{x}$ equals $\vec{b}$. Multiply $A\vec{x}$:

$$\begin{aligned} A\vec{x} &= A(\vec{x_p} + x_2\vec{v}) \\ &= A\vec{x_p} + A(x_2\vec{v}) \\ &= A\vec{x_p} + x_2(A\vec{v}) \\ &= A\vec{x_p} + x_2\vec{0} \\ &= A\vec{x_p} + \vec{0} \\ &= A\vec{x_p} \\ &= \vec{b} \end{aligned}$$

We know that the last line is true, that $A\vec{x_p} = \vec{b}$, since we know that $\vec{x}$ was a solution to $A\vec{x} = \vec{b}$. The whole point is that $\vec{x_p}$ itself is a solution to $A\vec{x} = \vec{b}$, and we could find more solutions by adding vectors "that go to zero" when multiplied by A. (The subscript p of "$\vec{x_p}$" is used to denote that this vector is a "particular" solution.)

Stated in a different way, let's say that we know two things: that $A\vec{x_p} = \vec{b}$ and $A\vec{v} = \vec{0}$. What is $A(\vec{x_p} + \vec{v})$? We can multiply it out:

$$\begin{aligned} A(\vec{x_p} + \vec{v}) &= A\vec{x_p} + A\vec{v} \\ &= \vec{b} + \vec{0} \\ &= \vec{b} \end{aligned}$$

and see that $A(\vec{x_p} + \vec{v})$ also equals $\vec{b}$.

So we wonder: does this mean that $A\vec{x} = \vec{b}$ will have infinite solutions? After all, if $\vec{x_p}$ and $\vec{x_p} + \vec{v}$ are both solutions, don't we have infinite solutions?

No. If $A\vec{x} = \vec{0}$ has exactly one solution, then $\vec{v} = \vec{0}$, and $\vec{x_p} = \vec{x_p} + \vec{v}$; we only have one solution.

So here is the culmination of all of our fun that started a few pages back. If $\vec{v}$ is a solution to $A\vec{x} = \vec{0}$ and $\vec{x_p}$ is a solution to $A\vec{x} = \vec{b}$, then $\vec{x_p} + \vec{v}$ is also a solution to $A\vec{x} = \vec{b}$. If $A\vec{x} = \vec{0}$ has infinite solutions, so does $A\vec{x} = \vec{b}$; if $A\vec{x} = \vec{0}$ has only one solution, so does $A\vec{x} = \vec{b}$. This culminating idea is of course important enough to be stated again.

Key Idea 8 **Solutions of Consistent Systems**

Let $A\vec{x} = \vec{b}$ be a consistent system of linear equations.

1. If $A\vec{x} = \vec{0}$ has exactly one solution ($\vec{x} = \vec{0}$), then $A\vec{x} = \vec{b}$ has exactly one solution.
2. If $A\vec{x} = \vec{0}$ has infinite solutions, then $A\vec{x} = \vec{b}$ has infinite solutions.

A key word in the above statement is *consistent*. If $A\vec{x} = \vec{b}$ is inconsistent (the linear system has no solution), then it doesn't matter how many solutions $A\vec{x} = \vec{0}$ has; $A\vec{x} = \vec{b}$ has no solution.

Enough fun, enough theory. We need to practice.

Example 47 Let

$$A = \begin{bmatrix} 1 & -1 & 1 & 3 \\ 4 & 2 & 4 & 6 \end{bmatrix} \quad \text{and} \quad \vec{b} = \begin{bmatrix} 1 \\ 10 \end{bmatrix}.$$

Solve the linear systems $A\vec{x} = \vec{0}$ and $A\vec{x} = \vec{b}$ for $\vec{x}$, and write the solutions in vector form.

Solution We'll tackle $A\vec{x} = \vec{0}$ first. We form the associated augmented matrix, put it into reduced row echelon form, and interpret the result.

$$\begin{bmatrix} 1 & -1 & 1 & 3 & 0 \\ 4 & 2 & 4 & 6 & 0 \end{bmatrix} \quad \xrightarrow{\text{rref}} \quad \begin{bmatrix} 1 & 0 & 1 & 2 & 0 \\ 0 & 1 & 0 & -1 & 0 \end{bmatrix}$$

$$\begin{aligned} x_1 &= -x_3 - 2x_4 \\ x_2 &= x_4 \\ x_3 &\text{ is free} \\ x_4 &\text{ is free} \end{aligned}$$

To write our solution in vector form, we rewrite x_1 and x_2 in $\vec{x}$ in terms of x_3 and x_4.

$$\vec{x} = \begin{bmatrix} x_1 \\ x_2 \\ x_3 \\ x_4 \end{bmatrix} = \begin{bmatrix} -x_3 - 2x_4 \\ x_4 \\ x_3 \\ x_4 \end{bmatrix}$$

Finally, we "pull apart" this vector into two vectors, one with the "x_3 stuff" and one with the "x_4 stuff."

$$\begin{aligned}
\vec{x} &= \begin{bmatrix} -x_3 - 2x_4 \\ x_4 \\ x_3 \\ x_4 \end{bmatrix} \\
&= \begin{bmatrix} -x_3 \\ 0 \\ x_3 \\ 0 \end{bmatrix} + \begin{bmatrix} -2x_4 \\ x_4 \\ 0 \\ x_4 \end{bmatrix} \\
&= x_3 \begin{bmatrix} -1 \\ 0 \\ 1 \\ 0 \end{bmatrix} + x_4 \begin{bmatrix} -2 \\ 1 \\ 0 \\ 1 \end{bmatrix} \\
&= x_3\vec{u} + x_4\vec{v}
\end{aligned}$$

We use $\vec{u}$ and $\vec{v}$ simply to give these vectors names (and save some space).

It is easy to confirm that both $\vec{u}$ and $\vec{v}$ are solutions to the linear system $A\vec{x} = \vec{0}$. (Just multiply $A\vec{u}$ and $A\vec{v}$ and see that both are $\vec{0}$.) Since both are solutions to a homogeneous system of linear equations, any linear combination of $\vec{u}$ and $\vec{v}$ will be a solution, too.

Now let's tackle $A\vec{x} = \vec{b}$. Once again we put the associated augmented matrix into reduced row echelon form and interpret the results.

$$\begin{bmatrix} 1 & -1 & 1 & 3 & 1 \\ 4 & 2 & 4 & 6 & 10 \end{bmatrix} \quad \overrightarrow{\text{rref}} \quad \begin{bmatrix} 1 & 0 & 1 & 2 & 2 \\ 0 & 1 & 0 & -1 & 1 \end{bmatrix}$$

$$\begin{aligned}
x_1 &= 2 - x_3 - 2x_4 \\
x_2 &= 1 + x_4 \\
x_3 &\text{ is free} \\
x_4 &\text{ is free}
\end{aligned}$$

Writing this solution in vector form gives

$$\vec{x} = \begin{bmatrix} x_1 \\ x_2 \\ x_3 \\ x_4 \end{bmatrix} = \begin{bmatrix} 2 - x_3 - 2x_4 \\ 1 + x_4 \\ x_3 \\ x_4 \end{bmatrix}.$$

Again, we pull apart this vector, but this time we break it into three vectors: one with

"x_3" stuff, one with "x_4" stuff, and one with just constants.

$$\vec{x} = \begin{bmatrix} 2 - x_3 - 2x_4 \\ 1 + x_4 \\ x_3 \\ x_4 \end{bmatrix}$$

$$= \begin{bmatrix} 2 \\ 1 \\ 0 \\ 0 \end{bmatrix} + \begin{bmatrix} -x_3 \\ 0 \\ x_3 \\ 0 \end{bmatrix} + \begin{bmatrix} -2x_4 \\ x_4 \\ 0 \\ x_4 \end{bmatrix}$$

$$= \begin{bmatrix} 2 \\ 1 \\ 0 \\ 0 \end{bmatrix} + x_3 \begin{bmatrix} -1 \\ 0 \\ 1 \\ 0 \end{bmatrix} + x_4 \begin{bmatrix} -2 \\ 1 \\ 0 \\ 1 \end{bmatrix}$$

$$= \underbrace{\vec{x_p}}_{\text{particular solution}} + \underbrace{x_3\vec{u} + x_4\vec{v}}_{\text{solution to homogeneous equations } A\vec{x} = \vec{0}}$$

Note that $A\vec{x_p} = \vec{b}$; by itself, $\vec{x_p}$ is a solution. To get infinite solutions, we add a bunch of stuff that "goes to zero" when we multiply by A; we add the solution to the homogeneous equations.

Why don't we graph this solution as we did in the past? Before we had only two variables, meaning the solution could be graphed in 2D. Here we have four variables, meaning that our solution "lives" in 4D. You *can* draw this on paper, but it is *very* confusing.

Example 48 Rewrite the linear system

$$\begin{array}{ccccccccccc} x_1 & + & 2x_2 & - & 3x_3 & + & 2x_4 & + & 7x_5 & = & 2 \\ 3x_1 & + & 4x_2 & + & 5x_3 & + & 2x_4 & + & 3x_5 & = & -4 \end{array}$$

as a matrix–vector equation, solve the system using vector notation, and give the solution to the related homogeneous equations.

Solution Rewriting the linear system in the form of $A\vec{x} = \vec{b}$, we have that

$$A = \begin{bmatrix} 1 & 2 & -3 & 2 & 7 \\ 3 & 4 & 5 & 2 & 3 \end{bmatrix}, \quad \vec{x} = \begin{bmatrix} x_1 \\ x_2 \\ x_3 \\ x_4 \\ x_5 \end{bmatrix} \quad \text{and} \quad \vec{b} = \begin{bmatrix} 2 \\ -4 \end{bmatrix}.$$

To solve the system, we put the associated augmented matrix into reduced row echelon form and interpret the results.

$$\begin{bmatrix} 1 & 2 & -3 & 2 & 7 & 2 \\ 3 & 4 & 5 & 2 & 3 & -4 \end{bmatrix} \xrightarrow{\text{rref}} \begin{bmatrix} 1 & 0 & 11 & -2 & -11 & -8 \\ 0 & 1 & -7 & 2 & 9 & 5 \end{bmatrix}$$

$$\begin{aligned} x_1 &= -8 - 11x_3 + 2x_4 + 11x_5 \\ x_2 &= 5 + 7x_3 - 2x_4 - 9x_5 \\ x_3 &\text{ is free} \\ x_4 &\text{ is free} \\ x_5 &\text{ is free} \end{aligned}$$

We use this information to write $\vec{x}$, again pulling it apart. Since we have three free variables and also constants, we'll need to pull $\vec{x}$ apart into four separate vectors.

$$\begin{aligned} \vec{x} &= \begin{bmatrix} x_1 \\ x_2 \\ x_3 \\ x_4 \\ x_5 \end{bmatrix} \\ &= \begin{bmatrix} -8 - 11x_3 + 2x_4 + 11x_5 \\ 5 + 7x_3 - 2x_4 - 9x_5 \\ x_3 \\ x_4 \\ x_5 \end{bmatrix} \\ &= \begin{bmatrix} -8 \\ 5 \\ 0 \\ 0 \\ 0 \end{bmatrix} + \begin{bmatrix} -11x_3 \\ 7x_3 \\ x_3 \\ 0 \\ 0 \end{bmatrix} + \begin{bmatrix} 2x_4 \\ -2x_4 \\ 0 \\ x_4 \\ 0 \end{bmatrix} + \begin{bmatrix} 11x_5 \\ -9x_5 \\ 0 \\ 0 \\ x_5 \end{bmatrix} \\ &= \begin{bmatrix} -8 \\ 5 \\ 0 \\ 0 \\ 0 \end{bmatrix} + x_3 \begin{bmatrix} -11 \\ 7 \\ 1 \\ 0 \\ 0 \end{bmatrix} + x_4 \begin{bmatrix} 2 \\ -2 \\ 0 \\ 1 \\ 0 \end{bmatrix} + x_5 \begin{bmatrix} 11 \\ -9 \\ 0 \\ 0 \\ 1 \end{bmatrix} \\ &= \underbrace{\vec{x_p}}_{\substack{\text{particular} \\ \text{solution}}} + \underbrace{x_3\vec{u} + x_4\vec{v} + x_5\vec{w}}_{\substack{\text{solution to homogeneous} \\ \text{equations } A\vec{x} = \vec{0}}} \end{aligned}$$

So $\vec{x_p}$ is a particular solution; $A\vec{x_p} = \vec{b}$. (Multiply it out to verify that this is true.) The other vectors, $\vec{u}$, $\vec{v}$ and $\vec{w}$, that are multiplied by our free variables x_3, x_4 and x_5, are each solutions to the homogeneous equations, $A\vec{x} = \vec{0}$. Any linear combination of these three vectors, i.e., any vector found by choosing values for x_3, x_4 and x_5 in $x_3\vec{u} + x_4\vec{v} + x_5\vec{w}$ is a solution to $A\vec{x} = \vec{0}$.

Example 49 Let

$$A = \begin{bmatrix} 1 & 2 \\ 4 & 5 \end{bmatrix} \quad \text{and} \quad \vec{b} = \begin{bmatrix} 3 \\ 6 \end{bmatrix}.$$

Find the solutions to $A\vec{x} = \vec{b}$ and $A\vec{x} = \vec{0}$.

Solution We go through the familiar work of finding the reduced row echelon form of the appropriate augmented matrix and interpreting the solution.

$$\begin{bmatrix} 1 & 2 & 3 \\ 4 & 5 & 6 \end{bmatrix} \quad \overrightarrow{\text{rref}} \quad \begin{bmatrix} 1 & 0 & -1 \\ 0 & 1 & 2 \end{bmatrix}$$

$$\begin{aligned} x_1 &= -1 \\ x_2 &= 2 \end{aligned}$$

Thus

$$\vec{x} = \begin{bmatrix} x_1 \\ x_2 \end{bmatrix} = \begin{bmatrix} -1 \\ 2 \end{bmatrix}.$$

This may strike us as a bit odd; we are used to having lots of different vectors in the solution. However, in this case, the linear system $A\vec{x} = \vec{b}$ has exactly one solution, and we've found it. What is the solution to $A\vec{x} = \vec{0}$? Since we've only found one solution to $A\vec{x} = \vec{b}$, we can conclude from Key Idea 8 the related homogeneous equations $A\vec{x} = \vec{0}$ have only one solution, namely $\vec{x} = \vec{0}$. We can write our solution vector $\vec{x}$ in a form similar to our previous examples to highlight this:

$$\begin{aligned} \vec{x} &= \begin{bmatrix} -1 \\ 2 \end{bmatrix} \\ &= \begin{bmatrix} -1 \\ 2 \end{bmatrix} + \begin{bmatrix} 0 \\ 0 \end{bmatrix} \\ &= \underbrace{\vec{x_p}}_{\substack{\text{particular} \\ \text{solution}}} + \underbrace{\vec{0}}_{\substack{\text{solution to} \\ A\vec{x}=\vec{0}}}. \end{aligned}$$

Example 50 Let

$$A = \begin{bmatrix} 1 & 1 \\ 2 & 2 \end{bmatrix} \quad \text{and} \quad \vec{b} = \begin{bmatrix} 1 \\ 1 \end{bmatrix}.$$

Find the solutions to $A\vec{x} = \vec{b}$ and $A\vec{x} = \vec{0}$.

Solution To solve $A\vec{x} = \vec{b}$, we put the appropriate augmented matrix into reduced row echelon form and interpret the results.

$$\begin{bmatrix} 1 & 1 & 1 \\ 2 & 2 & 1 \end{bmatrix} \quad \overrightarrow{\text{rref}} \quad \begin{bmatrix} 1 & 1 & 0 \\ 0 & 0 & 1 \end{bmatrix}$$

We immediately have a problem; we see that the second row tells us that $0x_1 + 0x_2 = 1$, the sign that our system does not have a solution. Thus $A\vec{x} = \vec{b}$ has no solution. Of course, this does not mean that $A\vec{x} = \vec{0}$ has no solution; it always has a solution.

To find the solution to $A\vec{x} = \vec{0}$, we interpret the reduced row echelon form of the appropriate augmented matrix.

$$\begin{bmatrix} 1 & 1 & 0 \\ 2 & 2 & 0 \end{bmatrix} \xrightarrow{\text{rref}} \begin{bmatrix} 1 & 1 & 0 \\ 0 & 0 & 0 \end{bmatrix}$$

$$\begin{aligned} x_1 &= -x_2 \\ x_2 &\text{ is free} \end{aligned}$$

Thus

$$\begin{aligned} \vec{x} &= \begin{bmatrix} x_1 \\ x_2 \end{bmatrix} \\ &= \begin{bmatrix} -x_2 \\ x_2 \end{bmatrix} \\ &= x_2 \begin{bmatrix} -1 \\ 1 \end{bmatrix} \\ &= x_2\vec{u}. \end{aligned}$$

We have no solution to $A\vec{x} = \vec{b}$, but infinite solutions to $A\vec{x} = \vec{0}$.

The previous example may seem to violate the principle of Key Idea 8. After all, it seems that having infinite solutions to $A\vec{x} = \vec{0}$ should imply infinite solutions to $A\vec{x} = \vec{b}$. However, we remind ourselves of the key word in the idea that we observed before: *consistent*. If $A\vec{x} = \vec{b}$ is consistent and $A\vec{x} = \vec{0}$ has infinite solutions, then so will $A\vec{x} = \vec{b}$. But if $A\vec{x} = \vec{b}$ is not consistent, it does not matter how many solutions $A\vec{x} = \vec{0}$ has; $A\vec{x} = \vec{b}$ is still inconsistent.

This whole section is highlighting a very important concept that we won't fully understand until after two sections, but we get a glimpse of it here. When solving any system of linear equations (which we can write as $A\vec{x} = \vec{b}$), whether we have exactly one solution, infinite solutions, or no solution depends on an intrinsic property of A. We'll find out what that property is soon; in the next section we solve a problem we introduced at the beginning of this section, how to solve matrix equations $AX = B$.

Exercises 2.4

In Exercises 1 – 6, a matrix A and vectors $\vec{b}$, $\vec{u}$ and $\vec{v}$ are given. Verify that $\vec{u}$ and $\vec{v}$ are both solutions to the equation $A\vec{x} = \vec{b}$; that is, show that $A\vec{u} = A\vec{v} = \vec{b}$.

1. $A = \begin{bmatrix} 1 & -2 \\ -3 & 6 \end{bmatrix}$,
$\vec{b} = \begin{bmatrix} 0 \\ 0 \end{bmatrix}, \vec{u} = \begin{bmatrix} 2 \\ 1 \end{bmatrix}, \vec{v} = \begin{bmatrix} -10 \\ -5 \end{bmatrix}$

2. $A = \begin{bmatrix} 1 & -2 \\ -3 & 6 \end{bmatrix}$,

$\vec{b} = \begin{bmatrix} 2 \\ -6 \end{bmatrix}, \vec{u} = \begin{bmatrix} 0 \\ -1 \end{bmatrix}, \vec{v} = \begin{bmatrix} 2 \\ 0 \end{bmatrix}$

3. $A = \begin{bmatrix} 1 & 0 \\ 2 & 0 \end{bmatrix}$,

$\vec{b} = \begin{bmatrix} 0 \\ 0 \end{bmatrix}, \vec{u} = \begin{bmatrix} 0 \\ -1 \end{bmatrix}, \vec{v} = \begin{bmatrix} 0 \\ 59 \end{bmatrix}$

4. $A = \begin{bmatrix} 1 & 0 \\ 2 & 0 \end{bmatrix}$,

$\vec{b} = \begin{bmatrix} -3 \\ -6 \end{bmatrix}, \vec{u} = \begin{bmatrix} -3 \\ -1 \end{bmatrix}, \vec{v} = \begin{bmatrix} -3 \\ 59 \end{bmatrix}$

5. $A = \begin{bmatrix} 0 & -3 & -1 & -3 \\ -4 & 2 & -3 & 5 \end{bmatrix}$,

$\vec{b} = \begin{bmatrix} 0 \\ 0 \end{bmatrix}, \vec{u} = \begin{bmatrix} 11 \\ 4 \\ -12 \\ 0 \end{bmatrix}$,

$\vec{v} = \begin{bmatrix} 9 \\ -12 \\ 0 \\ 12 \end{bmatrix}$

6. $A = \begin{bmatrix} 0 & -3 & -1 & -3 \\ -4 & 2 & -3 & 5 \end{bmatrix}$,

$\vec{b} = \begin{bmatrix} 48 \\ 36 \end{bmatrix}, \vec{u} = \begin{bmatrix} -17 \\ -16 \\ 0 \\ 0 \end{bmatrix}$,

$\vec{v} = \begin{bmatrix} -8 \\ -28 \\ 0 \\ 12 \end{bmatrix}$

In Exercises 7 – 9, a matrix A and vectors $\vec{b}$, $\vec{u}$ and $\vec{v}$ are given. Verify that $A\vec{u} = \vec{0}$, $A\vec{v} = \vec{b}$ and $A(\vec{u} + \vec{v}) = \vec{b}$.

7. $A = \begin{bmatrix} 2 & -2 & -1 \\ -1 & 1 & -1 \\ -2 & 2 & -1 \end{bmatrix}$,

$\vec{b} = \begin{bmatrix} 1 \\ 1 \\ 1 \end{bmatrix}, \vec{u} = \begin{bmatrix} 1 \\ 1 \\ 0 \end{bmatrix}, \vec{v} = \begin{bmatrix} 1 \\ 1 \\ -1 \end{bmatrix}$

8. $A = \begin{bmatrix} 1 & -1 & 3 \\ 3 & -3 & -3 \\ -1 & 1 & 1 \end{bmatrix}$,

$\vec{b} = \begin{bmatrix} -1 \\ -3 \\ 1 \end{bmatrix}, \vec{u} = \begin{bmatrix} 2 \\ 2 \\ 0 \end{bmatrix}, \vec{v} = \begin{bmatrix} 2 \\ 3 \\ 0 \end{bmatrix}$

9. $A = \begin{bmatrix} 2 & 0 & 0 \\ 0 & 1 & -3 \\ 3 & 1 & -3 \end{bmatrix}$,

$\vec{b} = \begin{bmatrix} 2 \\ -4 \\ -1 \end{bmatrix}, \vec{u} = \begin{bmatrix} 0 \\ 6 \\ 2 \end{bmatrix}, \vec{v} = \begin{bmatrix} 1 \\ -1 \\ 1 \end{bmatrix}$

In Exercises 10 – 24, a matrix A and vector $\vec{b}$ are given.

(a) Solve the equation $A\vec{x} = \vec{0}$.

(b) Solve the equation $A\vec{x} = \vec{b}$.

In each of the above, be sure to write your answer in vector format. Also, when possible, give 2 particular solutions to each equation.

10. $A = \begin{bmatrix} 0 & 2 \\ -1 & 3 \end{bmatrix}, \vec{b} = \begin{bmatrix} -2 \\ -1 \end{bmatrix}$

11. $A = \begin{bmatrix} -4 & -1 \\ -3 & -2 \end{bmatrix}, \vec{b} = \begin{bmatrix} 1 \\ 4 \end{bmatrix}$

12. $A = \begin{bmatrix} 1 & -2 \\ 0 & 1 \end{bmatrix}, \vec{b} = \begin{bmatrix} 0 \\ -5 \end{bmatrix}$

13. $A = \begin{bmatrix} 1 & 0 \\ 5 & -4 \end{bmatrix}, \vec{b} = \begin{bmatrix} -2 \\ -1 \end{bmatrix}$

14. $A = \begin{bmatrix} 2 & -3 \\ -4 & 6 \end{bmatrix}, \vec{b} = \begin{bmatrix} 1 \\ -1 \end{bmatrix}$

15. $A = \begin{bmatrix} -4 & 3 & 2 \\ -4 & 5 & 0 \end{bmatrix}, \vec{b} = \begin{bmatrix} -4 \\ -4 \end{bmatrix}$

16. $A = \begin{bmatrix} 1 & 5 & -2 \\ 1 & 4 & 5 \end{bmatrix}, \vec{b} = \begin{bmatrix} 0 \\ 1 \end{bmatrix}$

17. $A = \begin{bmatrix} -1 & -2 & -2 \\ 3 & 4 & -2 \end{bmatrix}, \vec{b} = \begin{bmatrix} -4 \\ -4 \end{bmatrix}$

18. $A = \begin{bmatrix} 2 & 2 & 2 \\ 5 & 5 & -3 \end{bmatrix}, \vec{b} = \begin{bmatrix} 3 \\ -3 \end{bmatrix}$

19. $A = \begin{bmatrix} 1 & 5 & -4 & -1 \\ 1 & 0 & -2 & 1 \end{bmatrix}$,

$\vec{b} = \begin{bmatrix} 0 \\ -2 \end{bmatrix}$

20. $A = \begin{bmatrix} -4 & 2 & -5 & 4 \\ 0 & 1 & -1 & 5 \end{bmatrix}$,

$\vec{b} = \begin{bmatrix} -3 \\ -2 \end{bmatrix}$

21. $A = \begin{bmatrix} 0 & 0 & 2 & 1 & 4 \\ -2 & -1 & -4 & -1 & 5 \end{bmatrix}$,

$\vec{b} = \begin{bmatrix} 3 \\ 4 \end{bmatrix}$

22. $A = \begin{bmatrix} 3 & 0 & -2 & -4 & 5 \\ 2 & 3 & 2 & 0 & 2 \\ -5 & 0 & 4 & 0 & 5 \end{bmatrix}$,

$\vec{b} = \begin{bmatrix} -1 \\ -5 \\ 4 \end{bmatrix}$

23. $A = \begin{bmatrix} -1 & 3 & 1 & -3 & 4 \\ 3 & -3 & -1 & 1 & -4 \\ -2 & 3 & -2 & -3 & 1 \end{bmatrix}$,

$\vec{b} = \begin{bmatrix} 1 \\ 1 \\ -5 \end{bmatrix}$

24. $A = \begin{bmatrix} -4 & -2 & -1 & 4 & 0 \\ 5 & -4 & 3 & -1 & 1 \\ 4 & -5 & 3 & 1 & -4 \end{bmatrix}$,

$\vec{b} = \begin{bmatrix} 3 \\ 2 \\ 1 \end{bmatrix}$

In Exercises 25 – 28, a matrix A and vector $\vec{b}$ are given. Solve the equation $A\vec{x} = \vec{b}$, write the solution in vector format, and sketch the solution as the appropriate line on the Cartesian plane.

25. $A = \begin{bmatrix} 2 & 4 \\ -1 & -2 \end{bmatrix}, \vec{b} = \begin{bmatrix} 0 \\ 0 \end{bmatrix}$

26. $A = \begin{bmatrix} 2 & 4 \\ -1 & -2 \end{bmatrix}, \vec{b} = \begin{bmatrix} -6 \\ 3 \end{bmatrix}$

27. $A = \begin{bmatrix} 2 & -5 \\ -4 & -10 \end{bmatrix}, \vec{b} = \begin{bmatrix} 1 \\ 2 \end{bmatrix}$

28. $A = \begin{bmatrix} 2 & -5 \\ -4 & -10 \end{bmatrix}, \vec{b} = \begin{bmatrix} 0 \\ 0 \end{bmatrix}$

2.5 Solving Matrix Equations $AX = B$

AS YOU READ . . .

1. T/F: To solve the matrix equation $AX = B$, put the matrix $\begin{bmatrix} A & X \end{bmatrix}$ into reduced row echelon form and interpret the result properly.

2. T/F: The first column of a matrix product AB is A times the first column of B.

3. Give two reasons why one might solve for the columns of X in the equation AX=B separately.

We began last section talking about solving numerical equations like $ax = b$ for x. We mentioned that solving matrix equations of the form $AX = B$ is of interest, but we first learned how to solve the related, but simpler, equations $A\vec{x} = \vec{b}$. In this section we will learn how to solve the general matrix equation $AX = B$ for X.

We will start by considering the best case scenario when solving $A\vec{x} = \vec{b}$; that is, when A is square and we have exactly one solution. For instance, suppose we want to solve $A\vec{x} = \vec{b}$ where

$$A = \begin{bmatrix} 1 & 1 \\ 2 & 1 \end{bmatrix} \quad \text{and} \quad \vec{b} = \begin{bmatrix} 0 \\ 1 \end{bmatrix}.$$

We know how to solve this; put the appropriate matrix into reduced row echelon form and interpret the result.

$$\begin{bmatrix} 1 & 1 & 0 \\ 2 & 1 & 1 \end{bmatrix} \quad \overrightarrow{\text{rref}} \quad \begin{bmatrix} 1 & 0 & 1 \\ 0 & 1 & -1 \end{bmatrix}$$

We read from this that

$$\vec{x} = \begin{bmatrix} 1 \\ -1 \end{bmatrix}.$$

Written in a more general form, we found our solution by forming the augmented matrix

$$\begin{bmatrix} A & \vec{b} \end{bmatrix}$$

and interpreting its reduced row echelon form:

$$\begin{bmatrix} A & \vec{b} \end{bmatrix} \quad \overrightarrow{\text{rref}} \quad \begin{bmatrix} I & \vec{x} \end{bmatrix}$$

Notice that when the reduced row echelon form of A is the identity matrix I we have exactly one solution. This, again, is the best case scenario.

We apply the same general technique to solving the matrix equation $AX = B$ for X. We'll assume that A is a square matrix (B need not be) and we'll form the augmented matrix

$$\begin{bmatrix} A & B \end{bmatrix}.$$

Putting this matrix into reduced row echelon form will give us X, much like we found $\vec{x}$ before.

$$\begin{bmatrix} A & B \end{bmatrix} \quad \overrightarrow{\text{rref}} \quad \begin{bmatrix} I & X \end{bmatrix}$$

As long as the reduced row echelon form of A is the identity matrix, this technique works great. After a few examples, we'll discuss why this technique works, and we'll also talk just a little bit about what happens when the reduced row echelon form of A is not the identity matrix.

First, some examples.

Example 51 Solve the matrix equation $AX = B$ where

$$A = \begin{bmatrix} 1 & -1 \\ 5 & 3 \end{bmatrix} \quad \text{and} \quad B = \begin{bmatrix} -8 & -13 & 1 \\ 32 & -17 & 21 \end{bmatrix}.$$

Solution To solve $AX = B$ for X, we form the proper augmented matrix, put it into reduced row echelon form, and interpret the result.

$$\begin{bmatrix} 1 & -1 & -8 & -13 & 1 \\ 5 & 3 & 32 & -17 & 21 \end{bmatrix} \quad \overrightarrow{\text{rref}} \quad \begin{bmatrix} 1 & 0 & 1 & -7 & 3 \\ 0 & 1 & 9 & 6 & 2 \end{bmatrix}$$

We read from the reduced row echelon form of the matrix that

$$X = \begin{bmatrix} 1 & -7 & 3 \\ 9 & 6 & 2 \end{bmatrix}.$$

We can easily check to see if our answer is correct by multiplying AX.

Example 52 Solve the matrix equation $AX = B$ where

$$A = \begin{bmatrix} 1 & 0 & 2 \\ 0 & -1 & -2 \\ 2 & -1 & 0 \end{bmatrix} \quad \text{and} \quad B = \begin{bmatrix} -1 & 2 \\ 2 & -6 \\ 2 & -4 \end{bmatrix}.$$

SOLUTION To solve, let's again form the augmented matrix

$$\begin{bmatrix} A & B \end{bmatrix},$$

put it into reduced row echelon form, and interpret the result.

$$\begin{bmatrix} 1 & 0 & 2 & -1 & 2 \\ 0 & -1 & -2 & 2 & -6 \\ 2 & -1 & 0 & 2 & -4 \end{bmatrix} \quad \overrightarrow{\text{rref}} \quad \begin{bmatrix} 1 & 0 & 0 & 1 & 0 \\ 0 & 1 & 0 & 0 & 4 \\ 0 & 0 & 1 & -1 & 1 \end{bmatrix}$$

We see from this that

$$X = \begin{bmatrix} 1 & 0 \\ 0 & 4 \\ -1 & 1 \end{bmatrix}.$$

Why does this work? To see the answer, let's define five matrices.

$$A = \begin{bmatrix} 1 & 2 \\ 3 & 4 \end{bmatrix}, \quad \vec{u} = \begin{bmatrix} 1 \\ 1 \end{bmatrix}, \quad \vec{v} = \begin{bmatrix} -1 \\ 1 \end{bmatrix}, \quad \vec{w} = \begin{bmatrix} 5 \\ 6 \end{bmatrix} \quad \text{and} \quad X = \begin{bmatrix} 1 & -1 & 5 \\ 1 & 1 & 6 \end{bmatrix}$$

Notice that $\vec{u}$, $\vec{v}$ and $\vec{w}$ are the first, second and third columns of X, respectively. Now consider this list of matrix products: $A\vec{u}$, $A\vec{v}$, $A\vec{w}$ and AX.

$$A\vec{u} = \begin{bmatrix} 1 & 2 \\ 3 & 4 \end{bmatrix} \begin{bmatrix} 1 \\ 1 \end{bmatrix} = \begin{bmatrix} 3 \\ 7 \end{bmatrix} \qquad A\vec{v} = \begin{bmatrix} 1 & 2 \\ 3 & 4 \end{bmatrix} \begin{bmatrix} -1 \\ 1 \end{bmatrix} = \begin{bmatrix} 1 \\ 1 \end{bmatrix}$$

$$A\vec{w} = \begin{bmatrix} 1 & 2 \\ 3 & 4 \end{bmatrix} \begin{bmatrix} 5 \\ 6 \end{bmatrix} = \begin{bmatrix} 17 \\ 39 \end{bmatrix} \qquad AX = \begin{bmatrix} 1 & 2 \\ 3 & 4 \end{bmatrix} \begin{bmatrix} 1 & -1 & 5 \\ 1 & 1 & 6 \end{bmatrix} = \begin{bmatrix} 3 & 1 & 17 \\ 7 & 1 & 39 \end{bmatrix}$$

So again note that the columns of X are $\vec{u}$, $\vec{v}$ and $\vec{w}$; that is, we can write

$$X = \begin{bmatrix} \vec{u} & \vec{v} & \vec{w} \end{bmatrix}.$$

Notice also that the columns of AX are $A\vec{u}$, $A\vec{v}$ and $A\vec{w}$, respectively. Thus we can write

$$\begin{aligned} AX &= A\begin{bmatrix} \vec{u} & \vec{v} & \vec{w} \end{bmatrix} \\ &= \begin{bmatrix} A\vec{u} & A\vec{v} & A\vec{w} \end{bmatrix} \\ &= \begin{bmatrix} \begin{bmatrix} 3 \\ 7 \end{bmatrix} & \begin{bmatrix} 1 \\ 1 \end{bmatrix} & \begin{bmatrix} 17 \\ 39 \end{bmatrix} \end{bmatrix} \\ &= \begin{bmatrix} 3 & 1 & 17 \\ 7 & 1 & 39 \end{bmatrix} \end{aligned}$$

We summarize what we saw above in the following statement:

The columns of a matrix product AX are A times the columns of X.

How does this help us solve the matrix equation $AX = B$ for X? Assume that A is a square matrix (that forces X and B to be the same size). We'll let $\vec{x_1}, \vec{x_2}, \cdots \vec{x_n}$ denote the columns of the (unknown) matrix X, and we'll let $\vec{b_1}, \vec{b_2}, \cdots \vec{b_n}$ denote the columns of B. We want to solve $AX = B$ for X. That is, we want X where

$$\begin{aligned} AX &= B \\ A\begin{bmatrix} \vec{x_1} & \vec{x_2} & \cdots & \vec{x_n} \end{bmatrix} &= \begin{bmatrix} \vec{b_1} & \vec{b_2} & \cdots & \vec{b_n} \end{bmatrix} \\ \begin{bmatrix} A\vec{x_1} & A\vec{x_2} & \cdots & A\vec{x_n} \end{bmatrix} &= \begin{bmatrix} \vec{b_1} & \vec{b_2} & \cdots & \vec{b_n} \end{bmatrix} \end{aligned}$$

If the matrix on the left hand side is equal to the matrix on the right, then their respective columns must be equal. This means we need to solve n equations:

$$\begin{aligned} A\vec{x_1} &= \vec{b_1} \\ A\vec{x_2} &= \vec{b_2} \\ \vdots &= \vdots \\ A\vec{x_n} &= \vec{b_n} \end{aligned}$$

We already know how to do this; this is what we learned in the previous section. Let's do this in a concrete example. In our above work we defined matrices A and X, and looked at the product AX. Let's call the product B; that is, set B= AX. Now, let's pretend that we don't know what X is, and let's try to find the matrix X that satisfies the equation $AX = B$. As a refresher, recall that

$$A = \begin{bmatrix} 1 & 2 \\ 3 & 4 \end{bmatrix} \quad \text{and} \quad B = \begin{bmatrix} 3 & 1 & 17 \\ 7 & 1 & 39 \end{bmatrix}.$$

Since A is a 2×2 matrix and B is a 2×3 matrix, what dimensions must X be in the equation $AX = B$? The number of rows of X must match the number of columns of A; the number of columns of X must match the number of columns of B. Therefore we know that X must be a 2×3 matrix.

We'll call the three columns of X $\vec{x_1}$, $\vec{x_2}$ and $\vec{x_3}$. Our previous explanation tells us that if $AX = B$, then:

$$\begin{aligned} AX &= B \\ A\begin{bmatrix} \vec{x_1} & \vec{x_2} & \vec{x_3} \end{bmatrix} &= \begin{bmatrix} 3 & 1 & 17 \\ 7 & 1 & 39 \end{bmatrix} \\ \begin{bmatrix} A\vec{x_1} & A\vec{x_2} & A\vec{x_3} \end{bmatrix} &= \begin{bmatrix} 3 & 1 & 17 \\ 7 & 1 & 39 \end{bmatrix}. \end{aligned}$$

Hence

$$\begin{aligned} A\vec{x_1} &= \begin{bmatrix} 3 \\ 7 \end{bmatrix} \\ A\vec{x_2} &= \begin{bmatrix} 1 \\ 1 \end{bmatrix} \\ A\vec{x_3} &= \begin{bmatrix} 17 \\ 39 \end{bmatrix} \end{aligned}$$

To find $\vec{x_1}$, we form the proper augmented matrix and put it into reduced row echelon form and interpret the results.

$$\begin{bmatrix} 1 & 2 & 3 \\ 3 & 4 & 7 \end{bmatrix} \quad \overrightarrow{\text{rref}} \quad \begin{bmatrix} 1 & 0 & 1 \\ 0 & 1 & 1 \end{bmatrix}$$

This shows us that

$$\vec{x_1} = \begin{bmatrix} 1 \\ 1 \end{bmatrix}.$$

To find $\vec{x_2}$, we again form an augmented matrix and interpret its reduced row echelon form.

$$\begin{bmatrix} 1 & 2 & 1 \\ 3 & 4 & 1 \end{bmatrix} \quad \overrightarrow{\text{rref}} \quad \begin{bmatrix} 1 & 0 & -1 \\ 0 & 1 & 1 \end{bmatrix}$$

Thus

$$\vec{x_2} = \begin{bmatrix} -1 \\ 1 \end{bmatrix}$$

which matches with what we already knew from above.

Before continuing on in this manner to find $\vec{x_3}$, we should stop and think. If the matrix vector equation $A\vec{x} = \vec{b}$ is consistent, then the steps involved in putting

$$\begin{bmatrix} A & \vec{b} \end{bmatrix}$$

into reduced row echelon form depend only on A; it does not matter what $\vec{b}$ is. So when we put the two matrices

$$\begin{bmatrix} 1 & 2 & 3 \\ 3 & 4 & 7 \end{bmatrix} \quad \text{and} \quad \begin{bmatrix} 1 & 2 & 1 \\ 3 & 4 & 1 \end{bmatrix}$$

from above into reduced row echelon form, we performed exactly the same steps! (In fact, those steps are: $-3R_1 + R_2 \rightarrow R_2$; $-\frac{1}{2}R_2 \rightarrow R_2$; $-2R_2 + R_1 \rightarrow R_1$.)

Instead of solving for each column of X separately, performing the same steps to put the necessary matrices into reduced row echelon form three different times, why don't we just do it all at once?[16] Instead of individually putting

$$\begin{bmatrix} 1 & 2 & 3 \\ 3 & 4 & 7 \end{bmatrix}, \quad \begin{bmatrix} 1 & 2 & 1 \\ 3 & 4 & 1 \end{bmatrix} \quad \text{and} \quad \begin{bmatrix} 1 & 2 & 17 \\ 3 & 4 & 39 \end{bmatrix}$$

into reduced row echelon form, let's just put

$$\begin{bmatrix} 1 & 2 & 3 & 1 & 17 \\ 3 & 4 & 7 & 1 & 39 \end{bmatrix}$$

into reduced row echelon form.

$$\begin{bmatrix} 1 & 2 & 3 & 1 & 17 \\ 3 & 4 & 7 & 1 & 39 \end{bmatrix} \quad \xrightarrow{\text{rref}} \quad \begin{bmatrix} 1 & 0 & 1 & -1 & 5 \\ 0 & 1 & 1 & 1 & 6 \end{bmatrix}$$

By looking at the last three columns, we see X:

$$X = \begin{bmatrix} 1 & -1 & 5 \\ 1 & 1 & 6 \end{bmatrix}.$$

Now that we've justified the technique we've been using in this section to solve $AX = B$ for X, we reinfornce its importance by restating it as a Key Idea.

Key Idea 9

Solving $AX = B$

Let A be an $n \times n$ matrix, where the reduced row echelon form of A is I. To solve the matrix equation $AX = B$ for X,

1. Form the augmented matrix $\begin{bmatrix} A & B \end{bmatrix}$.
2. Put this matrix into reduced row echelon form. It will be of the form $\begin{bmatrix} I & X \end{bmatrix}$, where X appears in the columns where B once was.

These simple steps cause us to ask certain questions. First, we specify above that A should be a square matrix. What happens if A isn't square? Is a solution still possible? Secondly, we only considered cases where the reduced row echelon form of A was I (and stated that as a requirement in our Key Idea). What if the reduced row echelon form of A isn't I? Would we still be able to find a solution? (Instead of having exactly one solution, could we have no solution? Infinite solutions? How would we be able to tell?)

[16] One reason to do it three different times is that we enjoy doing unnecessary work. Another reason could be that we are stupid.

These questions are good to ask, and we leave it to the reader to discover their answers. Instead of tackling these questions, we instead tackle the problem of "Why do we care about solving $AX = B$?" The simple answer is that, for now, we only care about the special case when $B = I$. By solving $AX = I$ for X, we find a matrix X that, when multiplied by A, gives the identity I. That will be very useful.

Exercises 2.5

In Exercises 1 – 12, matrices A and B are given. Solve the matrix equation $AX = B$.

1. $A = \begin{bmatrix} 4 & -1 \\ -7 & 5 \end{bmatrix}$,
   $B = \begin{bmatrix} 8 & -31 \\ -27 & 38 \end{bmatrix}$

2. $A = \begin{bmatrix} 1 & -3 \\ -3 & 6 \end{bmatrix}$,
   $B = \begin{bmatrix} 12 & -10 \\ -27 & 27 \end{bmatrix}$

3. $A = \begin{bmatrix} 3 & 3 \\ 6 & 4 \end{bmatrix}$,
   $B = \begin{bmatrix} 15 & -39 \\ 16 & -66 \end{bmatrix}$

4. $A = \begin{bmatrix} -3 & -6 \\ 4 & 0 \end{bmatrix}$,
   $B = \begin{bmatrix} 48 & -30 \\ 0 & -8 \end{bmatrix}$

5. $A = \begin{bmatrix} -1 & -2 \\ -2 & -3 \end{bmatrix}$,
   $B = \begin{bmatrix} 13 & 4 & 7 \\ 22 & 5 & 12 \end{bmatrix}$

6. $A = \begin{bmatrix} -4 & 1 \\ -1 & -2 \end{bmatrix}$,
   $B = \begin{bmatrix} -2 & -10 & 19 \\ 13 & 2 & -2 \end{bmatrix}$

7. $A = \begin{bmatrix} 1 & 0 \\ 3 & -1 \end{bmatrix}$, $B = I_2$

8. $A = \begin{bmatrix} 2 & 2 \\ 3 & 1 \end{bmatrix}$, $B = I_2$

9. $A = \begin{bmatrix} -2 & 0 & 4 \\ -5 & -4 & 5 \\ -3 & 5 & -3 \end{bmatrix}$,
   $B = \begin{bmatrix} -18 & 2 & -14 \\ -38 & 18 & -13 \\ 10 & 2 & -18 \end{bmatrix}$

10. $A = \begin{bmatrix} -5 & -4 & -1 \\ 8 & -2 & -3 \\ 6 & 1 & -8 \end{bmatrix}$,
    $B = \begin{bmatrix} -21 & -8 & -19 \\ 65 & -11 & -10 \\ 75 & -51 & 33 \end{bmatrix}$

11. $A = \begin{bmatrix} 0 & -2 & 1 \\ 0 & 2 & 2 \\ 1 & 2 & -3 \end{bmatrix}$, $B = I_3$

12. $A = \begin{bmatrix} -3 & 3 & -2 \\ 1 & -3 & 2 \\ -1 & -1 & 2 \end{bmatrix}$, $B = I_3$

2.6 The Matrix Inverse

1. T/F: If A and B are square matrices where $AB = I$, then $BA = I$.

2. T/F: A matrix A has exactly one inverse, infinite inverses, or no inverse.

3. T/F: Everyone is special.

4. T/F: If A is invertible, then $A\vec{x} = \vec{0}$ has exactly 1 solution.
5. What is a corollary?
6. Fill in the blanks: ________ a matrix is invertible is useful; computing the inverse is ________.

Once again we visit the old algebra equation, $ax = b$. How do we solve for x? We know that, as long as $a \neq 0$,

$$x = \frac{b}{a}, \text{ or, stated in another way, } x = a^{-1}b.$$

What is a^{-1}? It is the number that, when multiplied by a, returns 1. That is,

$$a^{-1}a = 1.$$

Let us now think in terms of matrices. We have learned of the identity matrix I that "acts like the number 1." That is, if A is a square matrix, then

$$IA = AI = A.$$

If we had a matrix, which we'll call A^{-1}, where $A^{-1}A = I$, then by analogy to our algebra example above it seems like we might be able to solve the linear system $A\vec{x} = \vec{b}$ for $\vec{x}$ by multiplying both sides of the equation by A^{-1}. That is, perhaps

$$\vec{x} = A^{-1}\vec{b}.$$

Of course, there is a lot of speculation here. We don't know that such a matrix like A^{-1} exists. However, we do know how to solve the matrix equation $AX = B$, so we can use that technique to solve the equation $AX = I$ for X. This seems like it will get us close to what we want. Let's practice this once and then study our results.

Example 53 Let

$$A = \begin{bmatrix} 2 & 1 \\ 1 & 1 \end{bmatrix}.$$

Find a matrix X such that $AX = I$.

Solution We know how to solve this from the previous section: we form the proper augmented matrix, put it into reduced row echelon form and interpret the results.

$$\begin{bmatrix} 2 & 1 & 1 & 0 \\ 1 & 1 & 0 & 1 \end{bmatrix} \quad \overrightarrow{\text{rref}} \quad \begin{bmatrix} 1 & 0 & 1 & -1 \\ 0 & 1 & -1 & 2 \end{bmatrix}$$

We read from our matrix that

$$X = \begin{bmatrix} 1 & -1 \\ -1 & 2 \end{bmatrix}.$$

Let's check our work:

$$
\begin{aligned}
AX &= \begin{bmatrix} 2 & 1 \\ 1 & 1 \end{bmatrix} \begin{bmatrix} 1 & -1 \\ -1 & 2 \end{bmatrix} \\
&= \begin{bmatrix} 1 & 0 \\ 0 & 1 \end{bmatrix} \\
&= I
\end{aligned}
$$

Sure enough, it works.

Looking at our previous example, we are tempted to jump in and call the matrix X that we found "A^{-1}." However, there are two obstacles in the way of us doing this.

First, we know that in general $AB \neq BA$. So while we found that $AX = I$, we can't automatically assume that $XA = I$.

Secondly, we have seen examples of matrices where $AB = AC$, but $B \neq C$. So just because $AX = I$, it is possible that another matrix Y exists where $AY = I$. If this is the case, using the notation A^{-1} would be misleading, since it could refer to more than one matrix.

These obstacles that we face are not insurmountable. The first obstacle was that we know that $AX = I$ but didn't know that $XA = I$. That's easy enough to check, though. Let's look at A and X from our previous example.

$$
\begin{aligned}
XA &= \begin{bmatrix} 1 & -1 \\ -1 & 2 \end{bmatrix} \begin{bmatrix} 2 & 1 \\ 1 & 1 \end{bmatrix} \\
&= \begin{bmatrix} 1 & 0 \\ 0 & 1 \end{bmatrix} \\
&= I
\end{aligned}
$$

Perhaps this first obstacle isn't much of an obstacle after all. Of course, we only have one example where it worked, so this doesn't mean that it always works. We have good news, though: it always does work. The only "bad" news to come with this is that this is a bit harder to prove. We won't worry about proving it always works, but state formally that it does in the following theorem.

Theorem 5

Special Commuting Matrix Products

Let A be an $n \times n$ matrix.

1. If there is a matrix X such that $AX = I_n$, then $XA = I_n$.
2. If there is a matrix X such that $XA = I_n$, then $AX = I_n$.

The second obstacle is easier to address. We want to know if another matrix Y exists where $AY = I = YA$. Let's suppose that it does. Consider the expression XAY.

Since matrix multiplication is associative, we can group this any way we choose. We could group this as $(XA)Y$; this results in

$$\begin{aligned}(XA)Y &= IY \\ &= Y.\end{aligned}$$

We could also group XAY as $X(AY)$. This tells us

$$\begin{aligned}X(AY) &= XI \\ &= X\end{aligned}$$

Combining the two ideas above, we see that $X = XAY = Y$; that is, $X = Y$. We conclude that there is only one matrix X where $XA = I = AX$. (Even if we think we have two, we can do the above exercise and see that we really just have one.)

We have just proved the following theorem.

Theorem 6 **Uniqueness of Solutions to $AX = I_n$**

Let A be an $n \times n$ matrix and let X be a matrix where $AX = I_n$. Then X is unique; it is the only matrix that satisfies this equation.

So given a square matrix A, if we can find a matrix X where $AX = I$, then we know that $XA = I$ and that X is the only matrix that does this. This makes X special, so we give it a special name.

Definition 18 **Invertible Matrices and the Inverse of A**

Let A and X be $n \times n$ matrices where $AX = I = XA$. Then:

1. A is *invertible*.
2. X is the *inverse* of A, denoted by A^{-1}.

Let's do an example.

Example 54 Find the inverse of $A = \begin{bmatrix} 1 & 2 \\ 2 & 4 \end{bmatrix}$.

Solution By solving the equation $AX = I$ for X will give us the inverse of A. Forming the appropriate augmented matrix and finding its reduced row echelon form gives us

$$\left[\begin{array}{cc|cc} 1 & 2 & 1 & 0 \\ 2 & 4 & 0 & 1 \end{array}\right] \quad \overrightarrow{\text{rref}} \quad \begin{bmatrix} 1 & 2 & 0 & 1/2 \\ 0 & 0 & 1 & -1/2 \end{bmatrix}$$

Yikes! We were expecting to find that the reduced row echelon form of this matrix would look like

$$\begin{bmatrix} I & A^{-1} \end{bmatrix}.$$

However, we don't have the identity on the left hand side. Our conclusion: A is not invertible.

We have just seen that not all matrices are invertible.[17] With this thought in mind, let's complete the array of boxes we started before the example. We've discovered that if a matrix has an inverse, it has only one. Therefore, we gave that special matrix a name, "*the* inverse." Finally, we describe the most general way to find the inverse of a matrix, and a way to tell if it does not have one.

Key Idea 10

Finding A^{-1}

Let A be an $n \times n$ matrix. To find A^{-1}, put the augmented matrix

$$\begin{bmatrix} A & I_n \end{bmatrix}$$

into reduced row echelon form. If the result is of the form

$$\begin{bmatrix} I_n & X \end{bmatrix},$$

then $A^{-1} = X$. If not, (that is, if the first n columns of the reduced row echelon form are not I_n), then A is not invertible.

Let's try again.

Example 55 Find the inverse, if it exists, of $A = \begin{bmatrix} 1 & 1 & -1 \\ 1 & -1 & 1 \\ 1 & 2 & 3 \end{bmatrix}$.

SOLUTION We'll try to solve $AX = I$ for X and see what happens.

$$\left[\begin{array}{ccc|ccc} 1 & 1 & -1 & 1 & 0 & 0 \\ 1 & -1 & 1 & 0 & 1 & 0 \\ 1 & 2 & 3 & 0 & 0 & 1 \end{array}\right] \quad \overrightarrow{\text{rref}} \quad \begin{bmatrix} 1 & 0 & 0 & 0.5 & 0.5 & 0 \\ 0 & 1 & 0 & 0.2 & -0.4 & 0.2 \\ 0 & 0 & 1 & -0.3 & 0.1 & 0.2 \end{bmatrix}$$

[17] Hence our previous definition; why bother calling A "invertible" if every square matrix is? If everyone is special, then no one is. Then again, everyone *is* special.

We have a solution, so

$$A = \begin{bmatrix} 0.5 & 0.5 & 0 \\ 0.2 & -0.4 & 0.2 \\ -0.3 & 0.1 & 0.2 \end{bmatrix}.$$

Multiply AA^{-1} to verify that it is indeed the inverse of A.

In general, given a matrix A, to find A^{-1} we need to form the augmented matrix $\begin{bmatrix} A & I \end{bmatrix}$ and put it into reduced row echelon form and interpret the result. In the case of a 2×2 matrix, though, there is a shortcut. We give the shortcut in terms of a theorem.[18]

Theorem 7 **The Inverse of a 2×2 Matrix**

Let

$$A = \begin{bmatrix} a & b \\ c & d \end{bmatrix}.$$

A is invertible if and only if $ad - bc \neq 0$.

If $ad - bc \neq 0$, then

$$A^{-1} = \frac{1}{ad - bc} \begin{bmatrix} d & -b \\ -c & a \end{bmatrix}.$$

We can't divide by 0, so if $ad - bc = 0$, we don't have an inverse. Recall Example 54, where

$$A = \begin{bmatrix} 1 & 2 \\ 2 & 4 \end{bmatrix}.$$

Here, $ad - bc = 1(4) - 2(2) = 0$, which is why A didn't have an inverse.

Although this idea is simple, we should practice it.

Example 56 Use Theorem 7 to find the inverse of

$$A = \begin{bmatrix} 3 & 2 \\ -1 & 9 \end{bmatrix}$$

if it exists.

---

[18] We don't prove this theorem here, but it really isn't hard to do. Put the matrix

$$\begin{bmatrix} a & b & 1 & 0 \\ c & d & 0 & 1 \end{bmatrix}$$

into reduced row echelon form and you'll discover the result of the theorem. Alternatively, multiply A by what we propose is the inverse and see that we indeed get I.

Solution Since $ad - bc = 29 \neq 0$, A^{-1} exists. By the Theorem,

$$A^{-1} = \frac{1}{3(9) - 2(-1)} \begin{bmatrix} 9 & -2 \\ 1 & 3 \end{bmatrix} = \frac{1}{29} \begin{bmatrix} 9 & -2 \\ 1 & 3 \end{bmatrix}$$

We can leave our answer in this form, or we could "simplify" it as

$$A^{-1} = \frac{1}{29} \begin{bmatrix} 9 & -2 \\ 1 & 3 \end{bmatrix} = \begin{bmatrix} 9/29 & -2/29 \\ 1/29 & 3/29 \end{bmatrix}.$$

We started this section out by speculating that just as we solved algebraic equations of the form $ax = b$ by computing $x = a^{-1}b$, we might be able to solve matrix equations of the form $A\vec{x} = \vec{b}$ by computing $\vec{x} = A^{-1}\vec{b}$. If A^{-1} does exist, then we *can* solve the equation $A\vec{x} = \vec{b}$ this way. Consider:

$$\begin{aligned} A\vec{x} &= \vec{b} && \text{(original equation)} \\ A^{-1}A\vec{x} &= A^{-1}\vec{b} && \text{(multiply both sides \textit{on the left} by } A^{-1}\text{)} \\ I\vec{x} &= A^{-1}\vec{b} && \text{(since } A^{-1}A = I\text{)} \\ \vec{x} &= A^{-1}\vec{b} && \text{(since } I\vec{x} = \vec{x}\text{)} \end{aligned}$$

Let's step back and think about this for a moment. The only thing we know about the equation $A\vec{x} = \vec{b}$ is that A is invertible. We also know that solutions to $A\vec{x} = \vec{b}$ come in three forms: exactly one solution, infinite solutions, and no solution. We just showed that if A is invertible, then $A\vec{x} = \vec{b}$ has *at least* one solution. We showed that by setting $\vec{x}$ equal to $A^{-1}\vec{b}$, we have a solution. Is it possible that more solutions exist?

No. Suppose we are told that a known vector $\vec{v}$ is a solution to the equation $A\vec{x} = \vec{b}$; that is, we know that $A\vec{v} = \vec{b}$. We can repeat the above steps:

$$\begin{aligned} A\vec{v} &= \vec{b} \\ A^{-1}A\vec{v} &= A^{-1}\vec{b} \\ I\vec{v} &= A^{-1}\vec{b} \\ \vec{v} &= A^{-1}\vec{b}. \end{aligned}$$

This shows that *all* solutions to $A\vec{x} = \vec{b}$ are exactly $\vec{x} = A^{-1}\vec{b}$ when A is invertible. We have just proved the following theorem.

Theorem 8

Invertible Matrices and Solutions to $A\vec{x} = \vec{b}$

Let A be an invertible $n \times n$ matrix, and let $\vec{b}$ be any $n \times 1$ column vector. Then the equation $A\vec{x} = \vec{b}$ has exactly one solution, namely

$$\vec{x} = A^{-1}\vec{b}.$$

A corollary[19] to this theorem is: If A is not invertible, then $A\vec{x} = \vec{b}$ does not have exactly one solution. It may have infinite solutions and it may have no solution, and we would need to examine the reduced row echelon form of the augmented matrix $\begin{bmatrix} A & \vec{b} \end{bmatrix}$ to see which case applies.

We demonstrate our theorem with an example.

Example 57 Solve $A\vec{x} = \vec{b}$ by computing $\vec{x} = A^{-1}\vec{b}$, where

$$A = \begin{bmatrix} 1 & 0 & -3 \\ -3 & -4 & 10 \\ 4 & -5 & -11 \end{bmatrix} \text{ and } \vec{b} = \begin{bmatrix} -15 \\ 57 \\ -46 \end{bmatrix}.$$

SOLUTION Without showing our steps, we compute

$$A^{-1} = \begin{bmatrix} 94 & 15 & -12 \\ 7 & 1 & -1 \\ 31 & 5 & -4 \end{bmatrix}.$$

We then find the solution to $A\vec{x} = \vec{b}$ by computing $A^{-1}\vec{b}$:

$$\begin{aligned} \vec{x} &= A^{-1}\vec{b} \\ &= \begin{bmatrix} 94 & 15 & -12 \\ 7 & 1 & -1 \\ 31 & 5 & -4 \end{bmatrix} \begin{bmatrix} -15 \\ 57 \\ -46 \end{bmatrix} \\ &= \begin{bmatrix} -3 \\ -2 \\ 4 \end{bmatrix}. \end{aligned}$$

We can easily check our answer:

$$\begin{bmatrix} 1 & 0 & -3 \\ -3 & -4 & 10 \\ 4 & -5 & -11 \end{bmatrix} \begin{bmatrix} -3 \\ -2 \\ 4 \end{bmatrix} = \begin{bmatrix} -15 \\ 57 \\ -46 \end{bmatrix}.$$

[19] a *corollary* is an idea that follows directly from a theorem

Knowing a matrix is invertible is incredibly useful.[20] Among many other reasons, if you know A is invertible, then you know for sure that $A\vec{x} = \vec{b}$ has a solution (as we just stated in Theorem 8). In the next section we'll demonstrate many different properties of invertible matrices, including stating several different ways in which we know that a matrix is invertible.

Exercises 2.6

In Exercises 1 – 8, A matrix A is given. Find A^{-1} using Theorem 7, if it exists.

1. $\begin{bmatrix} 1 & 5 \\ -5 & -24 \end{bmatrix}$

2. $\begin{bmatrix} 1 & -4 \\ 1 & -3 \end{bmatrix}$

3. $\begin{bmatrix} 3 & 0 \\ 0 & 7 \end{bmatrix}$

4. $\begin{bmatrix} 2 & 5 \\ 3 & 4 \end{bmatrix}$

5. $\begin{bmatrix} 1 & -3 \\ -2 & 6 \end{bmatrix}$

6. $\begin{bmatrix} 3 & 7 \\ 2 & 4 \end{bmatrix}$

7. $\begin{bmatrix} 1 & 0 \\ 0 & 1 \end{bmatrix}$

8. $\begin{bmatrix} 0 & 1 \\ 1 & 0 \end{bmatrix}$

In Exercises 9 – 28, a matrix A is given. Find A^{-1} using Key Idea 10, if it exists.

9. $\begin{bmatrix} -2 & 3 \\ 1 & 5 \end{bmatrix}$

10. $\begin{bmatrix} -5 & -2 \\ 9 & 2 \end{bmatrix}$

11. $\begin{bmatrix} 1 & 2 \\ 3 & 4 \end{bmatrix}$

12. $\begin{bmatrix} 5 & 7 \\ 5/3 & 7/3 \end{bmatrix}$

13. $\begin{bmatrix} 25 & -10 & -4 \\ -18 & 7 & 3 \\ -6 & 2 & 1 \end{bmatrix}$

14. $\begin{bmatrix} 2 & 3 & 4 \\ -3 & 6 & 9 \\ -1 & 9 & 13 \end{bmatrix}$

15. $\begin{bmatrix} 1 & 0 & 0 \\ 4 & 1 & -7 \\ 20 & 7 & -48 \end{bmatrix}$

16. $\begin{bmatrix} -4 & 1 & 5 \\ -5 & 1 & 9 \\ -10 & 2 & 19 \end{bmatrix}$

17. $\begin{bmatrix} 5 & -1 & 0 \\ 7 & 7 & 1 \\ -2 & -8 & -1 \end{bmatrix}$

18. $\begin{bmatrix} 1 & -5 & 0 \\ -2 & 15 & 4 \\ 4 & -19 & 1 \end{bmatrix}$

19. $\begin{bmatrix} 25 & -8 & 0 \\ -78 & 25 & 0 \\ 48 & -15 & 1 \end{bmatrix}$

20. $\begin{bmatrix} 1 & 0 & 0 \\ 7 & 5 & 8 \\ -2 & -2 & -3 \end{bmatrix}$

21. $\begin{bmatrix} 0 & 0 & 1 \\ 1 & 0 & 0 \\ 0 & 1 & 0 \end{bmatrix}$

22. $\begin{bmatrix} 0 & 1 & 0 \\ 1 & 0 & 0 \\ 0 & 0 & 1 \end{bmatrix}$

23. $\begin{bmatrix} 1 & 0 & 0 & 0 \\ -19 & -9 & 0 & 4 \\ 33 & 4 & 1 & -7 \\ 4 & 2 & 0 & -1 \end{bmatrix}$

24. $\begin{bmatrix} 1 & 0 & 0 & 0 \\ 27 & 1 & 0 & 4 \\ 18 & 0 & 1 & 4 \\ 4 & 0 & 0 & 1 \end{bmatrix}$

[20] As odd as it may sound, *knowing* a matrix is invertible is useful; actually computing the inverse isn't. This is discussed at the end of the next section.

25. $\begin{bmatrix} -15 & 45 & -3 & 4 \\ 55 & -164 & 15 & -15 \\ -215 & 640 & -62 & 59 \\ -4 & 12 & 0 & 1 \end{bmatrix}$

26. $\begin{bmatrix} 1 & 0 & 2 & 8 \\ 0 & 1 & 0 & 0 \\ 0 & -4 & -29 & -110 \\ 0 & -3 & -5 & -19 \end{bmatrix}$

27. $\begin{bmatrix} 0 & 0 & 1 & 0 \\ 0 & 0 & 0 & 1 \\ 1 & 0 & 0 & 0 \\ 0 & 1 & 0 & 0 \end{bmatrix}$

28. $\begin{bmatrix} 1 & 0 & 0 & 0 \\ 0 & 2 & 0 & 0 \\ 0 & 0 & 3 & 0 \\ 0 & 0 & 0 & -4 \end{bmatrix}$

In Exercises 29 – 36, a matrix A and a vector $\vec{b}$ are given. Solve the equation $A\vec{x} = \vec{b}$ using Theorem 8.

29. $A = \begin{bmatrix} 3 & 5 \\ 2 & 3 \end{bmatrix}$, $\vec{b} = \begin{bmatrix} 21 \\ 13 \end{bmatrix}$

30. $A = \begin{bmatrix} 1 & -4 \\ 4 & -15 \end{bmatrix}$, $\vec{b} = \begin{bmatrix} 21 \\ 77 \end{bmatrix}$

31. $A = \begin{bmatrix} 9 & 70 \\ -4 & -31 \end{bmatrix}$, $\vec{b} = \begin{bmatrix} -2 \\ 1 \end{bmatrix}$

32. $A = \begin{bmatrix} 10 & -57 \\ 3 & -17 \end{bmatrix}$, $\vec{b} = \begin{bmatrix} -14 \\ -4 \end{bmatrix}$

33. $A = \begin{bmatrix} 1 & 2 & 12 \\ 0 & 1 & 6 \\ -3 & 0 & 1 \end{bmatrix}$,

$\vec{b} = \begin{bmatrix} -17 \\ -5 \\ 20 \end{bmatrix}$

34. $A = \begin{bmatrix} 1 & 0 & -3 \\ 8 & -2 & -13 \\ 12 & -3 & -20 \end{bmatrix}$,

$\vec{b} = \begin{bmatrix} -34 \\ -159 \\ -243 \end{bmatrix}$

35. $A = \begin{bmatrix} 5 & 0 & -2 \\ -8 & 1 & 5 \\ -2 & 0 & 1 \end{bmatrix}$,

$\vec{b} = \begin{bmatrix} 33 \\ -70 \\ -15 \end{bmatrix}$

36. $A = \begin{bmatrix} 1 & -6 & 0 \\ 0 & 1 & 0 \\ 2 & -8 & 1 \end{bmatrix}$,

$\vec{b} = \begin{bmatrix} -69 \\ 10 \\ -102 \end{bmatrix}$

2.7 Properties of the Matrix Inverse

AS YOU READ . . .

1. What does it mean to say that two statements are "equivalent?"
2. T/F: If A is not invertible, then $A\vec{x} = \vec{0}$ could have no solutions.
3. T/F: If A is not invertible, then $A\vec{x} = \vec{b}$ could have infinite solutions.
4. What is the inverse of the inverse of A?
5. T/F: Solving $A\vec{x} = \vec{b}$ using Gaussian elimination is faster than using the inverse of A.

We ended the previous section by stating that invertible matrices are important. Since they are, in this section we study invertible matrices in two ways. First, we look at ways to tell whether or not a matrix is invertible, and second, we study properties of invertible matrices (that is, how they interact with other matrix operations).

We start with collecting ways in which we know that a matrix is invertible. We actually already know the truth of this theorem from our work in the previous section, but it is good to list the following statements in one place. As we move through other sections, we'll add on to this theorem.

Theorem 9 **Invertible Matrix Theorem**

Let A be an $n \times n$ matrix. The following statements are equivalent.

(a) A is invertible.

(b) There exists a matrix B such that $BA = I$.

(c) There exists a matrix C such that $AC = I$.

(d) The reduced row echelon form of A is I.

(e) The equation $A\vec{x} = \vec{b}$ has exactly one solution for every $n \times 1$ vector $\vec{b}$.

(f) The equation $A\vec{x} = \vec{0}$ has exactly one solution (namely, $\vec{x} = \vec{0}$).

Let's make note of a few things about the Invertible Matrix Theorem.

1. First, note that the theorem uses the phrase "the following statements are *equivalent*." When two or more statements are equivalent, it means that the truth of any one of them implies that the rest are also true; if any one of the statements is false, then they are all false. So, for example, if we determined that the equation $A\vec{x} = \vec{0}$ had exactly one solution (and A was an $n \times n$ matrix) then we would know that A was invertible, that $A\vec{x} = \vec{b}$ had only one solution, that the reduced row echelon form of A was I, etc.

2. Let's go through each of the statements and see why we already knew they all said essentially the same thing.

   (a) This simply states that A is invertible – that is, that there exists a matrix A^{-1} such that $A^{-1}A = AA^{-1} = I$. We'll go on to show why all the other statements basically tell us "A is invertible."

   (b) If we know that A is invertible, then we already know that there is a matrix B where $BA = I$. That is part of the definition of invertible. However, we can also "go the other way." Recall from Theorem 5 that even if all we know is that there is a matrix B where $BA = I$, then we also know that $AB = I$. That is, we know that B is the inverse of A (and hence A is invertible).

(c) We use the same logic as in the previous statement to show why this is the same as "A is invertible."

(d) If A is invertible, we can find the inverse by using Key Idea 10 (which in turn depends on Theorem 5). The crux of Key Idea 10 is that the reduced row echelon form of A is I; if it is something else, we can't find A^{-1} (it doesn't exist). Knowing that A is invertible means that the reduced row echelon form of A is I. We can go the other way; if we know that the reduced row echelon form of A is I, then we can employ Key Idea 10 to find A^{-1}, so A is invertible.

(e) We know from Theorem 8 that if A is invertible, then given any vector $\vec{b}$, $A\vec{x} = \vec{b}$ has always has exactly one solution, namely $\vec{x} = A^{-1}\vec{b}$. However, we can go the other way; let's say we know that $A\vec{x} = \vec{b}$ always has exactly solution. How can we conclude that A is invertible?

Think about how we, up to this point, determined the solution to $A\vec{x} = \vec{b}$. We set up the augmented matrix $\begin{bmatrix} A & \vec{b} \end{bmatrix}$ and put it into reduced row echelon form. We know that getting the identity matrix on the left means that we had a unique solution (and not getting the identity means we either have no solution or infinite solutions). So getting I on the left means having a unique solution; having I on the left means that the reduced row echelon form of A is I, which we know from above is the same as A being invertible.

(f) This is the same as the above; simply replace the vector $\vec{b}$ with the vector $\vec{0}$.

So we came up with a list of statements that are all *equivalent* to the statement "A is invertible." Again, if we know that if any one of them is true (or false), then they are all true (or all false).

Theorem 9 states formally that if A is invertible, then $A\vec{x} = \vec{b}$ has exactly one solution, namely $A^{-1}\vec{b}$. What if A is not invertible? What are the possibilities for solutions to $A\vec{x} = \vec{b}$?

We know that $A\vec{x} = \vec{b}$ *cannot* have exactly one solution; if it did, then by our theorem it would be invertible. Recalling that linear equations have either one solution, infinite solutions, or no solution, we are left with the latter options when A is not invertible. This idea is important and so we'll state it again as a Key Idea.

Key Idea 11

Solutions to $A\vec{x} = \vec{b}$ and the Invertibility of A

Consider the system of linear equations $A\vec{x} = \vec{b}$.

1. If A is invertible, then $A\vec{x} = \vec{b}$ has exactly one solution, namely $A^{-1}\vec{b}$.

2. If A is not invertible, then $A\vec{x} = \vec{b}$ has either infinite solutions or no solution.

In Theorem 9 we've come up with a list of ways in which we can tell whether or not a matrix is invertible. At the same time, we have come up with a list of properties of invertible matrices – things we know that are true about them. (For instance, if we know that A is invertible, then we know that $A\vec{x} = \vec{b}$ has only one solution.)

We now go on to discover other properties of invertible matrices. Specifically, we want to find out how invertibility interacts with other matrix operations. For instance, if we know that A and B are invertible, what is the inverse of $A+B$? What is the inverse of AB? What is "the inverse of the inverse?" We'll explore these questions through an example.

Example 58 Let

$$A = \begin{bmatrix} 3 & 2 \\ 0 & 1 \end{bmatrix} \text{ and } B = \begin{bmatrix} -2 & 0 \\ 1 & 1 \end{bmatrix}.$$

Find:

1. A^{-1}
2. B^{-1}
3. $(AB)^{-1}$
4. $(A^{-1})^{-1}$
5. $(A+B)^{-1}$
6. $(5A)^{-1}$

In addition, try to find connections between each of the above.

Solution

1. Computing A^{-1} is straightforward; we'll use Theorem 7.

$$A^{-1} = \frac{1}{3}\begin{bmatrix} 1 & -2 \\ 0 & 3 \end{bmatrix} = \begin{bmatrix} 1/3 & -2/3 \\ 0 & 1 \end{bmatrix}$$

2. We compute B^{-1} in the same way as above.

$$B^{-1} = \frac{1}{-2}\begin{bmatrix} 1 & 0 \\ -1 & -2 \end{bmatrix} = \begin{bmatrix} -1/2 & 0 \\ 1/2 & 1 \end{bmatrix}$$

3. To compute $(AB)^{-1}$, we first compute AB:

$$AB = \begin{bmatrix} 3 & 2 \\ 0 & 1 \end{bmatrix}\begin{bmatrix} -2 & 0 \\ 1 & 1 \end{bmatrix} = \begin{bmatrix} -4 & 2 \\ 1 & 1 \end{bmatrix}$$

We now apply Theorem 7 to find $(AB)^{-1}$.

$$(AB)^{-1} = \frac{1}{-6}\begin{bmatrix} 1 & -2 \\ -1 & -4 \end{bmatrix} = \begin{bmatrix} -1/6 & 1/3 \\ 1/6 & 2/3 \end{bmatrix}$$

4. To compute $(A^{-1})^{-1}$, we simply apply Theorem 7 to A^{-1}:

$$(A^{-1})^{-1} = \frac{1}{1/3}\begin{bmatrix} 1 & 2/3 \\ 0 & 1/3 \end{bmatrix} = \begin{bmatrix} 3 & 2 \\ 0 & 1 \end{bmatrix}.$$

5. To compute $(A+B)^{-1}$, we first compute $A+B$ then apply Theorem 7:

$$A+B = \begin{bmatrix} 3 & 2 \\ 0 & 1 \end{bmatrix} + \begin{bmatrix} -2 & 0 \\ 1 & 1 \end{bmatrix} = \begin{bmatrix} 1 & 2 \\ 1 & 2 \end{bmatrix}.$$

Hence

$$(A+B)^{-1} = \frac{1}{0}\begin{bmatrix} 2 & -2 \\ -1 & 1 \end{bmatrix} = !$$

Our last expression is really nonsense; we know that if $ad - bc = 0$, then the given matrix is not invertible. That is the case with $A+B$, so we conclude that $A+B$ is not invertible.

6. To compute $(5A)^{-1}$, we compute $5A$ and then apply Theorem 7.

$$(5A)^{-1} = \left(\begin{bmatrix} 15 & 10 \\ 0 & 5 \end{bmatrix}\right)^{-1} = \frac{1}{75}\begin{bmatrix} 5 & -10 \\ 0 & 15 \end{bmatrix} = \begin{bmatrix} 1/15 & -2/15 \\ 0 & 1/5 \end{bmatrix}$$

We now look for connections between A^{-1}, B^{-1}, $(AB)^{-1}$, $(A^{-1})^{-1}$ and $(A+B)^{-1}$.

3. Is there some sort of relationship between $(AB)^{-1}$ and A^{-1} and B^{-1}? A first guess that seems plausible is $(AB)^{-1} = A^{-1}B^{-1}$. Is this true? Using our work from above, we have

$$A^{-1}B^{-1} = \begin{bmatrix} 1/3 & -2/3 \\ 0 & 1 \end{bmatrix}\begin{bmatrix} -1/2 & 0 \\ 1/2 & 1 \end{bmatrix} = \begin{bmatrix} -1/2 & -2/3 \\ 1/2 & 1 \end{bmatrix}.$$

Obviously, this is not equal to $(AB)^{-1}$. Before we do some further guessing, let's think about what the inverse of AB is supposed to do. The inverse – let's call it C – is supposed to be a matrix such that

$$(AB)C = C(AB) = I.$$

In examining the expression $(AB)C$, we see that we want B to somehow "cancel" with C. What "cancels" B? An obvious answer is B^{-1}. This gives us a thought:

perhaps we got the order of A^{-1} and B^{-1} wrong before. After all, we were hoping to find that

$$ABA^{-1}B^{-1} \stackrel{?}{=} I,$$

but algebraically speaking, it is hard to cancel out these terms.[21] However, switching the order of A^{-1} and B^{-1} gives us some hope. Is $(AB)^{-1} = B^{-1}A^{-1}$? Let's see.

$$\begin{aligned}
(AB)(B^{-1}A^{-1}) &= A(BB^{-1})A^{-1} && \text{(regrouping by the associative property)}\\
&= AIA^{-1} && (BB^{-1} = I)\\
&= AA^{-1} && (AI = A)\\
&= I && (AA^{-1} = I)
\end{aligned}$$

Thus it seems that $(AB)^{-1} = B^{-1}A^{-1}$. Let's confirm this with our example matrices.

$$B^{-1}A^{-1} = \begin{bmatrix} -1/2 & 0 \\ 1/2 & 1 \end{bmatrix} \begin{bmatrix} 1/3 & -2/3 \\ 0 & 1 \end{bmatrix} = \begin{bmatrix} -1/6 & 1/3 \\ 1/6 & 2/3 \end{bmatrix} = (AB)^{-1}.$$

It worked!

4. Is there some sort of connection between $(A^{-1})^{-1}$ and A? The answer is pretty obvious: they are equal. The "inverse of the inverse" returns one to the original matrix.

5. Is there some sort of relationship between $(A+B)^{-1}$, A^{-1} and B^{-1}? Certainly, if we were forced to make a guess without working any examples, we would guess that

$$(A+B)^{-1} \stackrel{?}{=} A^{-1} + B^{-1}.$$

However, we saw that in our example, the matrix $(A+B)$ isn't even invertible. This pretty much kills any hope of a connection.

6. Is there a connection between $(5A)^{-1}$ and A^{-1}? Consider:

$$\begin{aligned}
(5A)^{-1} &= \begin{bmatrix} 1/15 & -2/15 \\ 0 & 1/5 \end{bmatrix}\\
&= \frac{1}{5}\begin{bmatrix} 1/3 & -2/3 \\ 0 & 1/5 \end{bmatrix}\\
&= \frac{1}{5}A^{-1}
\end{aligned}$$

Yes, there is a connection!

Let's summarize the results of this example. If A and B are both invertible matrices, then so is their product, AB. We demonstrated this with our example, and there is

[21] Recall that matrix multiplication is not commutative.

more to be said. Let's suppose that A and B are $n \times n$ matrices, but we don't yet know if they are invertible. If AB is invertible, then each of A and B are; if AB is not invertible, then A or B is also not invertible.

In short, invertibility "works well" with matrix multiplication. However, we saw that it doesn't work well with matrix addition. Knowing that A and B are invertible does not help us find the inverse of $(A + B)$; in fact, the latter matrix may not even be invertible.[22]

Let's do one more example, then we'll summarize the results of this section in a theorem.

Example 59 Find the inverse of $A = \begin{bmatrix} 2 & 0 & 0 \\ 0 & 3 & 0 \\ 0 & 0 & -7 \end{bmatrix}$.

SOLUTION We'll find A^{-1} using Key Idea 10.

$$\begin{bmatrix} 2 & 0 & 0 & 1 & 0 & 0 \\ 0 & 3 & 0 & 0 & 1 & 0 \\ 0 & 0 & -7 & 0 & 0 & 1 \end{bmatrix} \quad \overrightarrow{\text{rref}} \quad \begin{bmatrix} 1 & 0 & 0 & 1/2 & 0 & 0 \\ 0 & 1 & 0 & 0 & 1/3 & 0 \\ 0 & 0 & 1 & 0 & 0 & -1/7 \end{bmatrix}$$

Therefore

$$A^{-1} = \begin{bmatrix} 1/2 & 0 & 0 \\ 0 & 1/3 & 0 \\ 0 & 0 & -1/7 \end{bmatrix}.$$

The matrix A in the previous example is a *diagonal* matrix: the only nonzero entries of A lie on the *diagonal*.[23] The relationship between A and A^{-1} in the above example seems pretty strong, and it holds true in general. We'll state this and summarize the results of this section with the following theorem.

---

[22] The fact that invertibility works well with matrix multiplication should not come as a surprise. After all, saying that A is invertible makes a statement about the mulitiplicative properties of A. It says that I can multiply A with a special matrix to get I. Invertibility, in and of itself, says nothing about matrix addition, therefore we should not be too surprised that it doesn't work well with it.

[23] We still haven't formally defined *diagonal*, but the definition is rather visual so we risk it. See Definition 20 on page 123 for more details.

Theorem 10

Properties of Invertible Matrices

Let A and B be $n \times n$ invertible matrices. Then:

1. AB is invertible; $(AB)^{-1} = B^{-1}A^{-1}$.
2. A^{-1} is invertible; $(A^{-1})^{-1} = A$.
3. nA is invertible for any nonzero scalar n; $(nA)^{-1} = \frac{1}{n}A^{-1}$.
4. If A is a diagonal matrix, with diagonal entries $d_1, d_2, \cdots, d_n$, where none of the diagonal entries are 0, then A^{-1} exists and is a diagonal matrix. Furthermore, the diagonal entries of A^{-1} are $1/d_1, 1/d_2, \cdots, 1/d_n$.

Furthermore,

1. If a product AB is not invertible, then A or B is not invertible.
2. If A or B are not invertible, then AB is not invertible.

We end this section with a comment about solving systems of equations "in real life."[24] Solving a system $A\vec{x} = \vec{b}$ by computing $A^{-1}\vec{b}$ seems pretty slick, so it would make sense that this is the way it is normally done. However, in practice, this is rarely done. There are two main reasons why this is the case.

First, computing A^{-1} and $A^{-1}\vec{b}$ is "expensive" in the sense that it takes up a lot of computing time. Certainly, our calculators have no trouble dealing with the 3×3 cases we often consider in this textbook, but in real life the matrices being considered are very large (as in, hundreds of thousand rows and columns). Computing A^{-1} alone is rather impractical, and we waste a lot of time if we come to find out that A^{-1} does not exist. Even if we already know what A^{-1} is, computing $A^{-1}\vec{b}$ is computationally expensive – Gaussian elimination is faster.

Secondly, computing A^{-1} using the method we've described often gives rise to numerical roundoff errors. Even though computers often do computations with an accuracy to more than 8 decimal places, after thousands of computations, roundoffs

[24]Yes, real people do solve linear equations in real life. Not just mathematicians, but economists, engineers, and scientists of all flavors regularly need to solve linear equations, and the matrices they use are often *huge*.

Most people see matrices at work without thinking about it. Digital pictures are simply "rectangular arrays" of numbers representing colors – they are matrices of colors. Many of the standard image processing operations involve matrix operations. The author's wife has a "7 megapixel" camera which creates pictures that are 3072×2304 in size, giving over 7 million pixels, and that isn't even considered a "large" picture these days.

can cause big errors. (A "small" $1,000 \times 1,000$ matrix has $1,000,000$ entries! That's a lot of places to have roundoff errors accumulate!) It is not unheard of to have a computer compute A^{-1} for a large matrix, and then immediately have it compute AA^{-1} and *not* get the identity matrix.[25]

Therefore, in real life, solutions to $A\vec{x} = \vec{b}$ are usually found using the methods we learned in Section 2.4. It turns out that even with all of our advances in mathematics, it is hard to beat the basic method that Gauss introduced a long time ago.

Exercises 2.7

In Exercises 1 – 4, matrices A and B are given. Compute $(AB)^{-1}$ and $B^{-1}A^{-1}$.

1. $A = \begin{bmatrix} 1 & 2 \\ 1 & 1 \end{bmatrix}$, $B = \begin{bmatrix} 3 & 5 \\ 2 & 5 \end{bmatrix}$

2. $A = \begin{bmatrix} 1 & 2 \\ 3 & 4 \end{bmatrix}$, $B = \begin{bmatrix} 7 & 1 \\ 2 & 1 \end{bmatrix}$

3. $A = \begin{bmatrix} 2 & 5 \\ 3 & 8 \end{bmatrix}$, $B = \begin{bmatrix} 1 & -1 \\ 1 & 4 \end{bmatrix}$

4. $A = \begin{bmatrix} 2 & 4 \\ 2 & 5 \end{bmatrix}$, $B = \begin{bmatrix} 2 & 2 \\ 6 & 5 \end{bmatrix}$

In Exercises 5 – 8, a 2×2 matrix A is given. Compute A^{-1} and $(A^{-1})^{-1}$ using Theorem 7.

5. $A = \begin{bmatrix} -3 & 5 \\ 1 & -2 \end{bmatrix}$

6. $A = \begin{bmatrix} 3 & 5 \\ 2 & 4 \end{bmatrix}$

7. $A = \begin{bmatrix} 2 & 7 \\ 1 & 3 \end{bmatrix}$

8. $A = \begin{bmatrix} 9 & 0 \\ 7 & 9 \end{bmatrix}$

9. Find 2×2 matrices A and B that are each invertible, but $A + B$ is not.

10. Create a random 6×6 matrix A, then have a calculator or computer compute AA^{-1}. Was the identity matrix returned exactly? Comment on your results.

11. Use a calculator or computer to compute AA^{-1}, where

$$A = \begin{bmatrix} 1 & 2 & 3 & 4 \\ 1 & 4 & 9 & 16 \\ 1 & 8 & 27 & 64 \\ 1 & 16 & 81 & 256 \end{bmatrix}.$$

Was the identity matrix returned exactly? Comment on your results.

[25] The result is usually very close, with the numbers on the diagonal close to 1 and the other entries near 0. But it isn't exactly the identity matrix.

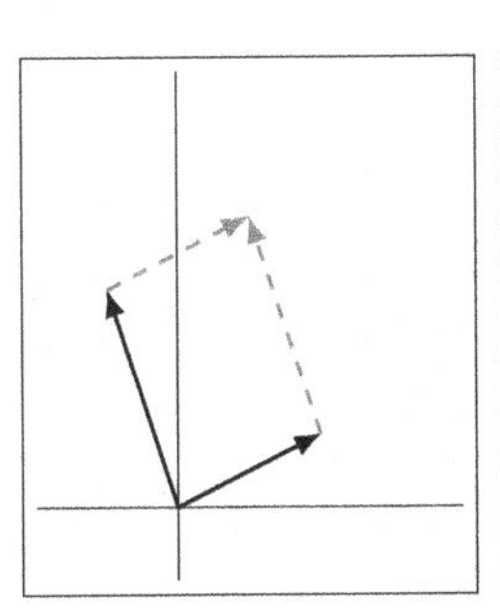

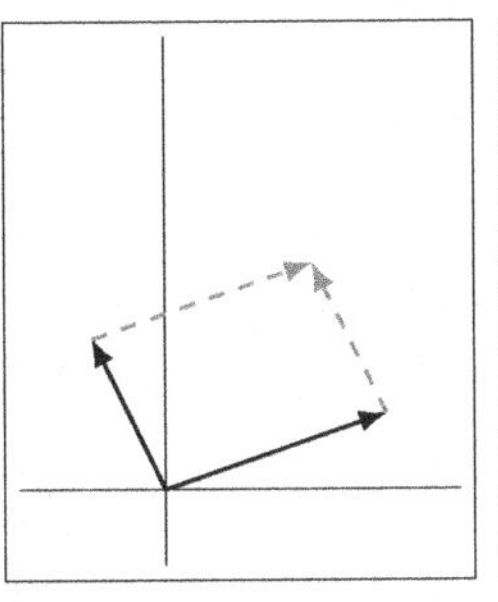

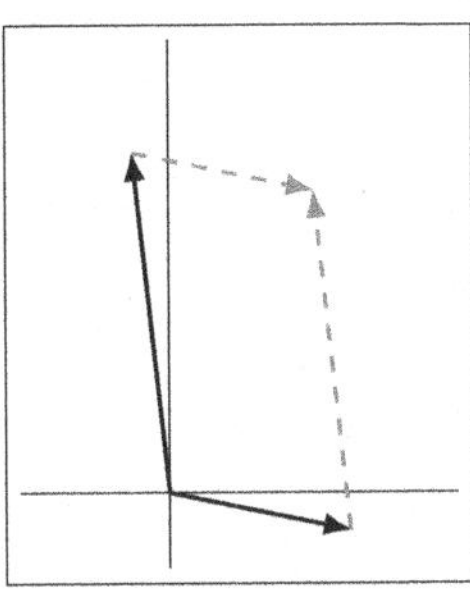

OPERATIONS ON MATRICES

In the previous chapter we learned about matrix arithmetic: adding, subtracting, and multiplying matrices, finding inverses, and multiplying by scalars. In this chapter we learn about some operations that we perform *on* matrices. We can think of them as functions: you input a matrix, and you get something back. One of these operations, the transpose, will return another matrix. With the other operations, the trace and the determinant, we input matrices and get numbers in return, an idea that is different than what we have seen before.

3.1 The Matrix Transpose

1. T/F: If A is a 3×5 matrix, then A^T will be a 5×3 matrix.

2. Where are there zeros in an upper triangular matrix?

3. T/F: A matrix is symmetric if it doesn't change when you take its transpose.

4. What is the transpose of the transpose of A?

5. Give 2 other terms to describe symmetric matrices besides "interesting."

We jump right in with a definition.

Definition 19

Transpose

Let A be an $m \times n$ matrix. The *tranpsose of* A, denoted A^T, is the $n \times m$ matrix whose columns are the respective rows of A.

Examples will make this definition clear.

Example 60 Find the transpose of $A = \begin{bmatrix} 1 & 2 & 3 \\ 4 & 5 & 6 \end{bmatrix}$.

Solution Note that A is a 2×3 matrix, so A^T will be a 3×2 matrix. By the definition, the first column of A^T is the first row of A; the second column of A^T is the second row of A. Therefore,

$$A^T = \begin{bmatrix} 1 & 4 \\ 2 & 5 \\ 3 & 6 \end{bmatrix}.$$

Example 61 Find the transpose of the following matrices.

$$A = \begin{bmatrix} 7 & 2 & 9 & 1 \\ 2 & -1 & 3 & 0 \\ -5 & 3 & 0 & 11 \end{bmatrix} \quad B = \begin{bmatrix} 1 & 10 & -2 \\ 3 & -5 & 7 \\ 4 & 2 & -3 \end{bmatrix} \quad C = \begin{bmatrix} 1 & -1 & 7 & 8 & 3 \end{bmatrix}$$

Solution We find each transpose using the definition without explanation. Make note of the dimensions of the original matrix and the dimensions of its transpose.

$$A^T = \begin{bmatrix} 7 & 2 & -5 \\ 2 & -1 & 3 \\ 9 & 3 & 0 \\ 1 & 0 & 11 \end{bmatrix} \quad B^T = \begin{bmatrix} 1 & 3 & 4 \\ 10 & -5 & 2 \\ -2 & 7 & -3 \end{bmatrix} \quad C^T = \begin{bmatrix} 1 \\ -1 \\ 7 \\ 8 \\ 3 \end{bmatrix}$$

Notice that with matrix B, when we took the transpose, the *diagonal* did not change. We can see what the diagonal is below where we rewrite B and B^T with the diagonal in bold. We'll follow this by a definition of what we mean by "the diagonal of a matrix," along with a few other related definitions.

$$B = \begin{bmatrix} \mathbf{1} & 10 & -2 \\ 3 & \mathbf{-5} & 7 \\ 4 & 2 & \mathbf{-3} \end{bmatrix} \quad B^T = \begin{bmatrix} \mathbf{1} & 3 & 4 \\ 10 & \mathbf{-5} & 2 \\ -2 & 7 & \mathbf{-3} \end{bmatrix}$$

It is probably pretty clear why we call those entries "the diagonal." Here is the formal definition.

Definition 20

The Diagonal, a Diagonal Matrix, Triangular Matrices

Let A be an $m \times n$ matrix. The *diagonal of A* consists of the entries $a_{11}, a_{22}, \ldots$ of A.

A *diagonal matrix* is an $n \times n$ matrix in which the only nonzero entries lie on the diagonal.

An *upper (lower) triangular* matrix is a matrix in which any nonzero entries lie on or above (below) the diagonal.

Example 62 Consider the matrices A, B, C and I_4, as well as their transposes, where

$$A = \begin{bmatrix} 1 & 2 & 3 \\ 0 & 4 & 5 \\ 0 & 0 & 6 \end{bmatrix} \quad B = \begin{bmatrix} 3 & 0 & 0 \\ 0 & 7 & 0 \\ 0 & 0 & -1 \end{bmatrix} \quad C = \begin{bmatrix} 1 & 2 & 3 \\ 0 & 4 & 5 \\ 0 & 0 & 6 \\ 0 & 0 & 0 \end{bmatrix}.$$

Identify the diagonal of each matrix, and state whether each matrix is diagonal, upper triangular, lower triangular, or none of the above.

Solution We first compute the transpose of each matrix.

$$A^T = \begin{bmatrix} 1 & 0 & 0 \\ 2 & 4 & 0 \\ 3 & 5 & 6 \end{bmatrix} \quad B^T = \begin{bmatrix} 3 & 0 & 0 \\ 0 & 7 & 0 \\ 0 & 0 & -1 \end{bmatrix} \quad C^T = \begin{bmatrix} 1 & 0 & 0 & 0 \\ 2 & 4 & 0 & 0 \\ 3 & 5 & 6 & 0 \end{bmatrix}$$

Note that $I_4^T = I_4$.

The diagonals of A and A^T are the same, consisting of the entries 1, 4 and 6. The diagonals of B and B^T are also the same, consisting of the entries 3, 7 and -1. Finally, the diagonals of C and C^T are the same, consisting of the entries 1, 4 and 6.

The matrix A is upper triangular; the only nonzero entries lie on or above the diagonal. Likewise, A^T is lower triangular.

The matrix B is diagonal. By their definitions, we can also see that B is both upper and lower triangular. Likewise, I_4 is diagonal, as well as upper and lower triangular.

Finally, C is upper triangular, with C^T being lower triangular.

Make note of the definitions of diagonal and triangular matrices. We specify that a diagonal matrix must be square, but triangular matrices don't have to be. ("Most" of the time, however, the ones we study are.) Also, as we mentioned before in the example, by definition a diagonal matrix is also both upper and lower triangular. Finally, notice that by definition, the transpose of an upper triangular matrix is a lower triangular matrix, and vice-versa.

There are many questions to probe concerning the transpose operations.[1] The first set of questions we'll investigate involve the matrix arithmetic we learned from last chapter. We do this investigation by way of examples, and then summarize what we have learned at the end.

Example 63 Let

$$A = \begin{bmatrix} 1 & 2 & 3 \\ 4 & 5 & 6 \end{bmatrix} \text{ and } B = \begin{bmatrix} 1 & 2 & 1 \\ 3 & -1 & 0 \end{bmatrix}.$$

Find $A^T + B^T$ and $(A + B)^T$.

SOLUTION We note that

$$A^T = \begin{bmatrix} 1 & 4 \\ 2 & 5 \\ 3 & 6 \end{bmatrix} \text{ and } B^T = \begin{bmatrix} 1 & 3 \\ 2 & -1 \\ 1 & 0 \end{bmatrix}.$$

Therefore

$$\begin{aligned} A^T + B^T &= \begin{bmatrix} 1 & 4 \\ 2 & 5 \\ 3 & 6 \end{bmatrix} + \begin{bmatrix} 1 & 3 \\ 2 & -1 \\ 1 & 0 \end{bmatrix} \\ &= \begin{bmatrix} 2 & 7 \\ 4 & 4 \\ 4 & 6 \end{bmatrix}. \end{aligned}$$

Also,

$$\begin{aligned} (A + B)^T &= \left(\begin{bmatrix} 1 & 2 & 3 \\ 4 & 5 & 6 \end{bmatrix} + \begin{bmatrix} 1 & 2 & 1 \\ 3 & -1 & 0 \end{bmatrix} \right)^T \\ &= \left(\begin{bmatrix} 2 & 4 & 4 \\ 7 & 4 & 6 \end{bmatrix} \right)^T \\ &= \begin{bmatrix} 2 & 7 \\ 4 & 4 \\ 4 & 6 \end{bmatrix}. \end{aligned}$$

It looks like "the sum of the transposes is the transpose of the sum."[2] This should lead us to wonder how the transpose works with multiplication.

Example 64 Let

$$A = \begin{bmatrix} 1 & 2 \\ 3 & 4 \end{bmatrix} \text{ and } B = \begin{bmatrix} 1 & 2 & -1 \\ 1 & 0 & 1 \end{bmatrix}.$$

[1]Remember, this is what mathematicians do. We learn something new, and then we ask lots of questions about it. Often the first questions we ask are along the lines of "How does this new thing relate to the old things I already know about?"

[2]This is kind of fun to say, especially when said fast. Regardless of how fast we say it, we should think about this statement. The "is" represents "equals." The stuff before "is" equals the stuff afterwards.

Find $(AB)^T$, A^TB^T and B^TA^T.

SOLUTION We first note that

$$A^T = \begin{bmatrix} 1 & 3 \\ 2 & 4 \end{bmatrix} \text{ and } B^T = \begin{bmatrix} 1 & 1 \\ 2 & 0 \\ -1 & 1 \end{bmatrix}.$$

Find $(AB)^T$:

$$\begin{aligned} (AB)^T &= \left(\begin{bmatrix} 1 & 2 \\ 3 & 4 \end{bmatrix} \begin{bmatrix} 1 & 2 & -1 \\ 1 & 0 & 1 \end{bmatrix} \right)^T \\ &= \left(\begin{bmatrix} 3 & 2 & 1 \\ 7 & 6 & 1 \end{bmatrix} \right)^T \\ &= \begin{bmatrix} 3 & 7 \\ 2 & 6 \\ 1 & 1 \end{bmatrix} \end{aligned}$$

Now find A^TB^T:

$$\begin{aligned} A^TB^T &= \begin{bmatrix} 1 & 3 \\ 2 & 4 \end{bmatrix} \begin{bmatrix} 1 & 1 \\ 2 & 0 \\ -1 & 1 \end{bmatrix} \\ &= \text{Not defined!} \end{aligned}$$

So we can't compute A^TB^T. Let's finish by computing B^TA^T:

$$\begin{aligned} B^TA^T &= \begin{bmatrix} 1 & 1 \\ 2 & 0 \\ -1 & 1 \end{bmatrix} \begin{bmatrix} 1 & 3 \\ 2 & 4 \end{bmatrix} \\ &= \begin{bmatrix} 3 & 7 \\ 2 & 6 \\ 1 & 1 \end{bmatrix} \end{aligned}$$

We may have suspected that $(AB)^T = A^TB^T$. We saw that this wasn't the case, though – and not only was it not equal, the second product wasn't even defined! Oddly enough, though, we saw that $(AB)^T = B^TA^T$. [3] To help understand why this is true, look back at the work above and confirm the steps of each multiplication.

We have one more arithmetic operation to look at: the inverse.

Example 65 Let

$$A = \begin{bmatrix} 2 & 7 \\ 1 & 4 \end{bmatrix}.$$

[3] Then again, maybe this isn't all that "odd." It is reminiscent of the fact that, when invertible, $(AB)^{-1} = B^{-1}A^{-1}$.

Find $(A^{-1})^T$ and $(A^T)^{-1}$.

SOLUTION We first find A^{-1} and A^T:

$$A^{-1} = \begin{bmatrix} 4 & -7 \\ -1 & 2 \end{bmatrix} \text{ and } A^T = \begin{bmatrix} 2 & 1 \\ 7 & 4 \end{bmatrix}.$$

Finding $(A^{-1})^T$:

$$\begin{aligned} (A^{-1})^T &= \begin{bmatrix} 4 & -7 \\ -1 & 2 \end{bmatrix}^T \\ &= \begin{bmatrix} 4 & -1 \\ -7 & 2 \end{bmatrix} \end{aligned}$$

Finding $(A^T)^{-1}$:

$$\begin{aligned} (A^T)^{-1} &= \begin{bmatrix} 2 & 1 \\ 7 & 4 \end{bmatrix}^{-1} \\ &= \begin{bmatrix} 4 & -1 \\ -7 & 2 \end{bmatrix} \end{aligned}$$

It seems that "the inverse of the transpose is the transpose of the inverse."[4]

We have just looked at some examples of how the transpose operation interacts with matrix arithmetic operations.[5] We now give a theorem that tells us that what we saw wasn't a coincidence, but rather is always true.

Theorem 11

Properties of the Matrix Transpose

Let A and B be matrices where the following operations are defined. Then:

1. $(A+B)^T = A^T + B^T$ and $(A-B)^T = A^T - B^T$
2. $(kA)^T = kA^T$
3. $(AB)^T = B^TA^T$
4. $(A^{-1})^T = (A^T)^{-1}$
5. $(A^T)^T = A$

We included in the theorem two ideas we didn't discuss already. First, that $(kA)^T =$

[4]Again, we should think about this statement. The part before "is" states that we take the transpose of a matrix, then find the inverse. The part after "is" states that we find the inverse of the matrix, then take the transpose. Since these two statements are linked by an "is," they are equal.

[5]These examples don't *prove* anything, other than it worked in specific examples.

kA^T. This is probably obvious. It doesn't matter when you multiply a matrix by a scalar when dealing with transposes.

The second "new" item is that $(A^T)^T = A$. That is, if we take the transpose of a matrix, then take its transpose again, what do we have? The original matrix.

Now that we know some properties of the transpose operation, we are tempted to play around with it and see what happens. For instance, if A is an $m \times n$ matrix, we know that A^T is an $n \times m$ matrix. So no matter what matrix A we start with, we can always perform the multiplication AA^T (and also A^TA) and the result is a square matrix!

Another thing to ask ourselves as we "play around" with the transpose: suppose A is a square matrix. Is there anything special about $A + A^T$? The following example has us try out these ideas.

Example 66 Let

$$A = \begin{bmatrix} 2 & 1 & 3 \\ 2 & -1 & 1 \\ 1 & 0 & 1 \end{bmatrix}.$$

Find AA^T, $A + A^T$ and $A - A^T$.

SOLUTION Finding AA^T:

$$AA^T = \begin{bmatrix} 2 & 1 & 3 \\ 2 & -1 & 1 \\ 1 & 0 & 1 \end{bmatrix} \begin{bmatrix} 2 & 2 & 1 \\ 1 & -1 & 0 \\ 3 & 1 & 1 \end{bmatrix} = \begin{bmatrix} 14 & 6 & 5 \\ 6 & 4 & 3 \\ 5 & 3 & 2 \end{bmatrix}$$

Finding $A + A^T$:

$$A + A^T = \begin{bmatrix} 2 & 1 & 3 \\ 2 & -1 & 1 \\ 1 & 0 & 1 \end{bmatrix} + \begin{bmatrix} 2 & 2 & 1 \\ 1 & -1 & 0 \\ 3 & 1 & 1 \end{bmatrix} = \begin{bmatrix} 2 & 3 & 4 \\ 3 & -2 & 1 \\ 4 & 1 & 2 \end{bmatrix}$$

Finding $A - A^T$:

$$A - A^T = \begin{bmatrix} 2 & 1 & 3 \\ 2 & -1 & 1 \\ 1 & 0 & 1 \end{bmatrix} - \begin{bmatrix} 2 & 2 & 1 \\ 1 & -1 & 0 \\ 3 & 1 & 1 \end{bmatrix} = \begin{bmatrix} 0 & -1 & 2 \\ 1 & 0 & 1 \\ -2 & -1 & 0 \end{bmatrix}$$

Let's look at the matrices we've formed in this example. First, consider AA^T. Something seems to be nice about this matrix – look at the location of the 6's, the 5's and the 3's. More precisely, let's look at the transpose of AA^T. We should notice that if we take the transpose of this matrix, we have the very same matrix. That is,

$$\left(\begin{bmatrix} 14 & 6 & 5 \\ 6 & 4 & 3 \\ 5 & 3 & 2 \end{bmatrix}\right)^T = \begin{bmatrix} 14 & 6 & 5 \\ 6 & 4 & 3 \\ 5 & 3 & 2 \end{bmatrix}!$$

We'll formally define this in a moment, but a matrix that is equal to its transpose is called *symmetric*.

Look at the next part of the example; what do we notice about $A + A^T$? We should see that it, too, is symmetric. Finally, consider the last part of the example: do we notice anything about $A - A^T$?

We should immediately notice that it is not symmetric, although it does seem "close." Instead of it being equal to its transpose, we notice that this matrix is the *opposite* of its transpose. We call this type of matrix *skew symmetric.*[6] We formally define these matrices here.

Definition 21

Symmetric and Skew Symmetric Matrices

A matrix A is *symmetric* if $A^T = A$.

A matrix A is *skew symmetric* if $A^T = -A$.

Note that in order for a matrix to be either symmetric or skew symmetric, it must be square.

So why was AA^T symmetric in our previous example? Did we just luck out?[7] Let's take the transpose of AA^T and see what happens.

$$\begin{aligned} (AA^T)^T &= (A^T)^T(A)^T && \text{transpose multiplication rule} \\ &= AA^T && (A^T)^T = A \end{aligned}$$

We have just *proved* that no matter what matrix A we start with, the matrix AA^T will be symmetric. Nothing in our string of equalities even demanded that A be a square matrix; it is always true.

We can do a similar proof to show that as long as A is square, $A + A^T$ is a symmetric matrix.[8] We'll instead show here that if A is a square matrix, then $A - A^T$ is skew

[6]Some mathematicians use the term *antisymmetric*

[7]Of course not.

[8]Why do we say that A has to be square?

symmetric.

$$\begin{aligned}(A - A^T)^T &= A^T - (A^T)^T && \text{transpose subtraction rule}\\ &= A^T - A\\ &= -(A - A^T)\end{aligned}$$

So we took the transpose of $A - A^T$ and we got $-(A - A^T)$; this is the definition of being skew symmetric.

We'll take what we learned from Example 66 and put it in a box. (We've already proved most of this is true; the rest we leave to solve in the Exercises.)

Theorem 12 **Symmetric and Skew Symmetric Matrices**

1. Given any matrix A, the matrices AA^T and A^TA are symmetric.
2. Let A be a square matrix. The matrix $A + A^T$ is symmetric.
3. Let A be a square matrix. The matrix $A - A^T$ is skew symmetric.

Why do we care about the transpose of a matrix? Why do we care about symmetric matrices?

There are two answers that each answer both of these questions. First, we are interested in the tranpose of a matrix and symmetric matrices because they are interesting.[9] One particularly interesting thing about symmetric and skew symmetric matrices is this: consider the sum of $(A + A^T)$ and $(A - A^T)$:

$$(A + A^T) + (A - A^T) = 2A.$$

This gives us an idea: if we were to multiply both sides of this equation by $\frac{1}{2}$, then the right hand side would just be A. This means that

$$A = \underbrace{\frac{1}{2}(A + A^T)}_{\text{symmetric}} + \underbrace{\frac{1}{2}(A - A^T)}_{\text{skew symmetric}}.$$

That is, any matrix A can be written as the sum of a symmetric and skew symmetric matrix. That's interesting.

The second reason we care about them is that they are very useful and important in various areas of mathematics. The transpose of a matrix turns out to be an important

[9] Or: "neat," "cool," "bad," "wicked," "phat," "fo-shizzle."

operation; symmetric matrices have many nice properties that make solving certain types of problems possible.

Most of this text focuses on the preliminaries of matrix algebra, and the actual uses are beyond our current scope. One easy to describe example is curve fitting. Suppose we are given a large set of data points that, when plotted, look roughly quadratic. How do we find the quadratic that "best fits" this data? The solution can be found using matrix algebra, and specifically a matrix called the *pseudoinverse*. If A is a matrix, the pseudoinverse of A is the matrix $A^{\dagger} = (A^TA)^{-1}A^T$ (assuming that the inverse exists). We aren't going to worry about what all the above means; just notice that it has a cool sounding name and the transpose appears twice.

In the next section we'll learn about the trace, another operation that can be performed on a matrix that is relatively simple to compute but can lead to some deep results.

Exercises 3.1

In Exercises 1 – 24, a matrix A is given. Find A^T; make note if A is upper/lower triangular, diagonal, symmetric and/or skew symmetric.

1. $\begin{bmatrix} -7 & 4 \\ 4 & -6 \end{bmatrix}$

2. $\begin{bmatrix} 3 & 1 \\ -7 & 8 \end{bmatrix}$

3. $\begin{bmatrix} 1 & 0 \\ 0 & 9 \end{bmatrix}$

4. $\begin{bmatrix} 13 & -3 \\ -3 & 1 \end{bmatrix}$

5. $\begin{bmatrix} -5 & -9 \\ 3 & 1 \\ -10 & -8 \end{bmatrix}$

6. $\begin{bmatrix} -2 & 10 \\ 1 & -7 \\ 9 & -2 \end{bmatrix}$

7. $\begin{bmatrix} 4 & -7 & -4 & -9 \\ -9 & 6 & 3 & -9 \end{bmatrix}$

8. $\begin{bmatrix} 3 & -10 & 0 & 6 \\ -10 & -2 & -3 & 1 \end{bmatrix}$

9. $\begin{bmatrix} -7 & -8 & 2 & -3 \end{bmatrix}$

10. $\begin{bmatrix} -9 & 8 & 2 & -7 \end{bmatrix}$

11. $\begin{bmatrix} -9 & 4 & 10 \\ 6 & -3 & -7 \\ -8 & 1 & -1 \end{bmatrix}$

12. $\begin{bmatrix} 4 & -5 & 2 \\ 1 & 5 & 9 \\ 9 & 2 & 3 \end{bmatrix}$

13. $\begin{bmatrix} 4 & 0 & -2 \\ 0 & 2 & 3 \\ -2 & 3 & 6 \end{bmatrix}$

14. $\begin{bmatrix} 0 & 3 & -2 \\ 3 & -4 & 1 \\ -2 & 1 & 0 \end{bmatrix}$

15. $\begin{bmatrix} 2 & -5 & -3 \\ 5 & 5 & -6 \\ 7 & -4 & -10 \end{bmatrix}$

16. $\begin{bmatrix} 0 & -6 & 1 \\ 6 & 0 & 4 \\ -1 & -4 & 0 \end{bmatrix}$

17. $\begin{bmatrix} 4 & 2 & -9 \\ 5 & -4 & -10 \\ -6 & 6 & 9 \end{bmatrix}$

18. $\begin{bmatrix} 4 & 0 & 0 \\ -2 & -7 & 0 \\ 4 & -2 & 5 \end{bmatrix}$

19. $\begin{bmatrix} -3 & -4 & -5 \\ 0 & -3 & 5 \\ 0 & 0 & -3 \end{bmatrix}$

20. $\begin{bmatrix} 6 & -7 & 2 & 6 \\ 0 & -8 & -1 & 0 \\ 0 & 0 & 1 & -7 \end{bmatrix}$

21. $\begin{bmatrix} 1 & 0 & 0 \\ 0 & 2 & 0 \\ 0 & 0 & -1 \end{bmatrix}$

22. $\begin{bmatrix} 6 & -4 & -5 \\ -4 & 0 & 2 \\ -5 & 2 & -2 \end{bmatrix}$

23. $\begin{bmatrix} 0 & 1 & -2 \\ -1 & 0 & 4 \\ 2 & -4 & 0 \end{bmatrix}$

24. $\begin{bmatrix} 0 & 0 & 0 \\ 0 & 0 & 0 \\ 0 & 0 & 0 \end{bmatrix}$

3.2 The Matrix Trace

AS YOU READ . . .

1. T/F: We only compute the trace of square matrices.
2. T/F: One can tell if a matrix is invertible by computing the trace.

In the previous section, we learned about an operation we can peform on matrices, namely the transpose. Given a matrix A, we can "find the transpose of A," which is another matrix. In this section we learn about a new operation called the *trace*. It is a different type of operation than the transpose. Given a matrix A, we can "find the trace of A," which is not a matrix but rather a number. We formally define it here.

Definition 22 **The Trace**

Let A be an $n \times n$ matrix. The *trace of A*, denoted tr(A), is the sum of the diagonal elements of A. That is,

$$\text{tr}(A) = a_{11} + a_{22} + \cdots + a_{nn}.$$

This seems like a simple definition, and it really is. Just to make sure it is clear, let's practice.

Example 67 Find the trace of A, B, C and I_4, where

$$A = \begin{bmatrix} 1 & 2 \\ 3 & 4 \end{bmatrix}, \quad B = \begin{bmatrix} 1 & 2 & 0 \\ 3 & 8 & 1 \\ -2 & 7 & -5 \end{bmatrix} \quad \text{and } C = \begin{bmatrix} 1 & 2 & 3 \\ 4 & 5 & 6 \end{bmatrix}.$$

SOLUTION To find the trace of A, note that the diagonal elements of A are 1 and 4. Therefore, $\text{tr}(A) = 1 + 4 = 5$.

We see that the diagonal elements of B are 1, 8 and -5, so $\text{tr}(B) = 1 + 8 - 5 = 4$.

The matrix C is not a square matrix, and our definition states that we must start with a square matrix. Therefore $\text{tr}(C)$ is not defined.

Finally, the diagonal of I_4 consists of four 1s. Therefore $\text{tr}(I_4) = 4$.

Now that we have defined the trace of a matrix, we should think like mathematicians and ask some questions. The first questions that should pop into our minds should be along the lines of "How does the trace work with other matrix operations?"[10] We should think about how the trace works with matrix addition, scalar multiplication, matrix multiplication, matrix inverses, and the transpose.

We'll give a theorem that will formally tell us what is true in a moment, but first let's play with two sample matrices and see if we can see what will happen. Let

$$A = \begin{bmatrix} 2 & 1 & 3 \\ 2 & 0 & -1 \\ 3 & -1 & 3 \end{bmatrix} \text{ and } B = \begin{bmatrix} 2 & 0 & 1 \\ -1 & 2 & 0 \\ 0 & 2 & -1 \end{bmatrix}.$$

It should be clear that $\text{tr}(A) = 5$ and $\text{tr}(B) = 3$. What is $\text{tr}(A+B)$?

$$\begin{aligned} \text{tr}(A+B) &= \text{tr}\left(\begin{bmatrix} 2 & 1 & 3 \\ 2 & 0 & -1 \\ 3 & -1 & 3 \end{bmatrix} + \begin{bmatrix} 2 & 0 & 1 \\ -1 & 2 & 0 \\ 0 & 2 & -1 \end{bmatrix}\right) \\ &= \text{tr}\left(\begin{bmatrix} 4 & 1 & 4 \\ 1 & 2 & -1 \\ 3 & 1 & 2 \end{bmatrix}\right) \\ &= 8 \end{aligned}$$

So we notice that $\text{tr}(A+B) = \text{tr}(A) + \text{tr}(B)$. This probably isn't a coincidence.

How does the trace work with scalar multiplication? If we multiply A by 4, then the diagonal elements will be 8, 0 and 12, so $\text{tr}(4A) = 20$. Is it a coincidence that this is 4 times the trace of A?

Let's move on to matrix multiplication. How will the trace of AB relate to the traces of A and B? Let's see:

$$\begin{aligned} \text{tr}(AB) &= \text{tr}\left(\begin{bmatrix} 2 & 1 & 3 \\ 2 & 0 & -1 \\ 3 & -1 & 3 \end{bmatrix} \begin{bmatrix} 2 & 0 & 1 \\ -1 & 2 & 0 \\ 0 & 2 & -1 \end{bmatrix}\right) \\ &= \text{tr}\left(\begin{bmatrix} 3 & 8 & -1 \\ 4 & -2 & 3 \\ 7 & 4 & 0 \end{bmatrix}\right) \\ &= 1 \end{aligned}$$

[10] Recall that we asked a similar question once we learned about the transpose.

It isn't exactly clear what the relationship is among tr(A), tr(B) and tr(AB). Before moving on, let's find tr(BA):

$$\begin{aligned}
\text{tr}(BA) &= \text{tr}\left(\begin{bmatrix} 2 & 0 & 1 \\ -1 & 2 & 0 \\ 0 & 2 & -1 \end{bmatrix}\begin{bmatrix} 2 & 1 & 3 \\ 2 & 0 & -1 \\ 3 & -1 & 3 \end{bmatrix}\right) \\
&= \text{tr}\left(\begin{bmatrix} 7 & 1 & 9 \\ 2 & -1 & -5 \\ 1 & 1 & -5 \end{bmatrix}\right) \\
&= 1
\end{aligned}$$

We notice that $\text{tr}(AB) = \text{tr}(BA)$. Is this coincidental?
How are the traces of A and A^{-1} related? We compute A^{-1} and find that

$$A^{-1} = \begin{bmatrix} 1/17 & 6/17 & 1/17 \\ 9/17 & 3/17 & -8/17 \\ 2/17 & -5/17 & 2/17 \end{bmatrix}.$$

Therefore tr(A^{-1}) = 6/17. Again, the relationship isn't clear.[11]

Finally, let's see how the trace is related to the transpose. We actually don't have to formally compute anything. Recall from the previous section that the diagonals of A and A^T are identical; therefore, tr(A) = tr(A^T). That, we know for sure, isn't a coincidence.

We now formally state what equalities are true when considering the interaction of the trace with other matrix operations.

Theorem 13

Properties of the Matrix Trace

Let A and B be $n \times n$ matrices. Then:

1. tr($A + B$) = tr(A) + tr(B)
2. tr($A - B$) = tr(A) − tr(B)
3. tr(kA) = k·tr(A)
4. tr(AB) = tr(BA)
5. tr(A^T) = tr(A)

One of the key things to note here is what this theorem does *not* say. It says nothing about how the trace relates to inverses. The reason for the silence in these areas is that there simply is not a relationship.

[11]Something to think about: we know that not all square matrices are invertible. Would we be able to tell just by the trace? That seems unlikely.

We end this section by again wondering why anyone would care about the trace of matrix. One reason mathematicians are interested in it is that it can give a measurement of the "size"[12] of a matrix.

Consider the following 2×2 matrices:

$$A = \begin{bmatrix} 1 & -2 \\ 1 & 1 \end{bmatrix} \text{ and } B = \begin{bmatrix} 6 & 7 \\ 11 & -4 \end{bmatrix}.$$

These matrices have the same trace, yet B clearly has bigger elements in it. So how can we use the trace to determine a "size" of these matrices? We can consider $\text{tr}(A^TA)$ and $\text{tr}(B^TB)$.

$$\begin{aligned} \text{tr}(A^TA) &= \text{tr}\left(\begin{bmatrix} 1 & 1 \\ -2 & 1 \end{bmatrix}\begin{bmatrix} 1 & -2 \\ 1 & 1 \end{bmatrix}\right) \\ &= \text{tr}\left(\begin{bmatrix} 2 & -1 \\ -1 & 5 \end{bmatrix}\right) \\ &= 7 \end{aligned}$$

$$\begin{aligned} \text{tr}(B^TB) &= \text{tr}\left(\begin{bmatrix} 6 & 11 \\ 7 & -4 \end{bmatrix}\begin{bmatrix} 6 & 7 \\ 11 & -4 \end{bmatrix}\right) \\ &= \text{tr}\left(\begin{bmatrix} 157 & -2 \\ -2 & 65 \end{bmatrix}\right) \\ &= 222 \end{aligned}$$

Our concern is not how to interpret what this "size" measurement means, but rather to demonstrate that the trace (along with the transpose) can be used to give (perhaps useful) information about a matrix.[13]

---

[12]There are many different measurements of a matrix size. In this text, we just refer to its dimensions. Some measurements of size refer the magnitude of the elements in the matrix. The next section describes yet another measurement of matrix size.

[13]This example brings to light many interesting ideas that we'll flesh out just a little bit here.

1. Notice that the elements of A are $1, -2, 1$ and 1. Add the squares of these numbers: $1^2 + (-2)^2 + 1^2 + 1^2 = 7 = \text{tr}(A^TA)$.

   Notice that the elements of B are 6, 7, 11 and -4. Add the squares of these numbers: $6^2 + 7^2 + 11^2 + (-4)^2 = 222 = \text{tr}(B^TB)$.

   Can you see why this is true? When looking at multiplying A^TA, focus only on where the elements on the diagonal come from since they are the only ones that matter when taking the trace.

2. You can confirm on your own that regardless of the dimensions of A, $\text{tr}(A^TA) = \text{tr}(AA^T)$. To see why this is true, consider the previous point. (Recall also that A^TA and AA^T are always square, regardless of the dimensions of A.)

3. Mathematicians are actually more interested in $\sqrt{\text{tr}(A^TA)}$ than just $\text{tr}(A^TA)$. The reason for this is a bit complicated; the short answer is that "it works better." The reason "it works better" is related to the Pythagorean Theorem, all of all things. If we know that the legs of a right triangle have length a and b, we are more interested in $\sqrt{a^2 + b^2}$ than just $a^2 + b^2$. Of course, this explanation raises more questions than it answers; our goal here is just to whet your appetite and get you to do some more reading. A Numerical Linear Algebra book would be a good place to start.

Exercises 3.2

In Exercises 1 – 15, find the trace of the given matrix.

1. $\begin{bmatrix} 1 & -5 \\ 9 & 5 \end{bmatrix}$

2. $\begin{bmatrix} -3 & -10 \\ -6 & 4 \end{bmatrix}$

3. $\begin{bmatrix} 7 & 5 \\ -5 & -4 \end{bmatrix}$

4. $\begin{bmatrix} -6 & 0 \\ -10 & 9 \end{bmatrix}$

5. $\begin{bmatrix} -4 & 1 & 1 \\ -2 & 0 & 0 \\ -1 & -2 & -5 \end{bmatrix}$

6. $\begin{bmatrix} 0 & -3 & 1 \\ 5 & -5 & 5 \\ -4 & 1 & 0 \end{bmatrix}$

7. $\begin{bmatrix} -2 & -3 & 5 \\ 5 & 2 & 0 \\ -1 & -3 & 1 \end{bmatrix}$

8. $\begin{bmatrix} 4 & 2 & -1 \\ -4 & 1 & 4 \\ 0 & -5 & 5 \end{bmatrix}$

9. $\begin{bmatrix} 2 & 6 & 4 \\ -1 & 8 & -10 \end{bmatrix}$

10. $\begin{bmatrix} 6 & 5 \\ 2 & 10 \\ 3 & 3 \end{bmatrix}$

11. $\begin{bmatrix} -10 & 6 & -7 & -9 \\ -2 & 1 & 6 & -9 \\ 0 & 4 & -4 & 0 \\ -3 & -9 & 3 & -10 \end{bmatrix}$

12. $\begin{bmatrix} 5 & 2 & 2 & 2 \\ -7 & 4 & -7 & -3 \\ 9 & -9 & -7 & 2 \\ -4 & 8 & -8 & -2 \end{bmatrix}$

13. I_4

14. I_n

15. A matrix A that is skew symmetric.

In Exercises 16 – 19, verify Theorem 13 by:

1. **Showing that tr(A)+tr(B) = tr($A + B$) and**

2. **Showing that tr(AB) = tr(BA).**

16. $A = \begin{bmatrix} 1 & -1 \\ 9 & -6 \end{bmatrix}$, $B = \begin{bmatrix} -1 & 0 \\ -6 & 3 \end{bmatrix}$

17. $A = \begin{bmatrix} 0 & -8 \\ 1 & 8 \end{bmatrix}$, $B = \begin{bmatrix} -4 & 5 \\ -4 & 2 \end{bmatrix}$

18. $A = \begin{bmatrix} -8 & -10 & 10 \\ 10 & 5 & -6 \\ -10 & 1 & 3 \end{bmatrix}$

$B = \begin{bmatrix} -10 & -4 & -3 \\ -4 & -5 & 4 \\ 3 & 7 & 3 \end{bmatrix}$

19. $A = \begin{bmatrix} -10 & 7 & 5 \\ 7 & 7 & -5 \\ 8 & -9 & 2 \end{bmatrix}$

$B = \begin{bmatrix} -3 & -4 & 9 \\ 4 & -1 & -9 \\ -7 & -8 & 10 \end{bmatrix}$

3.3 The Determinant

1. T/F: The determinant of a matrix is always positive.

2. T/F: To compute the determinant of a 3×3 matrix, one needs to compute the determinants of 3 2×2 matrices.

3. Give an example of a 2×2 matrix with a determinant of 3.

In this chapter so far we've learned about the transpose (an operation on a matrix that returns another matrix) and the trace (an operation on a square matrix that returns a number). In this section we'll learn another operation on square matrices that returns a number, called the *determinant*. We give a pseudo-definition of the determinant here.

> The *determinant* of an $n \times n$ matrix A is a number, denoted $\det(A)$, that is determined by A.

That definition isn't meant to explain everything; it just gets us started by making us realize that the determinant is a number. The determinant is kind of a tricky thing to define. Once you know and understand it, it isn't that hard, but getting started is a bit complicated.[14] We start simply; we define the determinant for 2×2 matrices.

Definition 23

Determinant of 2×2 Matrices

Let

$$A = \begin{bmatrix} a & b \\ c & d \end{bmatrix}.$$

The *determinant of A*, denoted by

$$\det(A) \text{ or } \begin{vmatrix} a & b \\ c & d \end{vmatrix},$$

is $ad - bc$.

We've seen the expression $ad - bc$ before. In Section 2.6, we saw that a 2×2 matrix A has inverse

$$\frac{1}{ad-bc}\begin{bmatrix} d & -b \\ -c & a \end{bmatrix}$$

as long as $ad - bc \neq 0$; otherwise, the inverse does not exist. We can rephrase the above statement now: If $\det(A) \neq 0$, then

$$A^{-1} = \frac{1}{\det(A)}\begin{bmatrix} d & -b \\ -c & a \end{bmatrix}.$$

A brief word about the notation: notice that we can refer to the determinant by using what *looks like* absolute value bars around the entries of a matrix. We discussed at the end of the last section the idea of measuring the "size" of a matrix, and mentioned that there are many different ways to measure size. The determinant is one such way. Just as the absolute value of a number measures its size (and ignores its sign), the determinant of a matrix is a measurement of the size of the matrix. (Be careful, though: $\det(A)$ can be negative!)

Let's practice.

[14] It's similar to learning to ride a bike. The riding itself isn't hard, it is getting started that's difficult.

Example 68 Find the determinant of A, B and C where

$$A=\begin{bmatrix}1 & 2\\ 3 & 4\end{bmatrix},\ B=\begin{bmatrix}3 & -1\\ 2 & 7\end{bmatrix}\ \text{and}\ C=\begin{bmatrix}1 & -3\\ -2 & 6\end{bmatrix}.$$

SOLUTION Finding the determinant of A:

$$\begin{aligned}\det(A) &= \begin{vmatrix}1 & 2\\ 3 & 4\end{vmatrix}\\ &= 1(4)-2(3)\\ &= -2.\end{aligned}$$

Similar computations show that $\det(B) = 3(7)-(-1)(2) = 23$ and $\det(C) = 1(6)-(-3)(-2) = 0$.

Finding the determinant of a 2×2 matrix is pretty straightforward. It is natural to ask next "How do we compute the determinant of matrices that are not 2×2?" We first need to define some terms.[15]

Definition 24 **Matrix Minor, Cofactor**

Let A be an $n\times n$ matrix. The *i,j minor of A*, denoted $A_{i,j}$, is the determinant of the $(n-1)\times(n-1)$ matrix formed by deleting the ith row and j^{th} column of A.

The *i,j-cofactor of A* is the number

$$C_{ij} = (-1)^{i+j}A_{i,j}.$$

Notice that this definition makes reference to taking the determinant of a matrix, while we haven't yet defined what the determinant is beyond 2×2 matrices. We recognize this problem, and we'll see how far we can go before it becomes an issue.

Examples will help.

Example 69 Let

$$A=\begin{bmatrix}1 & 2 & 3\\ 4 & 5 & 6\\ 7 & 8 & 9\end{bmatrix}\ \text{and}\ B=\begin{bmatrix}1 & 2 & 0 & 8\\ -3 & 5 & 7 & 2\\ -1 & 9 & -4 & 6\\ 1 & 1 & 1 & 1\end{bmatrix}.$$

Find $A_{1,3}$, $A_{3,2}$, $B_{2,1}$, $B_{4,3}$ and their respective cofactors.

[15] This is the standard definition of these two terms, although slight variations exist.

Solution To compute the minor $A_{1,3}$, we remove the first row and third column of A then take the determinant.

$$A = \begin{bmatrix} 1 & 2 & 3 \\ 4 & 5 & 6 \\ 7 & 8 & 9 \end{bmatrix} \Rightarrow \begin{bmatrix} \cancel{1} & \cancel{2} & \cancel{3} \\ 4 & 5 & \cancel{6} \\ 7 & 8 & \cancel{9} \end{bmatrix} \Rightarrow \begin{bmatrix} 4 & 5 \\ 7 & 8 \end{bmatrix}$$

$$A_{1,3} = \begin{vmatrix} 4 & 5 \\ 7 & 8 \end{vmatrix} = 32 - 35 = -3.$$

The corresponding cofactor, $C_{1,3}$, is

$$C_{1,3} = (-1)^{1+3} A_{1,3} = (-1)^4(-3) = -3.$$

The minor $A_{3,2}$ is found by removing the third row and second column of A then taking the determinant.

$$A = \begin{bmatrix} 1 & 2 & 3 \\ 4 & 5 & 6 \\ 7 & 8 & 9 \end{bmatrix} \Rightarrow \begin{bmatrix} 1 & \cancel{2} & 3 \\ 4 & \cancel{5} & 6 \\ \cancel{7} & \cancel{8} & \cancel{9} \end{bmatrix} \Rightarrow \begin{bmatrix} 1 & 3 \\ 4 & 6 \end{bmatrix}$$

$$A_{3,2} = \begin{vmatrix} 1 & 3 \\ 4 & 6 \end{vmatrix} = 6 - 12 = -6.$$

The corresponding cofactor, $C_{3,2}$, is

$$C_{3,2} = (-1)^{3+2} A_{3,2} = (-1)^5(-6) = 6.$$

The minor $B_{2,1}$ is found by removing the second row and first column of B then taking the determinant.

$$B = \begin{bmatrix} 1 & 2 & 0 & 8 \\ -3 & 5 & 7 & 2 \\ -1 & 9 & -4 & 6 \\ 1 & 1 & 1 & 1 \end{bmatrix} \Rightarrow \begin{bmatrix} \cancel{1} & 2 & 0 & 8 \\ \cancel{-3} & \cancel{5} & \cancel{7} & \cancel{2} \\ \cancel{-1} & 9 & -4 & 6 \\ \cancel{1} & 1 & 1 & 1 \end{bmatrix} \Rightarrow \begin{bmatrix} 2 & 0 & 8 \\ 9 & -4 & 6 \\ 1 & 1 & 1 \end{bmatrix}$$

$$B_{2,1} = \begin{vmatrix} 2 & 0 & 8 \\ 9 & -4 & 6 \\ 1 & 1 & 1 \end{vmatrix} \stackrel{!}{=} ?$$

We're a bit stuck. We don't know how to find the determinate of this 3×3 matrix. We'll come back to this later. The corresponding cofactor is

$$C_{2,1} = (-1)^{2+1} B_{2,1} = -B_{2,1},$$

whatever this number happens to be.

The minor $B_{4,3}$ is found by removing the fourth row and third column of B then taking the determinant.

$$B = \begin{bmatrix} 1 & 2 & 0 & 8 \\ -3 & 5 & 7 & 2 \\ -1 & 9 & -4 & 6 \\ 1 & 1 & 1 & 1 \end{bmatrix} \Rightarrow \begin{bmatrix} 1 & 2 & \cancel{0} & 8 \\ -3 & 5 & \cancel{7} & 2 \\ -1 & 9 & \cancel{-4} & 6 \\ \cancel{1} & \cancel{1} & \cancel{1} & \cancel{1} \end{bmatrix} \Rightarrow \begin{bmatrix} 1 & 2 & 8 \\ -3 & 5 & 2 \\ -1 & 9 & 6 \end{bmatrix}$$

$$B_{4,3} = \begin{vmatrix} 1 & 2 & 8 \\ -3 & 5 & 2 \\ -1 & 9 & 6 \end{vmatrix} \stackrel{!}{=} ?$$

Again, we're stuck. We won't be able to fully compute $C_{4,3}$; all we know so far is that

$$C_{4,3} = (-1)^{4+3}B_{4,3} = (-1)B_{4,3}.$$

Once we learn how to compute determinates for matrices larger than 2×2 we can come back and finish this exercise.

In our previous example we ran into a bit of trouble. By our definition, in order to compute a minor of an $n \times n$ matrix we needed to compute the determinant of a $(n-1) \times (n-1)$ matrix. This was fine when we started with a 3×3 matrix, but when we got up to a 4×4 matrix (and larger) we run into trouble.

We are almost ready to define the determinant for any square matrix; we need one last definition.

Definition 25 **Cofactor Expansion**

Let A be an $n \times n$ matrix.

The *cofactor expansion of A along the ith row* is the sum

$$a_{i,1}C_{i,1} + a_{i,2}C_{i,2} + \cdots + a_{i,n}C_{i,n}.$$

The *cofactor expansion of A down the jth column* is the sum

$$a_{1,j}C_{1,j} + a_{2,j}C_{2,j} + \cdots + a_{n,j}C_{n,j}.$$

The notation of this definition might be a little intimidating, so let's look at an example.

Example 70 Let

$$A = \begin{bmatrix} 1 & 2 & 3 \\ 4 & 5 & 6 \\ 7 & 8 & 9 \end{bmatrix}.$$

Find the cofactor expansions along the second row and down the first column.

SOLUTION By the definition, the cofactor expansion along the second row is the sum

$$a_{2,1}C_{2,1} + a_{2,2}C_{2,2} + a_{2,3}C_{2,3}.$$

(Be sure to compare the above line to the definition of cofactor expansion, and see how the "i" in the definition is replaced by "2" here.)

We'll find each cofactor and then compute the sum.

$$C_{2,1} = (-1)^{2+1} \begin{vmatrix} 2 & 3 \\ 8 & 9 \end{vmatrix} = (-1)(-6) = 6 \qquad \left(\begin{array}{c}\text{we removed the second row and} \\ \text{first column of } A \text{ to compute the} \\ \text{minor}\end{array}\right)$$

$$C_{2,2} = (-1)^{2+2} \begin{vmatrix} 1 & 3 \\ 7 & 9 \end{vmatrix} = (1)(-12) = -12 \qquad \left(\begin{array}{c}\text{we removed the second row and} \\ \text{second column of } A \text{ to compute} \\ \text{the minor}\end{array}\right)$$

$$C_{2,3} = (-1)^{2+3} \begin{vmatrix} 1 & 2 \\ 7 & 8 \end{vmatrix} = (-1)(-6) = 6 \qquad \left(\begin{array}{c}\text{we removed the second row and} \\ \text{third column of } A \text{ to compute the} \\ \text{minor}\end{array}\right)$$

Thus the cofactor expansion along the second row is

$$\begin{aligned} a_{2,1}C_{2,1} + a_{2,2}C_{2,2} + a_{2,3}C_{2,3} &= 4(6) + 5(-12) + 6(6) \\ &= 24 - 60 + 36 \\ &= 0 \end{aligned}$$

At the moment, we don't know what to do with this cofactor expansion; we've just successfully found it.

We move on to find the cofactor expansion down the first column. By the definition, this sum is

$$a_{1,1}C_{1,1} + a_{2,1}C_{2,1} + a_{3,1}C_{3,1}.$$

(Again, compare this to the above definition and see how we replaced the "j" with "1.")

We find each cofactor:

$$C_{1,1} = (-1)^{1+1} \begin{vmatrix} 5 & 6 \\ 8 & 9 \end{vmatrix} = (1)(-3) = -3 \qquad \left(\begin{array}{c}\text{we removed the first row and first} \\ \text{column of } A \text{ to compute the minor}\end{array}\right)$$

$$C_{2,1} = (-1)^{2+1} \begin{vmatrix} 2 & 3 \\ 8 & 9 \end{vmatrix} = (-1)(-6) = 6 \qquad \left(\text{ we computed this cofactor above }\right)$$

$$C_{3,1} = (-1)^{3+1} \begin{vmatrix} 2 & 3 \\ 5 & 6 \end{vmatrix} = (1)(-3) = -3 \qquad \left(\begin{array}{c}\text{we removed the third row and first} \\ \text{column of } A \text{ to compute the minor}\end{array}\right)$$

The cofactor expansion down the first column is

$$\begin{aligned} a_{1,1}C_{1,1} + a_{2,1}C_{2,1} + a_{3,1}C_{3,1} &= 1(-3) + 4(6) + 7(-3) \\ &= -3 + 24 - 21 \\ &= 0 \end{aligned}$$

Is it a coincidence that both cofactor expansions were 0? We'll answer that in a while.

This section is entitled "The Determinant," yet we don't know how to compute it yet except for 2×2 matrices. We finally define it now.

Definition 26

The Determinant

The *determinant* of an $n \times n$ matrix A, denoted $\det(A)$ or $|A|$, is a number given by the following:

- if A is a 1×1 matrix $A = [a]$, then $\det(A) = a$.
- if A is a 2×2 matrix

$$A = \begin{bmatrix} a & b \\ c & d \end{bmatrix},$$

  then $\det(A) = ad - bc$.
- if A is an $n \times n$ matrix, where $n \geq 2$, then $\det(A)$ is the number found by taking the cofactor expansion along the first row of A. That is,

$$\det(A) = a_{1,1}C_{1,1} + a_{1,2}C_{1,2} + \cdots + a_{1,n}C_{1,n}.$$

Notice that in order to compute the determinant of an $n \times n$ matrix, we need to compute the determinants of n $(n-1) \times (n-1)$ matrices. This can be a lot of work. We'll later learn how to shorten some of this. First, let's practice.

Example 71 Find the determinant of

$$A = \begin{bmatrix} 1 & 2 & 3 \\ 4 & 5 & 6 \\ 7 & 8 & 9 \end{bmatrix}.$$

SOLUTION Notice that this is the matrix from Example 70. The cofactor expansion along the first row is

$$\det(A) = a_{1,1}C_{1,1} + a_{1,2}C_{1,2} + a_{1,3}C_{1,3}.$$

We'll compute each cofactor first then take the appropriate sum.

$$C_{1,1} = (-1)^{1+1}A_{1,1} = 1 \cdot \begin{vmatrix} 5 & 6 \\ 8 & 9 \end{vmatrix} = 45 - 48 = -3$$

$$C_{1,2} = (-1)^{1+2}A_{1,2} = (-1) \cdot \begin{vmatrix} 4 & 6 \\ 7 & 9 \end{vmatrix} = (-1)(36 - 42) = 6$$

$$C_{1,3} = (-1)^{1+3}A_{1,3} = 1 \cdot \begin{vmatrix} 4 & 5 \\ 7 & 8 \end{vmatrix} = 32 - 35 = -3$$

Therefore the determinant of A is

$$\det(A) = 1(-3) + 2(6) + 3(-3) = 0.$$

Example 72 Find the determinant of

$$A = \begin{bmatrix} 3 & 6 & 7 \\ 0 & 2 & -1 \\ 3 & -1 & 1 \end{bmatrix}.$$

SOLUTION We'll compute each cofactor first then find the determinant.

$$C_{1,1} = (-1)^{1+1}A_{1,1} = 1 \cdot \begin{vmatrix} 2 & -1 \\ -1 & 1 \end{vmatrix} = 2 - 1 = 1$$

$$C_{1,2} = (-1)^{1+2}A_{1,2} = (-1) \cdot \begin{vmatrix} 0 & -1 \\ 3 & 1 \end{vmatrix} = (-1)(0 + 3) = -3$$

$$C_{1,3} = (-1)^{1+3}A_{1,3} = 1 \cdot \begin{vmatrix} 0 & 2 \\ 3 & -1 \end{vmatrix} = 0 - 6 = -6$$

Thus the determinant is

$$\det(A) = 3(1) + 6(-3) + 7(-6) = -57.$$

Example 73 Find the determinant of

$$A = \begin{bmatrix} 1 & 2 & 1 & 2 \\ -1 & 2 & 3 & 4 \\ 8 & 5 & -3 & 1 \\ 5 & 9 & -6 & 3 \end{bmatrix}.$$

SOLUTION This, quite frankly, will take quite a bit of work. In order to compute this determinant, we need to compute 4 minors, each of which requires finding the determinant of a 3×3 matrix! Complaining won't get us any closer to the solution,[16]

[16] But it might make us feel a little better. Glance ahead: do you see how much work we have to do?!?

so let's get started. We first compute the cofactors:

$$\begin{aligned}
C_{1,1} &= (-1)^{1+1}A_{1,1} \\
&= 1 \cdot \begin{vmatrix} 2 & 3 & 4 \\ 5 & -3 & 1 \\ 9 & -6 & 3 \end{vmatrix} \qquad \begin{pmatrix}\text{we must compute the determinant} \\ \text{of this } 3 \times 3 \text{ matrix}\end{pmatrix} \\
&= 2 \cdot (-1)^{1+1} \begin{vmatrix} -3 & 1 \\ -6 & 3 \end{vmatrix} + 3 \cdot (-1)^{1+2} \begin{vmatrix} 5 & 1 \\ 9 & 3 \end{vmatrix} + 4 \cdot (-1)^{1+3} \begin{vmatrix} 5 & -3 \\ 9 & -6 \end{vmatrix} \\
&= 2(-3) + 3(-6) + 4(-3) \\
&= -36
\end{aligned}$$

$$\begin{aligned}
C_{1,2} &= (-1)^{1+2}A_{1,2} \\
&= (-1) \cdot \begin{vmatrix} -1 & 3 & 4 \\ 8 & -3 & 1 \\ 5 & -6 & 3 \end{vmatrix} \qquad \begin{pmatrix}\text{we must compute the determinant} \\ \text{of this } 3 \times 3 \text{ matrix}\end{pmatrix} \\
&= (-1) \underbrace{\left[(-1) \cdot (-1)^{1+1} \begin{vmatrix} -3 & 1 \\ -6 & 3 \end{vmatrix} + 3 \cdot (-1)^{1+2} \begin{vmatrix} 8 & 1 \\ 5 & 3 \end{vmatrix} + 4 \cdot (-1)^{1+3} \begin{vmatrix} 8 & -3 \\ 5 & -6 \end{vmatrix} \right]}_{\text{the determinate of the } 3 \times 3 \text{ matrix}} \\
&= (-1) \left[(-1)(-3) + 3(-19) + 4(-33) \right] \\
&= 186
\end{aligned}$$

$$\begin{aligned}
C_{1,3} &= (-1)^{1+3}A_{1,3} \\
&= 1 \cdot \begin{vmatrix} -1 & 2 & 4 \\ 8 & 5 & 1 \\ 5 & 9 & 3 \end{vmatrix} \qquad \begin{pmatrix}\text{we must compute the determinant} \\ \text{of this } 3 \times 3 \text{ matrix}\end{pmatrix} \\
&= (-1) \cdot (-1)^{1+1} \begin{vmatrix} 5 & 1 \\ 9 & 3 \end{vmatrix} + 2 \cdot (-1)^{1+2} \begin{vmatrix} 8 & 1 \\ 5 & 3 \end{vmatrix} + 4 \cdot (-1)^{1+3} \begin{vmatrix} 8 & 5 \\ 5 & 9 \end{vmatrix} \\
&= (-1)(6) + 2(-19) + 4(47) \\
&= 144
\end{aligned}$$

$$
\begin{aligned}
C_{1,4} &= (-1)^{1+4}A_{1,4} \\
&= (-1)\cdot \begin{vmatrix} -1 & 2 & 3 \\ 8 & 5 & -3 \\ 5 & 9 & -6 \end{vmatrix} \qquad \begin{pmatrix}\text{we must compute the determinant} \\ \text{of this } 3\times 3 \text{ matrix}\end{pmatrix} \\
&= (-1)\underbrace{\left[(-1)\cdot(-1)^{1+1}\begin{vmatrix} 5 & -3 \\ 9 & -6 \end{vmatrix} + 2\cdot(-1)^{1+2}\begin{vmatrix} 8 & -3 \\ 5 & -6 \end{vmatrix} + 3\cdot(-1)^{1+3}\begin{vmatrix} 8 & 5 \\ 5 & 9 \end{vmatrix}\right]}_{\text{the determinate of the } 3\times 3 \text{ matrix}} \\
&= (-1)\left[(-1)(-3) + 2(33) + 3(47)\right] \\
&= -210
\end{aligned}
$$

We've computed our four cofactors. All that is left is to compute the cofactor expansion.

$$\det(A) = 1(-36) + 2(186) + 1(144) + 2(-210) = 60.$$

As a way of "visualizing" this, let's write out the cofactor expansion again but including the matrices in their place.

$$
\begin{aligned}
\det(A) &= a_{1,1}C_{1,1} + a_{1,2}C_{1,2} + a_{1,3}C_{1,3} + a_{1,4}C_{1,4} \\
&= 1(-1)^2\underbrace{\begin{vmatrix} 2 & 3 & 4 \\ 5 & -3 & 1 \\ 9 & -6 & 3 \end{vmatrix}}_{=-36} + 2(-1)^3\underbrace{\begin{vmatrix} -1 & 3 & 4 \\ 8 & -3 & 1 \\ 5 & -6 & 3 \end{vmatrix}}_{=-186} \\
&+ \\
&\quad 1(-1)^4\underbrace{\begin{vmatrix} -1 & 2 & 4 \\ 8 & 5 & 1 \\ 5 & 9 & 3 \end{vmatrix}}_{=144} + 2(-1)^5\underbrace{\begin{vmatrix} -1 & 2 & 3 \\ 8 & 5 & -3 \\ 5 & 9 & -6 \end{vmatrix}}_{=210} \\
&= 60
\end{aligned}
$$

That certainly took a while; it required more than 50 multiplications (we didn't count the additions). To compute the determinant of a 5×5 matrix, we'll need to compute the determinants of five 4×4 matrices, meaning that we'll need over 250 multiplications! Not only is this a lot of work, but there are just too many ways to make silly mistakes.[17] There are some tricks to make this job easier, but regardless we see the need to employ technology. Even then, technology quickly bogs down. A 25×25 matrix is considered "small" by today's standards,[18] but it is essentially impossible for a computer to compute its determinant by only using cofactor expansion; it too needs to employ "tricks."

[17] The author made three when the above example was originally typed.

[18] It is common for mathematicians, scientists and engineers to consider linear systems with thousands of equations and variables.

In the next section we will learn some of these tricks as we learn some of the properties of the determinant. Right now, let's review the essentials of what we have learned.

1. The determinant of a square matrix is a number that is determined by the matrix.
2. We find the determinant by computing the cofactor expansion along the first row.
3. To compute the determinant of an $n \times n$ matrix, we need to compute n determinants of $(n-1) \times (n-1)$ matrices.

Exercises 3.3

In Exercises 1 – 8, find the determinant of the 2×2 matrix.

1. $\begin{bmatrix} 10 & 7 \\ 8 & 9 \end{bmatrix}$

2. $\begin{bmatrix} 6 & -1 \\ -7 & 8 \end{bmatrix}$

3. $\begin{bmatrix} -1 & -7 \\ -5 & 9 \end{bmatrix}$

4. $\begin{bmatrix} -10 & -1 \\ -4 & 7 \end{bmatrix}$

5. $\begin{bmatrix} 8 & 10 \\ 2 & -3 \end{bmatrix}$

6. $\begin{bmatrix} 10 & -10 \\ -10 & 0 \end{bmatrix}$

7. $\begin{bmatrix} 1 & -3 \\ 7 & 7 \end{bmatrix}$

8. $\begin{bmatrix} -4 & -5 \\ -1 & -4 \end{bmatrix}$

In Exercises 9 – 12, a matrix A is given.

(a) Construct the submatrices used to compute the minors $A_{1,1}$, $A_{1,2}$ and $A_{1,3}$.

(b) Find the cofactors $C_{1,1}$, $C_{1,2}$, and $C_{1,3}$.

9. $\begin{bmatrix} -7 & -3 & 10 \\ 3 & 7 & 6 \\ 1 & 6 & 10 \end{bmatrix}$

10. $\begin{bmatrix} -2 & -9 & 6 \\ -10 & -6 & 8 \\ 0 & -3 & -2 \end{bmatrix}$

11. $\begin{bmatrix} -5 & -3 & 3 \\ -3 & 3 & 10 \\ -9 & 3 & 9 \end{bmatrix}$

12. $\begin{bmatrix} -6 & -4 & 6 \\ -8 & 0 & 0 \\ -10 & 8 & -1 \end{bmatrix}$

In Exercises 13 – 24, find the determinant of the given matrix using cofactor expansion along the first row.

13. $\begin{bmatrix} 3 & 2 & 3 \\ -6 & 1 & -10 \\ -8 & -9 & -9 \end{bmatrix}$

14. $\begin{bmatrix} 8 & -9 & -2 \\ -9 & 9 & -7 \\ 5 & -1 & 9 \end{bmatrix}$

15. $\begin{bmatrix} -4 & 3 & -4 \\ -4 & -5 & 3 \\ 3 & -4 & 5 \end{bmatrix}$

16. $\begin{bmatrix} 1 & -2 & 1 \\ 5 & 5 & 4 \\ 4 & 0 & 0 \end{bmatrix}$

17. $\begin{bmatrix} 1 & -4 & 1 \\ 0 & 3 & 0 \\ 1 & 2 & 2 \end{bmatrix}$

18. $\begin{bmatrix} 3 & -1 & 0 \\ -3 & 0 & -4 \\ 0 & -1 & -4 \end{bmatrix}$

19. $\begin{bmatrix} -5 & 0 & -4 \\ 2 & 4 & -1 \\ -5 & 0 & -4 \end{bmatrix}$

20. $\begin{bmatrix} 1 & 0 & 0 \\ 0 & 1 & 0 \\ -1 & 1 & 1 \end{bmatrix}$

21. $\begin{bmatrix} 0 & 0 & -1 & -1 \\ 1 & 1 & 0 & 1 \\ 1 & 1 & -1 & 0 \\ -1 & 0 & 1 & 0 \end{bmatrix}$

22. $\begin{bmatrix} -1 & 0 & 0 & -1 \\ -1 & 0 & 0 & 1 \\ 1 & 1 & 1 & 0 \\ 1 & 0 & -1 & -1 \end{bmatrix}$

23. $\begin{bmatrix} -5 & 1 & 0 & 0 \\ -3 & -5 & 2 & 5 \\ -2 & 4 & -3 & 4 \\ 5 & 4 & -3 & 3 \end{bmatrix}$

24. $\begin{bmatrix} 2 & -1 & 4 & 4 \\ 3 & -3 & 3 & 2 \\ 0 & 4 & -5 & 1 \\ -2 & -5 & -2 & -5 \end{bmatrix}$

25. Let A be a 2×2 matrix;

$$A = \begin{bmatrix} a & b \\ c & d \end{bmatrix}.$$

Show why $\det(A) = ad - bc$ by computing the cofactor expansion of A along the first row.

3.4 Properties of the Determinant

1. Having the choice to compute the determinant of a matrix using cofactor expansion along any row or column is most useful when there are lots of what in a row or column?
2. Which elementary row operation does not change the determinant of a matrix?
3. Why do mathematicians rarely smile?
4. T/F: When computers are used to compute the determinant of a matrix, cofactor expansion is rarely used.

In the previous section we learned how to compute the determinant. In this section we learn some of the properties of the determinant, and this will allow us to compute determinants more easily. In the next section we will see one application of determinants.

We start with a theorem that gives us more freedom when computing determinants.

Theorem 14 **Cofactor Expansion Along Any Row or Column**

Let A be an $n \times n$ matrix. The determinant of A can be computed using cofactor expansion along any row or column of A.

We alluded to this fact way back after Example 70. We had just learned what cofactor expansion was and we practiced along the second row and down the third column. Later, we found the determinant of this matrix by computing the cofactor expansion along the first row. In all three cases, we got the number 0. This wasn't a coincidence. The above theorem states that all three expansions were actually computing the determinant.

How does this help us? By giving us freedom to choose any row or column to use for the expansion, we can choose a row or column that looks "most appealing." This usually means "it has lots of zeros." We demonstrate this principle below.

Example 74 Find the determinant of

$$A = \begin{bmatrix} 1 & 2 & 0 & 9 \\ 2 & -3 & 0 & 5 \\ 7 & 2 & 3 & 8 \\ -4 & 1 & 0 & 2 \end{bmatrix}.$$

SOLUTION Our first reaction may well be "Oh no! Not another 4×4 determinant!" However, we can use cofactor expansion along any row or column that we choose. The third column looks great; it has lots of zeros in it. The cofactor expansion along this column is

$$\begin{aligned} \det(A) &= a_{1,3}C_{1,3} + a_{2,3}C_{2,3} + a_{3,3}C_{3,3} + a_{4,3}C_{4,3} \\ &= 0 \cdot C_{1,3} + 0 \cdot C_{2,3} + 3 \cdot C_{3,3} + 0 \cdot C_{4,3} \end{aligned}$$

The wonderful thing here is that three of our cofactors are multiplied by 0. We won't bother computing them since they will not contribute to the determinant. Thus

$$\begin{aligned} \det(A) &= 3 \cdot C_{3,3} \\ &= 3 \cdot (-1)^{3+3} \cdot \begin{vmatrix} 1 & 2 & 9 \\ 2 & -3 & 5 \\ -4 & 1 & 2 \end{vmatrix} \\ &= 3 \cdot (-147) \qquad \left(\begin{array}{c} \text{we computed the determinant of the } 3 \times 3 \text{ matrix} \\ \text{without showing our work; it is } -147 \end{array} \right) \\ &= -447 \end{aligned}$$

Wow. That was a lot simpler than computing all that we did in Example 73. Of course, in that example, we didn't really have any shortcuts that we could have employed.

Example 75 Find the determinant of

$$A = \begin{bmatrix} 1 & 2 & 3 & 4 & 5 \\ 0 & 6 & 7 & 8 & 9 \\ 0 & 0 & 10 & 11 & 12 \\ 0 & 0 & 0 & 13 & 14 \\ 0 & 0 & 0 & 0 & 15 \end{bmatrix}.$$

Solution At first glance, we think "I don't want to find the determinant of a 5×5 matrix!" However, using our newfound knowledge, we see that things are not that bad. In fact, this problem is very easy.

What row or column should we choose to find the determinant along? There are two obvious choices: the first column or the last row. Both have 4 zeros in them. We choose the first column.[19] We omit most of the cofactor expansion, since most of it is just 0:

$$\det(A) = 1 \cdot (-1)^{1+1} \cdot \begin{vmatrix} 6 & 7 & 8 & 9 \\ 0 & 10 & 11 & 12 \\ 0 & 0 & 13 & 14 \\ 0 & 0 & 0 & 15 \end{vmatrix}.$$

Similarly, this determinant is not bad to compute; we again choose to use cofactor expansion along the first column. Note: technically, this cofactor expansion is $6 \cdot (-1)^{1+1}A_{1,1}$; we are going to drop the $(-1)^{1+1}$ terms from here on out in this example (it will show up a lot...).

$$\det(A) = 1 \cdot 6 \cdot \begin{vmatrix} 10 & 11 & 12 \\ 0 & 13 & 14 \\ 0 & 0 & 15 \end{vmatrix}.$$

You can probably can see a trend. We'll finish out the steps without explaining each one.

$$\begin{aligned} \det(A) &= 1 \cdot 6 \cdot 10 \cdot \begin{vmatrix} 13 & 14 \\ 0 & 15 \end{vmatrix} \\ &= 1 \cdot 6 \cdot 10 \cdot 13 \cdot 15 \\ &= 11700 \end{aligned}$$

We see that the final determinant is the product of the diagonal entries. This works for any triangular matrix (and since diagonal matrices are triangular, it works for diagonal matrices as well). This is an important enough idea that we'll put it into a box.

Key Idea 12 **The Determinant of Triangular Matrices**

The determinant of a triangular matrix is the product of its diagonal elements.

It is now again time to start thinking like a mathematician. Remember, mathematicians see something new and often ask "How does this relate to things I already

[19] We do not choose this because it is the better choice; both options are good. We simply had to make a choice.

know?" So now we ask, "If we change a matrix in some way, how is it's determinant changed?"

The standard way that we change matrices is through elementary row operations. If we perform an elementary row operation on a matrix, how will the determinant of the new matrix compare to the determinant of the original matrix?

Let's experiment first and then we'll officially state what happens.

Example 76 Let

$$A = \begin{bmatrix} 1 & 2 \\ 3 & 4 \end{bmatrix}.$$

Let B be formed from A by doing one of the following elementary row operations:

1. $2R_1 + R_2 \rightarrow R_2$

2. $5R_1 \rightarrow R_1$

3. $R_1 \leftrightarrow R_2$

Find $\det(A)$ as well as $\det(B)$ for each of the row operations above.

Solution It is straightforward to compute $\det(A) = -2$.

Let B be formed by performing the row operation in 1) on A; thus

$$B = \begin{bmatrix} 1 & 2 \\ 5 & 8 \end{bmatrix}.$$

It is clear that $\det(B) = -2$, the same as $\det(A)$.

Now let B be formed by performing the elementary row operation in 2) on A; that is,

$$B = \begin{bmatrix} 5 & 10 \\ 3 & 4 \end{bmatrix}.$$

We can see that $\det(B) = -10$, which is $5 \cdot \det(A)$.

Finally, let B be formed by the third row operation given; swap the two rows of A. We see that

$$B = \begin{bmatrix} 3 & 4 \\ 1 & 2 \end{bmatrix}$$

and that $\det(B) = 2$, which is $(-1) \cdot \det(A)$.

We've seen in the above example that there seems to be a relationship between the determinants of matrices "before and after" being changed by elementary row operations. Certainly, one example isn't enough to base a theory on, and we have not proved anything yet. Regardless, the following theorem is true.

Theorem 15 **The Determinant and Elementary Row Operations**

Let A be an $n \times n$ matrix and let B be formed by performing one elementary row operation on A.

1. If B is formed from A by adding a scalar multiple of one row to another, then $\det(B) = \det(A)$.

2. If B is formed from A by multiplying one row of A by a scalar k, then $\det(B) = k \cdot \det(A)$.

3. If B is formed from A by interchanging two rows of A, then $\det(B) = -\det(A)$.

Let's put this theorem to use in an example.

Example 77 Let

$$A = \begin{bmatrix} 1 & 2 & 1 \\ 0 & 1 & 1 \\ 1 & 1 & 1 \end{bmatrix}.$$

Compute $\det(A)$, then find the determinants of the following matrices by inspection using Theorem 15.

$$B = \begin{bmatrix} 1 & 1 & 1 \\ 1 & 2 & 1 \\ 0 & 1 & 1 \end{bmatrix} \quad C = \begin{bmatrix} 1 & 2 & 1 \\ 0 & 1 & 1 \\ 7 & 7 & 7 \end{bmatrix} \quad D = \begin{bmatrix} 1 & -1 & -2 \\ 0 & 1 & 1 \\ 1 & 1 & 1 \end{bmatrix}$$

Solution Computing $\det(A)$ by cofactor expansion down the first column or along the second row seems like the best choice, utilizing the one zero in the matrix. We can quickly confirm that $\det(A) = 1$.

To compute $\det(B)$, notice that the rows of A were rearranged to form B. There are different ways to describe what happened; saying $R_1 \leftrightarrow R_2$ was followed by $R_1 \leftrightarrow R_3$ produces B from A. Since there were *two* row swaps, $\det(B) = (-1)(-1)\det(A) = \det(A) = 1$.

Notice that C is formed from A by multiplying the third row by 7. Thus $\det(C) = 7 \cdot \det(A) = 7$.

It takes a little thought, but we can form D from A by the operation $-3R_2 + R_1 \rightarrow R_1$. This type of elementary row operation does not change determinants, so $\det(D) = \det(A)$.

Let's continue to think like mathematicians; mathematicians tend to remember "problems" they've encountered in the past,[20] and when they learn something new, in the backs of their minds they try to apply their new knowledge to solve their old problem.

What "problem" did we recently uncover? We stated in the last chapter that even computers could not compute the determinant of large matrices with cofactor expansion. How then can we compute the determinant of large matrices?

We just learned two interesting and useful facts about matrix determinants. First, the determinant of a triangular matrix is easy to compute: just multiply the diagonal elements. Secondly, we know how elementary row operations affect the determinant. Put these two ideas together: given any square matrix, we can use elementary row operations to put the matrix in triangular form,[21] find the determinant of the new matrix (which is easy), and then adjust that number by recalling what elementary operations we performed. Let's practice this.

Example 78 Find the determinant of A by first putting A into a triangular form, where

$$A = \begin{bmatrix} 2 & 4 & -2 \\ -1 & -2 & 5 \\ 3 & 2 & 1 \end{bmatrix}.$$

Solution In putting A into a triangular form, we need not worry about getting leading 1s, but it does tend to make our life easier as we work out a problem by hand. So let's scale the first row by $1/2$:

$$\tfrac{1}{2}R_1 \rightarrow R_1 \qquad \begin{bmatrix} 1 & 2 & -1 \\ -1 & -2 & 5 \\ 3 & 2 & 1 \end{bmatrix}.$$

Now let's get 0s below this leading 1:

$$\begin{array}{l} R_1 + R_2 \rightarrow R_2 \\ -3R_1 + R_3 \rightarrow R_3 \end{array} \qquad \begin{bmatrix} 1 & 2 & -1 \\ 0 & 0 & 4 \\ 0 & -4 & 4 \end{bmatrix}.$$

We can finish in one step; by interchanging rows 2 and 3 we'll have our matrix in triangular form.

$$R_2 \leftrightarrow R_3 \qquad \begin{bmatrix} 1 & 2 & -1 \\ 0 & -4 & 4 \\ 0 & 0 & 4 \end{bmatrix}.$$

Let's name this last matrix B. The determinant of B is easy to compute as it is triangular; $\det(B) = -16$. We can use this to find $\det(A)$.

Recall the steps we used to transform A into B. They are:

[20] which is why mathematicians rarely smile: they are remembering their problems

[21] or *echelon* form

$$\begin{aligned} \tfrac{1}{2}R_1 &\to R_1 \\ R_1 + R_2 &\to R_2 \\ -3R_1 + R_3 &\to R_3 \\ R_2 &\leftrightarrow R_3 \end{aligned}$$

The first operation multiplied a row of A by $\frac{1}{2}$. This means that the resulting matrix had a determinant that was $\frac{1}{2}$ the determinant of A.

The next two operations did not affect the determinant at all. The last operation, the row swap, changed the sign. Combining these effects, we know that

$$-16 = \det(B) = (-1)\frac{1}{2}\det(A).$$

Solving for $\det(A)$ we have that $\det(A) = 32$.

In practice, we don't need to keep track of operations where we add multiples of one row to another; they simply do not affect the determinant. Also, in practice, these steps are carried out by a computer, and computers don't care about leading 1s. Therefore, row scaling operations are rarely used. The only things to keep track of are row swaps, and even then all we care about are the number of row swaps. An odd number of row swaps means that the original determinant has the opposite sign of the triangular form matrix; an even number of row swaps means they have the same determinant.

Let's practice this again.

Example 79 The matrix B was formed from A using the following elementary row operations, though not necessarily in this order. Find $\det(A)$.

$$B = \begin{bmatrix} 1 & 2 & 3 \\ 0 & 4 & 5 \\ 0 & 0 & 6 \end{bmatrix} \qquad \begin{aligned} &2R_1 \to R_1 \\ &\tfrac{1}{3}R_3 \to R_3 \\ &R_1 \leftrightarrow R_2 \\ &6R_1 + R_2 \to R_2 \end{aligned}$$

Solution It is easy to compute $\det(B) = 24$. In looking at our list of elementary row operations, we see that only the first three have an effect on the determinant. Therefore

$$24 = \det(B) = 2 \cdot \frac{1}{3} \cdot (-1) \cdot \det(A)$$

and hence

$$\det(A) = -36.$$

In the previous example, we may have been tempted to "rebuild" A using the elementary row operations and then computing the determinant. This can be done, but in general it is a bad idea; it takes too much work and it is too easy to make a mistake.

Let's think some more like a mathematician. How does the determinant work with other matrix operations that we know? Specifically, how does the determinant interact with matrix addition, scalar multiplication, matrix multiplication, the transpose and the trace? We'll again do an example to get an idea of what is going on, then give a theorem to state what is true.

Example 80 Let

$$A = \begin{bmatrix} 1 & 2 \\ 3 & 4 \end{bmatrix} \text{ and } B = \begin{bmatrix} 2 & 1 \\ 3 & 5 \end{bmatrix}.$$

Find the determinants of the matrices A, B, $A + B$, $3A$, AB, A^T, A^{-1}, and compare the determinant of these matrices to their trace.

SOLUTION We can quickly compute that $\det(A) = -2$ and that $\det(B) = 7$.

$$\begin{aligned} \det(A - B) &= \det\left(\begin{bmatrix} 1 & 2 \\ 3 & 4 \end{bmatrix} - \begin{bmatrix} 2 & 1 \\ 3 & 5 \end{bmatrix}\right) \\ &= \begin{vmatrix} -1 & 1 \\ 0 & -1 \end{vmatrix} \\ &= 1 \end{aligned}$$

It's tough to find a connection between $\det(A - B)$, $\det(A)$ and $\det(B)$.

$$\begin{aligned} \det(3A) &= \begin{vmatrix} 3 & 6 \\ 9 & 12 \end{vmatrix} \\ &= -18 \end{aligned}$$

We can figure this one out; multiplying one row of A by 3 increases the determinant by a factor of 3; doing it again (and hence multiplying both rows by 3) increases the determinant again by a factor of 3. Therefore $\det(3A) = 3 \cdot 3 \cdot \det(A)$, or $3^2 \cdot A$.

$$\begin{aligned} \det(AB) &= \det\left(\begin{bmatrix} 1 & 2 \\ 3 & 4 \end{bmatrix}\begin{bmatrix} 2 & 1 \\ 3 & 5 \end{bmatrix}\right) \\ &= \begin{vmatrix} 8 & 11 \\ 18 & 23 \end{vmatrix} \\ &= -14 \end{aligned}$$

This one seems clear; $\det(AB) = \det(A)\det(B)$.

$$\begin{aligned} \det\left(A^T\right) &= \begin{vmatrix} 1 & 3 \\ 2 & 4 \end{vmatrix} \\ &= -2 \end{aligned}$$

Obviously $\det\left(A^T\right) = \det(A)$; is this always going to be the case? If we think about it, we can see that the cofactor expansion along the first *row* of A will give us the same result as the cofactor expansion along the first *column* of A^T.[22]

$$\begin{aligned}\det\left(A^{-1}\right) &= \begin{vmatrix} -2 & 1 \\ 3/2 & -1/2 \end{vmatrix} \\ &= 1 - 3/2 \\ &= -1/2\end{aligned}$$

It seems as though

$$\det\left(A^{-1}\right) = \frac{1}{\det(A)}.$$

We end by remarking that there seems to be no connection whatsoever between the trace of a matrix and its determinant. We leave it to the reader to compute the trace for some of the above matrices and confirm this statement.

We now state a theorem which will confirm our conjectures from the previous example.

Theorem 16 **Determinant Properties**

Let A and B be $n \times n$ matrices and let k be a scalar. The following are true:

1. $\det(kA) = k^n \cdot \det(A)$
2. $\det\left(A^T\right) = \det(A)$
3. $\det(AB) = \det(A)\det(B)$
4. If A is invertible, then
$$\det\left(A^{-1}\right) = \frac{1}{\det(A)}.$$
5. A matrix A is invertible if and only if $\det(A) \neq 0$.

This last statement of the above theorem is significant: what happens if $\det(A) = 0$? It seems that $\det\left(A^{-1}\right) =$"$1/0$", which is undefined. There actually isn't a problem here; it turns out that if $\det(A) = 0$, then A is not invertible (hence part 5 of Theorem 16). This allows us to add on to our Invertible Matrix Theorem.

[22] This can be a bit tricky to think out in your head. Try it with a 3×3 matrix A and see how it works. All the 2×2 submatrices that are created in A^T are the transpose of those found in A; this doesn't matter since it is easy to see that the determinant isn't affected by the transpose in a 2×2 matrix.

Theorem 17

Invertible Matrix Theorem

Let A be an $n \times n$ matrix. The following statements are equivalent.

(a) A is invertible.

(g) $\det(A) \neq 0$.

This new addition to the Invertible Matrix Theorem is very useful; we'll refer back to it in Chapter 4 when we discuss eigenvalues.

We end this section with a shortcut for computing the determinants of 3×3 matrices. Consider the matrix A:

$$\begin{bmatrix} 1 & 2 & 3 \\ 4 & 5 & 6 \\ 7 & 8 & 9 \end{bmatrix}.$$

We can compute its determinant using cofactor expansion as we did in Example 71. Once one becomes proficient at this method, computing the determinant of a 3×3 isn't all that hard. A method many find easier, though, starts with rewriting the matrix without the brackets, and repeating the first and second columns at the end as shown below.

$$\begin{matrix} 1 & 2 & 3 & 1 & 2 \\ 4 & 5 & 6 & 4 & 5 \\ 7 & 8 & 9 & 7 & 8 \end{matrix}$$

In this 3×5 array of numbers, there are 3 full "upper left to lower right" diagonals, and 3 full "upper right to lower left" diagonals, as shown below with the arrows.

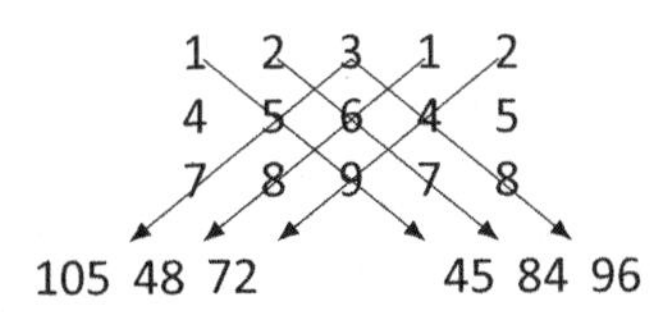

The numbers that appear at the ends of each of the arrows are computed by multiplying the numbers found along the arrows. For instance, the 105 comes from multiplying $3 \cdot 5 \cdot 7 = 105$. The determinant is found by adding the numbers on the right, and subtracting the sum of the numbers on the left. That is,

$$\det(A) = (45 + 84 + 96) - (105 + 48 + 72) = 0.$$

To help remind ourselves of this shortcut, we'll make it into a Key Idea.

Key Idea 13 **3 × 3 Determinant Shortcut**

Let A be a 3×3 matrix. Create a 3×5 array by repeating the first 2 columns and consider the products of the 3 "right hand" diagonals and 3 "left hand" diagonals as shown previously. Then

$$\begin{aligned}\det(A) &= \text{"(the sum of the right hand numbers)} \\ &\quad - \text{(the sum of the left hand numbers)".}\end{aligned}$$

We'll practice once more in the context of an example.

Example 81 Find the determinant of A using the previously described shortcut, where

$$A = \begin{bmatrix} 1 & 3 & 9 \\ -2 & 3 & 4 \\ -5 & 7 & 2 \end{bmatrix}.$$

SOLUTION Rewriting the first 2 columns, drawing the proper diagonals, and multiplying, we get:

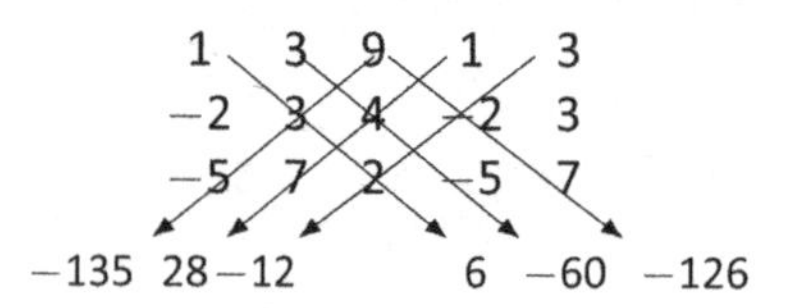

Summing the numbers on the right and subtracting the sum of the numbers on the left, we get

$$\det(A) = (6 - 60 - 126) - (-135 + 28 - 12) = -61.$$

In the next section we'll see how the determinant can be used to solve systems of linear equations.

Exercises 3.4

In Exercises 1 – 14, find the determinant of the given matrix using cofactor expansion along any row or column you choose.

1. $\begin{bmatrix} 1 & 2 & 3 \\ -5 & 0 & 3 \\ 4 & 0 & 6 \end{bmatrix}$

2. $\begin{bmatrix} -4 & 4 & -4 \\ 0 & 0 & -3 \\ -2 & -2 & -1 \end{bmatrix}$

3. $\begin{bmatrix} -4 & 1 & 1 \\ 0 & 0 & 0 \\ -1 & -2 & -5 \end{bmatrix}$

4. $\begin{bmatrix} 0 & -3 & 1 \\ 0 & 0 & 5 \\ -4 & 1 & 0 \end{bmatrix}$

5. $\begin{bmatrix} -2 & -3 & 5 \\ 5 & 2 & 0 \\ -1 & 0 & 0 \end{bmatrix}$

6. $\begin{bmatrix} -2 & -2 & 0 \\ 2 & -5 & -3 \\ -5 & 1 & 0 \end{bmatrix}$

7. $\begin{bmatrix} -3 & 0 & -5 \\ -2 & -3 & 3 \\ -1 & 0 & 1 \end{bmatrix}$

8. $\begin{bmatrix} 0 & 4 & -4 \\ 3 & 1 & -3 \\ -3 & -4 & 0 \end{bmatrix}$

9. $\begin{bmatrix} 5 & -5 & 0 & 1 \\ 2 & 4 & -1 & -1 \\ 5 & 0 & 0 & 4 \\ -1 & -2 & 0 & 5 \end{bmatrix}$

10. $\begin{bmatrix} -1 & 3 & 3 & 4 \\ 0 & 0 & 0 & 0 \\ 4 & -5 & -2 & 0 \\ 0 & 0 & 2 & 0 \end{bmatrix}$

11. $\begin{bmatrix} -5 & -5 & 0 & -2 \\ 0 & 0 & 5 & 0 \\ 1 & 3 & 3 & 1 \\ -4 & -2 & -1 & -5 \end{bmatrix}$

12. $\begin{bmatrix} -1 & 0 & -2 & 5 \\ 3 & -5 & 1 & -2 \\ -5 & -2 & -1 & -3 \\ -1 & 0 & 0 & 0 \end{bmatrix}$

13. $\begin{bmatrix} 4 & 0 & 5 & 1 & 0 \\ 1 & 0 & 3 & 1 & 5 \\ 2 & 2 & 0 & 2 & 2 \\ 1 & 0 & 0 & 0 & 0 \\ 4 & 4 & 2 & 5 & 3 \end{bmatrix}$

14. $\begin{bmatrix} 2 & 1 & 1 & 1 & 1 \\ 4 & 1 & 2 & 0 & 2 \\ 0 & 0 & 1 & 0 & 0 \\ 1 & 3 & 2 & 0 & 3 \\ 5 & 0 & 5 & 0 & 4 \end{bmatrix}$

In Exercises 15 – 18, a matrix M and det(M) are given. Matrices A, B and C are formed by performing operations on M. Determine the determinants of A, B and C using Theorems 15 and 16, and indicate the operations used to form A, B and C.

15. $M = \begin{bmatrix} 0 & 3 & 5 \\ 3 & 1 & 0 \\ -2 & -4 & -1 \end{bmatrix}$, $\det(M) = -41$.

(a) $A = \begin{bmatrix} 0 & 3 & 5 \\ -2 & -4 & -1 \\ 3 & 1 & 0 \end{bmatrix}$

(b) $B = \begin{bmatrix} 0 & 3 & 5 \\ 3 & 1 & 0 \\ 8 & 16 & 4 \end{bmatrix}$

(c) $C = \begin{bmatrix} 3 & 4 & 5 \\ 3 & 1 & 0 \\ -2 & -4 & -1 \end{bmatrix}$

16. $M = \begin{bmatrix} 9 & 7 & 8 \\ 1 & 3 & 7 \\ 6 & 3 & 3 \end{bmatrix}$, $\det(M) = 45$.

(a) $A = \begin{bmatrix} 18 & 14 & 16 \\ 1 & 3 & 7 \\ 6 & 3 & 3 \end{bmatrix}$

(b) $B = \begin{bmatrix} 9 & 7 & 8 \\ 1 & 3 & 7 \\ 96 & 73 & 83 \end{bmatrix}$

(c) $C = \begin{bmatrix} 9 & 1 & 6 \\ 7 & 3 & 3 \\ 8 & 7 & 3 \end{bmatrix}$

17. $M = \begin{bmatrix} 5 & 1 & 5 \\ 4 & 0 & 2 \\ 0 & 0 & 4 \end{bmatrix}$, $\det(M) = -16$.

(a) $A = \begin{bmatrix} 0 & 0 & 4 \\ 5 & 1 & 5 \\ 4 & 0 & 2 \end{bmatrix}$

(b) $B = \begin{bmatrix} -5 & -1 & -5 \\ -4 & 0 & -2 \\ 0 & 0 & 4 \end{bmatrix}$

(c) $C = \begin{bmatrix} 15 & 3 & 15 \\ 12 & 0 & 6 \\ 0 & 0 & 12 \end{bmatrix}$

18. $M = \begin{bmatrix} 5 & 4 & 0 \\ 7 & 9 & 3 \\ 1 & 3 & 9 \end{bmatrix}$,
$\det(M) = 120$.

(a) $A = \begin{bmatrix} 1 & 3 & 9 \\ 7 & 9 & 3 \\ 5 & 4 & 0 \end{bmatrix}$

(b) $B = \begin{bmatrix} 5 & 4 & 0 \\ 14 & 18 & 6 \\ 3 & 9 & 27 \end{bmatrix}$

(c) $C = \begin{bmatrix} -5 & -4 & 0 \\ -7 & -9 & -3 \\ -1 & -3 & -9 \end{bmatrix}$

In Exercises 19 – 22, matrices A and B are given. Verify part 3 of Theorem 16 by computing det(A), det(B) and det(AB).

19. $A = \begin{bmatrix} 2 & 0 \\ 1 & 2 \end{bmatrix}$,
$B = \begin{bmatrix} 0 & -4 \\ 1 & 3 \end{bmatrix}$

20. $A = \begin{bmatrix} 3 & -1 \\ 4 & 1 \end{bmatrix}$,
$B = \begin{bmatrix} -4 & -1 \\ -5 & 3 \end{bmatrix}$

21. $A = \begin{bmatrix} -4 & 4 \\ 5 & -2 \end{bmatrix}$,
$B = \begin{bmatrix} -3 & -4 \\ 5 & -3 \end{bmatrix}$

22. $A = \begin{bmatrix} -3 & -1 \\ 2 & -3 \end{bmatrix}$,
$B = \begin{bmatrix} 0 & 0 \\ 4 & -4 \end{bmatrix}$

In Exercises 23 – 30, find the determinant of the given matrix using Key Idea 13.

23. $\begin{bmatrix} 3 & 2 & 3 \\ -6 & 1 & -10 \\ -8 & -9 & -9 \end{bmatrix}$

24. $\begin{bmatrix} 8 & -9 & -2 \\ -9 & 9 & -7 \\ 5 & -1 & 9 \end{bmatrix}$

25. $\begin{bmatrix} -4 & 3 & -4 \\ -4 & -5 & 3 \\ 3 & -4 & 5 \end{bmatrix}$

26. $\begin{bmatrix} 1 & -2 & 1 \\ 5 & 5 & 4 \\ 4 & 0 & 0 \end{bmatrix}$

27. $\begin{bmatrix} 1 & -4 & 1 \\ 0 & 3 & 0 \\ 1 & 2 & 2 \end{bmatrix}$

28. $\begin{bmatrix} 3 & -1 & 0 \\ -3 & 0 & -4 \\ 0 & -1 & -4 \end{bmatrix}$

29. $\begin{bmatrix} -5 & 0 & -4 \\ 2 & 4 & -1 \\ -5 & 0 & -4 \end{bmatrix}$

30. $\begin{bmatrix} 1 & 0 & 0 \\ 0 & 1 & 0 \\ -1 & 1 & 1 \end{bmatrix}$

3.5 Cramer's Rule

1. T/F: Cramer's Rule is another method to compute the determinant of a matrix.
2. T/F: Cramer's Rule is often used because it is more efficient than Gaussian elimination.
3. Mathematicians use what word to describe the connections between seemingly unrelated ideas?

In the previous sections we have learned about the determinant, but we haven't given a really good reason *why* we would want to compute it.[23] This section shows one application of the determinant: solving systems of linear equations. We introduce this idea in terms of a theorem, then we will practice.

Theorem 18 **Cramer's Rule**

Let A be an $n \times n$ matrix with $\det(A) \neq 0$ and let $\vec{b}$ be an $n \times 1$ column vector. Then the linear system

$$A\vec{x} = \vec{b}$$

has solution

$$x_i = \frac{\det\left(A_i(\vec{b})\right)}{\det(A)},$$

where $A_i(\vec{b})$ is the matrix formed by replacing the ith column of A with $\vec{b}$.

Let's do an example.

Example 82 Use Cramer's Rule to solve the linear system $A\vec{x} = \vec{b}$ where

$$A = \begin{bmatrix} 1 & 5 & -3 \\ 1 & 4 & 2 \\ 2 & -1 & 0 \end{bmatrix} \quad \text{and} \quad \vec{b} = \begin{bmatrix} -36 \\ -11 \\ 7 \end{bmatrix}.$$

[23] The closest we came to motivation is that if $\det(A) = 0$, then we know that A is not invertible. But it seems that there may be easier ways to check.

Solution We first compute the determinant of A to see if we can apply Cramer's Rule.

$$\det(A) = \begin{vmatrix} 1 & 5 & -3 \\ 1 & 4 & 2 \\ 2 & -1 & 0 \end{vmatrix} = 49.$$

Since $\det(A) \neq 0$, we can apply Cramer's Rule. Following Theorem 18, we compute $\det\left(A_1(\vec{b})\right)$, $\det\left(A_2(\vec{b})\right)$ and $\det\left(A_3(\vec{b})\right)$.

$$\det\left(A_1(\vec{b})\right) = \begin{vmatrix} -36 & 5 & -3 \\ -11 & 4 & 2 \\ 7 & -1 & 0 \end{vmatrix} = 49.$$

(We used a bold font to show where $\vec{b}$ replaced the first column of A.)

$$\det\left(A_2(\vec{b})\right) = \begin{vmatrix} 1 & -36 & -3 \\ 1 & -11 & 2 \\ 2 & 7 & 0 \end{vmatrix} = -245.$$

$$\det\left(A_3(\vec{b})\right) = \begin{vmatrix} 1 & 5 & -36 \\ 1 & 4 & -11 \\ 2 & -1 & 7 \end{vmatrix} = 196.$$

Therefore we can compute $\vec{x}$:

$$x_1 = \frac{\det\left(A_1(\vec{b})\right)}{\det(A)} = \frac{49}{49} = 1$$

$$x_2 = \frac{\det\left(A_2(\vec{b})\right)}{\det(A)} = \frac{-245}{49} = -5$$

$$x_3 = \frac{\det\left(A_3(\vec{b})\right)}{\det(A)} = \frac{196}{49} = 4$$

Therefore

$$\vec{x} = \begin{bmatrix} x_1 \\ x_2 \\ x_3 \end{bmatrix} = \begin{bmatrix} 1 \\ -5 \\ 4 \end{bmatrix}.$$

Let's do another example.

Example 83 Use Cramer's Rule to solve the linear system $A\vec{x} = \vec{b}$ where

$$A = \begin{bmatrix} 1 & 2 \\ 3 & 4 \end{bmatrix} \text{ and } \vec{b} = \begin{bmatrix} -1 \\ 1 \end{bmatrix}.$$

SOLUTION The determinant of A is -2, so we can apply Cramer's Rule.

$$\det\left(A_1(\vec{b})\right) = \begin{vmatrix} -1 & 2 \\ 1 & 4 \end{vmatrix} = -6.$$

$$\det\left(A_2(\vec{b})\right) = \begin{vmatrix} 1 & -1 \\ 3 & 1 \end{vmatrix} = 4.$$

Therefore

$$x_1 = \frac{\det\left(A_1(\vec{b})\right)}{\det(A)} = \frac{-6}{-2} = 3$$

$$x_2 = \frac{\det\left(A_2(\vec{b})\right)}{\det(A)} = \frac{4}{-2} = -2$$

and

$$\vec{x} = \begin{bmatrix} x_1 \\ x_2 \end{bmatrix} = \begin{bmatrix} 3 \\ -2 \end{bmatrix}.$$

We learned in Section 3.4 that when considering a linear system $A\vec{x} = \vec{b}$ where A is square, if $\det(A) \neq 0$ then A is invertible and $A\vec{x} = \vec{b}$ has exactly one solution. We also stated in Key Idea 11 that if $\det(A) = 0$, then A is not invertible and so therefore either $A\vec{x} = \vec{b}$ has no solution or infinite solutions. Our method of figuring out which of these cases applied was to form the augmented matrix $\begin{bmatrix} A & \vec{b} \end{bmatrix}$, put it into reduced row echelon form, and then interpret the results.

Cramer's Rule specifies that $\det(A) \neq 0$ (so we are guaranteed a solution). When $\det(A) = 0$ we are not able to discern whether infinite solutions or no solution exists for a given vector $\vec{b}$. Cramer's Rule is only applicable to the case when exactly one solution exists.

We end this section with a practical consideration. We have mentioned before that finding determinants is a computationally intensive operation. To solve a linear system with 3 equations and 3 unknowns, we need to compute 4 determinants. Just think: with 10 equations and 10 unknowns, we'd need to compute 11 really hard determinants of 10×10 matrices! That is a lot of work!

The upshot of this is that Cramer's Rule makes for a poor choice in solving numerical linear systems. It simply is not done in practice; it is hard to beat Gaussian elimination.[24]

So why include it? *Because its truth is amazing.* The determinant is a very strange operation; it produces a number in a very odd way. It should seem incredible to the

[24] A version of Cramer's Rule is often taught in introductory differential equations courses as it can be used to find solutions to certain linear differential equations. In this situation, the entries of the matrices are functions, not numbers, and hence computing determinants is easier than using Gaussian elimination. Again, though, as the matrices get large, other solution methods are resorted to.

reader that by manipulating determinants in a particular way, we can solve linear systems.

In the next chapter we'll see another use for the determinant. Meanwhile, try to develop a deeper appreciation of math: odd, complicated things that seem completely unrelated often are intricately tied together. Mathematicians see these connections and describe them as "beautiful."

Exercises 3.5

In Exercises 1 – 12, matrices A and $\vec{b}$ are given.

(a) Give $\det(A)$ and $\det(A_i)$ for all i.

(b) Use Cramer's Rule to solve $A\vec{x} = \vec{b}$. If Cramer's Rule cannot be used to find the solution, then state whether or not a solution exists.

1. $A = \begin{bmatrix} 7 & -7 \\ -7 & 9 \end{bmatrix}, \quad \vec{b} = \begin{bmatrix} 28 \\ -26 \end{bmatrix}$

2. $A = \begin{bmatrix} 9 & 5 \\ -4 & -7 \end{bmatrix}, \quad \vec{b} = \begin{bmatrix} -45 \\ 20 \end{bmatrix}$

3. $A = \begin{bmatrix} -8 & 16 \\ 10 & -20 \end{bmatrix}, \quad \vec{b} = \begin{bmatrix} -48 \\ 60 \end{bmatrix}$

4. $A = \begin{bmatrix} 0 & -6 \\ 9 & -10 \end{bmatrix}, \quad \vec{b} = \begin{bmatrix} 6 \\ -17 \end{bmatrix}$

5. $A = \begin{bmatrix} 2 & 10 \\ -1 & 3 \end{bmatrix}, \quad \vec{b} = \begin{bmatrix} 42 \\ 19 \end{bmatrix}$

6. $A = \begin{bmatrix} 7 & 14 \\ -2 & -4 \end{bmatrix}, \quad \vec{b} = \begin{bmatrix} -1 \\ 4 \end{bmatrix}$

7. $A = \begin{bmatrix} 3 & 0 & -3 \\ 5 & 4 & 4 \\ 5 & 5 & -4 \end{bmatrix}, \quad \vec{b} = \begin{bmatrix} 24 \\ 0 \\ 31 \end{bmatrix}$

8. $A = \begin{bmatrix} 4 & 9 & 3 \\ -5 & -2 & -13 \\ -1 & 10 & -13 \end{bmatrix},$

$\vec{b} = \begin{bmatrix} -28 \\ 35 \\ 7 \end{bmatrix}$

9. $A = \begin{bmatrix} 4 & -4 & 0 \\ 5 & 1 & -1 \\ 3 & -1 & 2 \end{bmatrix}, \quad \vec{b} = \begin{bmatrix} 16 \\ 22 \\ 8 \end{bmatrix}$

10. $A = \begin{bmatrix} 1 & 0 & -10 \\ 4 & -3 & -10 \\ -9 & 6 & -2 \end{bmatrix},$

$\vec{b} = \begin{bmatrix} -40 \\ -94 \\ 132 \end{bmatrix}$

11. $A = \begin{bmatrix} 7 & -4 & 25 \\ -2 & 1 & -7 \\ 9 & -7 & 34 \end{bmatrix},$

$\vec{b} = \begin{bmatrix} -1 \\ -3 \\ 5 \end{bmatrix}$

12. $A = \begin{bmatrix} -6 & -7 & -7 \\ 5 & 4 & 1 \\ 5 & 4 & 8 \end{bmatrix},$

$\vec{b} = \begin{bmatrix} 58 \\ -35 \\ -49 \end{bmatrix}$

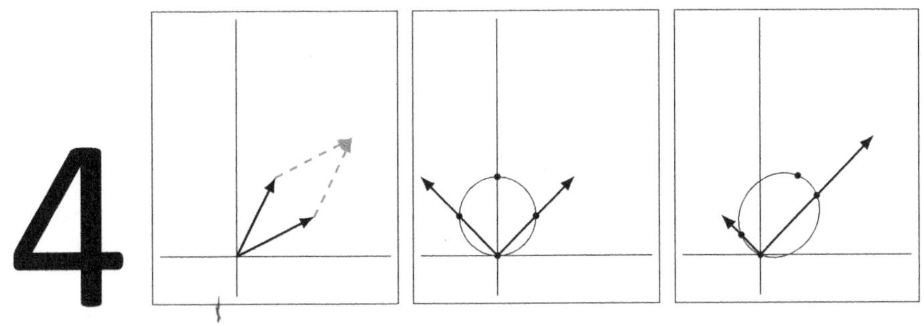

4 Eigenvalues and Eigenvectors

We have often explored new ideas in matrix algebra by making connections to our previous algebraic experience. Adding two numbers, $x + y$, led us to adding vectors $\vec{x} + \vec{y}$ and adding matrices $A + B$. We explored multiplication, which then led us to solving the matrix equation $A\vec{x} = \vec{b}$, which was reminiscent of solving the algebra equation $ax = b$.

This chapter is motivated by another analogy. Consider: when we multiply an unknown number x by another number such as 5, what do we know about the result? Unless, $x = 0$, we know that in some sense $5x$ will be "5 times bigger than x." Applying this to vectors, we would readily agree that $5\vec{x}$ gives a vector that is "5 times bigger than $\vec{x}$." Each entry in $\vec{x}$ is multiplied by 5.

Within the matrix algebra context, though, we have two types of multiplication: scalar and matrix multiplication. What happens to $\vec{x}$ when we multiply it by a matrix A? Our first response is likely along the lines of "You just get another vector. There is no definable relationship." We might wonder if there is ever the case where a matrix – vector multiplication is very similar to a scalar – vector multiplication. That is, do we ever have the case where $A\vec{x} = a\vec{x}$, where a is some scalar? That is the motivating question of this chapter.

4.1 Eigenvalues and Eigenvectors

1. T/F: Given any matrix A, we can always find a vector $\vec{x}$ where $A\vec{x} = \vec{x}$.

2. When is the zero vector an eigenvector for a matrix?

3. If $\vec{v}$ is an eigenvector of a matrix A with eigenvalue of 2, then what is $A\vec{v}$?

4. T/F: If A is a 5×5 matrix, to find the eigenvalues of A, we would need to find the roots of a 5th degree polynomial.

We start by considering the matrix A and vector $\vec{x}$ as given below.[1]

$$A = \begin{bmatrix} 1 & 4 \\ 2 & 3 \end{bmatrix} \qquad \vec{x} = \begin{bmatrix} 1 \\ 1 \end{bmatrix}$$

Multiplying $A\vec{x}$ gives:

$$\begin{aligned} A\vec{x} &= \begin{bmatrix} 1 & 4 \\ 2 & 3 \end{bmatrix} \begin{bmatrix} 1 \\ 1 \end{bmatrix} \\ &= \begin{bmatrix} 5 \\ 5 \end{bmatrix} \\ &= 5 \begin{bmatrix} 1 \\ 1 \end{bmatrix}! \end{aligned}$$

Wow! It looks like multiplying $A\vec{x}$ is the same as $5\vec{x}$! This makes us wonder lots of things: Is this the only case in the world where something like this happens?[2] Is A somehow a special matrix, and $A\vec{x} = 5\vec{x}$ for any vector $\vec{x}$ we pick?[3] Or maybe $\vec{x}$ was a special vector, and no matter what 2×2 matrix A we picked, we would have $A\vec{x} = 5\vec{x}$.[4]

A more likely explanation is this: given the matrix A, the number 5 and the vector $\vec{x}$ formed a special pair that happened to work together in a nice way. It is then natural to wonder if other "special" pairs exist. For instance, could we find a vector $\vec{x}$ where $A\vec{x} = 3\vec{x}$?

This equation is hard to solve *at first*; we are not used to matrix equations where $\vec{x}$ appears on both sides of "$=$." Therefore we put off solving this for just a moment to state a definition and make a few comments.

Definition 27

Eigenvalues and Eigenvectors

Let A be an $n \times n$ matrix, $\vec{x}$ a nonzero $n \times 1$ column vector and λ a scalar. If

$$A\vec{x} = \lambda\vec{x},$$

then $\vec{x}$ is an *eigenvector* of A and λ is an *eigenvalue* of A.

The word "eigen" is German for "proper" or "characteristic." Therefore, an *eigenvector* of A is a "characteristic vector of A." This vector tells us something about A.

Why do we use the Greek letter λ (lambda)? It is pure tradition. Above, we used a to represent the unknown scalar, since we are used to that notation. We now switch to λ because that is how everyone else does it.[5] Don't get hung up on this; λ is just a number.

[1] Recall this matrix and vector were used in Example 40 on page 75.

[2] Probably not.

[3] Probably not.

[4] See footnote 2.

[5] An example of mathematical peer pressure.

Note that our definition requires that A be a square matrix. If A isn't square then $A\vec{x}$ and $\lambda\vec{x}$ will have different sizes, and so they cannot be equal. Also note that $\vec{x}$ must be nonzero. Why? What if $\vec{x} = \vec{0}$? Then *no matter what* λ is, $A\vec{x} = \lambda\vec{x}$. This would then imply that *every number* is an eigenvalue; if every number is an eigenvalue, then we wouldn't need a definition for it.[6] Therefore we specify that $\vec{x} \neq \vec{0}$.

Our last comment before trying to find eigenvalues and eigenvectors for given matrices deals with "why we care." Did we stumble upon a mathematical curiosity, or does this somehow help us build better bridges, heal the sick, send astronauts into orbit, design optical equipment, and understand quantum mechanics? The answer, of course, is "Yes."[7] This is a wonderful topic in and of itself: we need no external application to appreciate its worth. At the same time, it has many, many applications to "the real world." A simple Internet seach on "applications of eigenvalues" with confirm this.

Back to our math. Given a square matrix A, we want to find a nonzero vector $\vec{x}$ and a scalar λ such that $A\vec{x} = \lambda\vec{x}$. We will solve this using the skills we developed in Chapter 2.

$$\begin{aligned} A\vec{x} &= \lambda\vec{x} && \text{original equation} \\ A\vec{x} - \lambda\vec{x} &= \vec{0} && \text{subtract } \lambda\vec{x} \text{ from both sides} \\ (A - \lambda I)\vec{x} &= \vec{0} && \text{factor out } \vec{x} \end{aligned}$$

Think about this last factorization. We are likely tempted to say

$$A\vec{x} - \lambda\vec{x} = (A - \lambda)\vec{x},$$

but this really doesn't make sense. After all, what does "a matrix minus a number" mean? We need the identity matrix in order for this to be logical.

Let us now think about the equation $(A - \lambda I)\vec{x} = \vec{0}$. While it looks complicated, it really is just matrix equation of the type we solved in Section 2.4. We are just trying to solve $B\vec{x} = \vec{0}$, where $B = (A - \lambda I)$.

We know from our previous work that this type of equation[8] always has a solution, namely, $\vec{x} = \vec{0}$. However, we want $\vec{x}$ to be an eigenvector and, by the definition, eigenvectors cannot be $\vec{0}$.

This means that we want solutions to $(A - \lambda I)\vec{x} = \vec{0}$ other than $\vec{x} = \vec{0}$. Recall that Theorem 8 says that if the matrix $(A - \lambda I)$ is invertible, then the *only* solution to $(A - \lambda I)\vec{x} = \vec{0}$ is $\vec{x} = \vec{0}$. Therefore, in order to have other solutions, we need $(A - \lambda I)$ to not be invertible.

Finally, recall from Theorem 16 that noninvertible matrices all have a determinant of 0. Therefore, if we want to find eigenvalues λ and eigenvectors $\vec{x}$, we need $\det(A - \lambda I) = 0$.

Let's start our practice of this theory by finding λ such that $\det(A - \lambda I) = 0$; that is, let's find the eigenvalues of a matrix.

---

[6] Recall footnote 17 on page 107.

[7] Except for the "understand quantum mechanics" part. Nobody truly understands that stuff; they just *probably* understand it.

[8] Recall this is a *homogeneous* system of equations.

Example 84 Find the eigenvalues of A, that is, find λ such that $\det(A - \lambda I) = 0$, where

$$A = \begin{bmatrix} 1 & 4 \\ 2 & 3 \end{bmatrix}.$$

Solution (Note that this is the matrix we used at the beginning of this section.) First, we write out what $A - \lambda I$ is:

$$\begin{aligned} A - \lambda I &= \begin{bmatrix} 1 & 4 \\ 2 & 3 \end{bmatrix} - \lambda \begin{bmatrix} 1 & 0 \\ 0 & 1 \end{bmatrix} \\ &= \begin{bmatrix} 1 & 4 \\ 2 & 3 \end{bmatrix} - \begin{bmatrix} \lambda & 0 \\ 0 & \lambda \end{bmatrix} \\ &= \begin{bmatrix} 1-\lambda & 4 \\ 2 & 3-\lambda \end{bmatrix} \end{aligned}$$

Therefore,

$$\begin{aligned} \det(A - \lambda I) &= \begin{vmatrix} 1-\lambda & 4 \\ 2 & 3-\lambda \end{vmatrix} \\ &= (1-\lambda)(3-\lambda) - 8 \\ &= \lambda^2 - 4\lambda - 5 \end{aligned}$$

Since we want $\det(A - \lambda I) = 0$, we want $\lambda^2 - 4\lambda - 5 = 0$. This is a simple quadratic equation that is easy to factor:

$$\begin{aligned} \lambda^2 - 4\lambda - 5 &= 0 \\ (\lambda - 5)(\lambda + 1) &= 0 \\ \lambda &= -1,\ 5 \end{aligned}$$

According to our above work, $\det(A - \lambda I) = 0$ when $\lambda = -1, 5$. Thus, the eigenvalues of A are -1 and 5.

Earlier, when looking at the same matrix as used in our example, we wondered if we could find a vector $\vec{x}$ such that $A\vec{x} = 3\vec{x}$. According to this example, the answer is "No." With this matrix A, the only values of λ that work are -1 and 5.

Let's restate the above in a different way: It is pointless to try to find $\vec{x}$ where $A\vec{x} = 3\vec{x}$, for there is no such $\vec{x}$. There are only 2 equations of this form that have a solution, namely

$$A\vec{x} = -\vec{x} \qquad \text{and} \qquad A\vec{x} = 5\vec{x}.$$

As we introduced this section, we gave a vector $\vec{x}$ such that $A\vec{x} = 5\vec{x}$. Is this the only one? Let's find out while calling our work an example; this will amount to finding the eigenvectors of A that correspond to the eigenvector of 5.

Example 85 Find $\vec{x}$ such that $A\vec{x} = 5\vec{x}$, where

$$A = \begin{bmatrix} 1 & 4 \\ 2 & 3 \end{bmatrix}.$$

SOLUTION Recall that our algebra from before showed that if

$$A\vec{x} = \lambda\vec{x} \quad \text{then} \quad (A - \lambda I)\vec{x} = \vec{0}.$$

Therefore, we need to solve the equation $(A - \lambda I)\vec{x} = \vec{0}$ for $\vec{x}$ when $\lambda = 5$.

$$\begin{aligned} A - 5I &= \begin{bmatrix} 1 & 4 \\ 2 & 3 \end{bmatrix} - 5\begin{bmatrix} 1 & 0 \\ 0 & 1 \end{bmatrix} \\ &= \begin{bmatrix} -4 & 4 \\ 2 & -2 \end{bmatrix} \end{aligned}$$

To solve $(A - 5I)\vec{x} = \vec{0}$, we form the augmented matrix and put it into reduced row echelon form:

$$\begin{bmatrix} -4 & 4 & 0 \\ 2 & -2 & 0 \end{bmatrix} \quad \overrightarrow{\text{rref}} \quad \begin{bmatrix} 1 & -1 & 0 \\ 0 & 0 & 0 \end{bmatrix}.$$

Thus

$$\begin{aligned} x_1 &= x_2 \\ x_2 &\text{ is free} \end{aligned}$$

and

$$\vec{x} = \begin{bmatrix} x_1 \\ x_2 \end{bmatrix} = x_2\begin{bmatrix} 1 \\ 1 \end{bmatrix}.$$

We have infinite solutions to the equation $A\vec{x} = 5\vec{x}$; any nonzero scalar multiple of the vector $\begin{bmatrix} 1 \\ 1 \end{bmatrix}$ is a solution. We can do a few examples to confirm this:

$$\begin{bmatrix} 1 & 4 \\ 2 & 3 \end{bmatrix}\begin{bmatrix} 2 \\ 2 \end{bmatrix} = \begin{bmatrix} 10 \\ 10 \end{bmatrix} = 5\begin{bmatrix} 2 \\ 2 \end{bmatrix};$$

$$\begin{bmatrix} 1 & 4 \\ 2 & 3 \end{bmatrix}\begin{bmatrix} 7 \\ 7 \end{bmatrix} = \begin{bmatrix} 35 \\ 35 \end{bmatrix} = 5\begin{bmatrix} 7 \\ 7 \end{bmatrix};$$

$$\begin{bmatrix} 1 & 4 \\ 2 & 3 \end{bmatrix}\begin{bmatrix} -3 \\ -3 \end{bmatrix} = \begin{bmatrix} -15 \\ -15 \end{bmatrix} = 5\begin{bmatrix} -3 \\ -3 \end{bmatrix}.$$

Our method of finding the eigenvalues of a matrix A boils down to determining which values of λ give the matrix $(A-\lambda I)$ a determinant of 0. In computing $\det(A - \lambda I)$, we get a polynomial in λ whose roots are the eigenvalues of A. This polynomial is important and so it gets its own name.

Definition 28

Characteristic Polynomial

Let A be an $n \times n$ matrix. The *characteristic polynomial* of A is the nth degree polynomial $p(\lambda) = \det(A - \lambda I)$.

Our definition just states *what* the characteristic polynomial is. We know from our work so far *why* we care: the roots of the characteristic polynomial of an $n \times n$ matrix A are the eigenvalues of A.

In Examples 84 and 85, we found eigenvalues and eigenvectors, respectively, of a given matrix. That is, given a matrix A, we found values λ and vectors $\vec{x}$ such that $A\vec{x} = \lambda\vec{x}$. The steps that follow outline the general procedure for finding eigenvalues and eigenvectors; we'll follow this up with some examples.

Key Idea 14

Finding Eigenvalues and Eigenvectors

Let A be an $n \times n$ matrix.

1. To find the eigenvalues of A, compute $p(\lambda)$, the characteristic polynomial of A, set it equal to 0, then solve for λ.
2. To find the eigenvectors of A, *for each eigenvalue* solve the homogeneous system $(A - \lambda I)\vec{x} = \vec{0}$.

Example 86 Find the eigenvalues of A, and for each eigenvalue, find an eigenvector where

$$A = \begin{bmatrix} -3 & 15 \\ 3 & 9 \end{bmatrix}.$$

Solution To find the eigenvalues, we must compute $\det(A - \lambda I)$ and set it equal to 0.

$$\begin{aligned} \det(A - \lambda I) &= \begin{vmatrix} -3-\lambda & 15 \\ 3 & 9-\lambda \end{vmatrix} \\ &= (-3-\lambda)(9-\lambda) - 45 \\ &= \lambda^2 - 6\lambda - 27 - 45 \\ &= \lambda^2 - 6\lambda - 72 \\ &= (\lambda - 12)(\lambda + 6) \end{aligned}$$

Therefore, $\det(A - \lambda I) = 0$ when $\lambda = -6$ and 12; these are our eigenvalues. (We should note that $p(\lambda) = \lambda^2 - 6\lambda - 72$ is our characteristic polynomial.) It sometimes helps to give them "names," so we'll say $\lambda_1 = -6$ and $\lambda_2 = 12$. Now we find eigenvectors.

For $\lambda_1 = -6$:
We need to solve the equation $(A - (-6)I)\vec{x} = \vec{0}$. To do this, we form the appropriate augmented matrix and put it into reduced row echelon form.

$$\begin{bmatrix} 3 & 15 & 0 \\ 3 & 15 & 0 \end{bmatrix} \quad \xrightarrow{\text{rref}} \quad \begin{bmatrix} 1 & 5 & 0 \\ 0 & 0 & 0 \end{bmatrix}.$$

Our solution is

$$x_1 = -5x_2$$
$$x_2 \text{ is free;}$$

in vector form, we have

$$\vec{x} = x_2 \begin{bmatrix} -5 \\ 1 \end{bmatrix}.$$

We may pick any nonzero value for x_2 to get an eigenvector; a simple option is $x_2 = 1$. Thus we have the eigenvector

$$\vec{x_1} = \begin{bmatrix} -5 \\ 1 \end{bmatrix}.$$

(We used the notation $\vec{x_1}$ to associate this eigenvector with the eigenvalue λ_1.)

We now repeat this process to find an eigenvector for $\lambda_2 = 12$:
In solving $(A - 12I)\vec{x} = \vec{0}$, we find

$$\begin{bmatrix} -15 & 15 & 0 \\ 3 & -3 & 0 \end{bmatrix} \quad \xrightarrow{\text{rref}} \quad \begin{bmatrix} 1 & -1 & 0 \\ 0 & 0 & 0 \end{bmatrix}.$$

In vector form, we have

$$\vec{x} = x_2 \begin{bmatrix} 1 \\ 1 \end{bmatrix}.$$

Again, we may pick any nonzero value for x_2, and so we choose $x_2 = 1$. Thus an eigenvector for λ_2 is

$$\vec{x_2} = \begin{bmatrix} 1 \\ 1 \end{bmatrix}.$$

To summarize, we have:

$$\text{eigenvalue } \lambda_1 = -6 \text{ with eigenvector } \vec{x_1} = \begin{bmatrix} -5 \\ 1 \end{bmatrix}$$

and

$$\text{eigenvalue } \lambda_2 = 12 \text{ with eigenvector } \vec{x_2} = \begin{bmatrix} 1 \\ 1 \end{bmatrix}.$$

We should take a moment and check our work: is it true that $A\vec{x_1} = \lambda_1\vec{x_1}$?

$$\begin{aligned} A\vec{x_1} &= \begin{bmatrix} -3 & 15 \\ 3 & 9 \end{bmatrix} \begin{bmatrix} -5 \\ 1 \end{bmatrix} \\ &= \begin{bmatrix} 30 \\ -6 \end{bmatrix} \\ &= (-6) \begin{bmatrix} -5 \\ 1 \end{bmatrix} \\ &= \lambda_1\vec{x_1}. \end{aligned}$$

Yes; it appears we have truly found an eigenvalue/eigenvector pair for the matrix A.

Let's do another example.

Example 87 Let $A = \begin{bmatrix} -3 & 0 \\ 5 & 1 \end{bmatrix}$. Find the eigenvalues of A and an eigenvector for each eigenvalue.

Solution We first compute the characteristic polynomial, set it equal to 0, then solve for λ.

$$\begin{aligned} \det(A - \lambda I) &= \begin{vmatrix} -3-\lambda & 0 \\ 5 & 1-\lambda \end{vmatrix} \\ &= (-3-\lambda)(1-\lambda) \end{aligned}$$

From this, we see that $\det(A - \lambda I) = 0$ when $\lambda = -3, 1$. We'll set $\lambda_1 = -3$ and $\lambda_2 = 1$.

Finding an eigenvector for λ_1:
We solve $(A - (-3)I)\vec{x} = \vec{0}$ for $\vec{x}$ by row reducing the appropriate matrix:

$$\begin{bmatrix} 0 & 0 & 0 \\ 5 & 4 & 0 \end{bmatrix} \quad \overrightarrow{\text{rref}} \quad \begin{bmatrix} 1 & 5/4 & 0 \\ 0 & 0 & 0 \end{bmatrix}.$$

Our solution, in vector form, is

$$\vec{x} = x_2 \begin{bmatrix} -5/4 \\ 1 \end{bmatrix}.$$

Again, we can pick any nonzero value for x_2; a nice choice would eliminate the fraction. Therefore we pick $x_2 = 4$, and find

$$\vec{x_1} = \begin{bmatrix} -5 \\ 4 \end{bmatrix}.$$

Finding an eigenvector for λ_2:

We solve $(A - (1)I)\vec{x} = \vec{0}$ for $\vec{x}$ by row reducing the appropriate matrix:

$$\begin{bmatrix} -4 & 0 & 0 \\ 5 & 0 & 0 \end{bmatrix} \quad \overrightarrow{\text{rref}} \quad \begin{bmatrix} 1 & 0 & 0 \\ 0 & 0 & 0 \end{bmatrix}.$$

We've seen a matrix like this before,[9] but we may need a bit of a refreshing. Our first row tells us that $x_1 = 0$, and we see that no rows/equations involve x_2. We conclude that x_2 is free. Therefore, our solution, in vector form, is

$$\vec{x} = x_2 \begin{bmatrix} 0 \\ 1 \end{bmatrix}.$$

We pick $x_2 = 1$, and find

$$\vec{x_2} = \begin{bmatrix} 0 \\ 1 \end{bmatrix}.$$

To summarize, we have:

$$\text{eigenvalue } \lambda_1 = -3 \text{ with eigenvector } \vec{x_1} = \begin{bmatrix} -5 \\ 4 \end{bmatrix}$$

and

$$\text{eigenvalue } \lambda_2 = 1 \text{ with eigenvector } \vec{x_2} = \begin{bmatrix} 0 \\ 1 \end{bmatrix}.$$

So far, our examples have involved 2×2 matrices. Let's do an example with a 3×3 matrix.

Example 88 Find the eigenvalues of A, and for each eigenvalue, give one eigenvector, where

$$A = \begin{bmatrix} -7 & -2 & 10 \\ -3 & 2 & 3 \\ -6 & -2 & 9 \end{bmatrix}.$$

SOLUTION We first compute the characteristic polynomial, set it equal to 0, then solve for λ. A warning: this process is rather long. We'll use cofactor expansion along the first row; don't get bogged down with the arithmetic that comes from each step; just try to get the basic idea of what was done from step to step.

[9] See page 31. Our future need of knowing how to handle this situation is foretold in footnote 5.

$$
\begin{aligned}
\det(A-\lambda I) &= \begin{vmatrix} -7-\lambda & -2 & 10 \\ -3 & 2-\lambda & 3 \\ -6 & -2 & 9-\lambda \end{vmatrix} \\
&= (-7-\lambda)\begin{vmatrix} 2-\lambda & 3 \\ -2 & 9-\lambda \end{vmatrix} - (-2)\begin{vmatrix} -3 & 3 \\ -6 & 9-\lambda \end{vmatrix} + 10\begin{vmatrix} -3 & 2-\lambda \\ -6 & -2 \end{vmatrix} \\
&= (-7-\lambda)(\lambda^2-11\lambda+24) + 2(3\lambda-9) + 10(-6\lambda+18) \\
&= -\lambda^3 + 4\lambda^2 - \lambda - 6 \\
&= -(\lambda+1)(\lambda-2)(\lambda-3)
\end{aligned}
$$

In the last step we factored the characteristic polynomial $-\lambda^3+4\lambda^2-\lambda-6$. Factoring polynomials of degree > 2 is not trivial; we'll assume the reader has access to methods for doing this accurately.[10]

Our eigenvalues are $\lambda_1 = -1$, $\lambda_2 = 2$ and $\lambda_3 = 3$. We now find corresponding eigenvectors.

For $\lambda_1 = -1$:

We need to solve the equation $(A-(-1)I)\vec{x} = \vec{0}$. To do this, we form the appropriate augmented matrix and put it into reduced row echelon form.

$$
\begin{bmatrix} -6 & -2 & 10 & 0 \\ -3 & 3 & 3 & 0 \\ -6 & -2 & 10 & 0 \end{bmatrix} \quad \overrightarrow{\text{rref}} \quad \begin{bmatrix} 1 & 0 & -1.5 & 0 \\ 0 & 1 & -.5 & 0 \\ 0 & 0 & 0 & 0 \end{bmatrix}
$$

Our solution, in vector form, is

$$
\vec{x} = x_3 \begin{bmatrix} 3/2 \\ 1/2 \\ 1 \end{bmatrix}.
$$

We can pick any nonzero value for x_3; a nice choice would get rid of the fractions. So we'll set $x_3 = 2$ and choose $\vec{x_1} = \begin{bmatrix} 3 \\ 1 \\ 2 \end{bmatrix}$ as our eigenvector.

For $\lambda_2 = 2$:

We need to solve the equation $(A-2I)\vec{x} = \vec{0}$. To do this, we form the appropriate augmented matrix and put it into reduced row echelon form.

---

[10] You probably learned how to do this in an algebra course. As a reminder, possible roots can be found by factoring the constant term (in this case, -6) of the polynomial. That is, the roots of this equation could be ±1, ±2, ±3 and ±6. That's 12 things to check.

One could also graph this polynomial to find the roots. Graphing will show us that $\lambda = 3$ *looks* like a root, and a simple calculation will confirm that it is.

$$\begin{bmatrix} -9 & -2 & 10 & 0 \\ -3 & 0 & 3 & 0 \\ -6 & -2 & 7 & 0 \end{bmatrix} \overrightarrow{\text{rref}} \begin{bmatrix} 1 & 0 & -1 & 0 \\ 0 & 1 & -.5 & 0 \\ 0 & 0 & 0 & 0 \end{bmatrix}$$

Our solution, in vector form, is

$$\vec{x} = x_3 \begin{bmatrix} 1 \\ 1/2 \\ 1 \end{bmatrix}.$$

We can pick any nonzero value for x_3; again, a nice choice would get rid of the fractions. So we'll set $x_3 = 2$ and choose $\vec{x_2} = \begin{bmatrix} 2 \\ 1 \\ 2 \end{bmatrix}$ as our eigenvector.

For $\lambda_3 = 3$:

We need to solve the equation $(A - 3I)\vec{x} = \vec{0}$. To do this, we form the appropriate augmented matrix and put it into reduced row echelon form.

$$\begin{bmatrix} -10 & -2 & 10 & 0 \\ -3 & -1 & 3 & 0 \\ -6 & -2 & 6 & 0 \end{bmatrix} \overrightarrow{\text{rref}} \begin{bmatrix} 1 & 0 & -1 & 0 \\ 0 & 1 & 0 & 0 \\ 0 & 0 & 0 & 0 \end{bmatrix}$$

Our solution, in vector form, is (note that $x_2 = 0$):

$$\vec{x} = x_3 \begin{bmatrix} 1 \\ 0 \\ 1 \end{bmatrix}.$$

We can pick any nonzero value for x_3; an easy choice is $x_3 = 1$, so $\vec{x_3} = \begin{bmatrix} 1 \\ 0 \\ 1 \end{bmatrix}$ as our eigenvector.

To summarize, we have the following eigenvalue/eigenvector pairs:

$$\text{eigenvalue } \lambda_1 = -1 \text{ with eigenvector } \vec{x_1} = \begin{bmatrix} 3 \\ 1 \\ 2 \end{bmatrix}$$

$$\text{eigenvalue } \lambda_2 = 2 \text{ with eigenvector } \vec{x_2} = \begin{bmatrix} 2 \\ 1 \\ 2 \end{bmatrix}$$

$$\text{eigenvalue } \lambda_3 = 3 \text{ with eigenvector } \vec{x_3} = \begin{bmatrix} 1 \\ 0 \\ 1 \end{bmatrix}$$

Let's practice once more.

Example 89 Find the eigenvalues of A, and for each eigenvalue, give one eigenvector, where

$$A = \begin{bmatrix} 2 & -1 & 1 \\ 0 & 1 & 6 \\ 0 & 3 & 4 \end{bmatrix}.$$

Solution We first compute the characteristic polynomial, set it equal to 0, then solve for λ. We'll use cofactor expansion down the first column (since it has lots of zeros).

$$\begin{aligned} \det(A - \lambda I) &= \begin{vmatrix} 2-\lambda & -1 & 1 \\ 0 & 1-\lambda & 6 \\ 0 & 3 & 4-\lambda \end{vmatrix} \\ &= (2-\lambda) \begin{vmatrix} 1-\lambda & 6 \\ 3 & 4-\lambda \end{vmatrix} \\ &= (2-\lambda)(\lambda^2 - 5\lambda - 14) \\ &= (2-\lambda)(\lambda - 7)(\lambda + 2) \end{aligned}$$

Notice that while the characteristic polynomial is cubic, we never actually saw a cubic; we never distributed the $(2-\lambda)$ across the quadratic. Instead, we realized that this was a factor of the cubic, and just factored the remaining quadratic. (This makes this example quite a bit simpler than the previous example.)

Our eigenvalues are $\lambda_1 = -2$, $\lambda_2 = 2$ and $\lambda_3 = 7$. We now find corresponding eigenvectors.

For $\lambda_1 = -2$:

We need to solve the equation $(A - (-2)I)\vec{x} = \vec{0}$. To do this, we form the appropriate augmented matrix and put it into reduced row echelon form.

$$\begin{bmatrix} 4 & -1 & 1 & 0 \\ 0 & 3 & 6 & 0 \\ 0 & 3 & 6 & 0 \end{bmatrix} \quad \overrightarrow{\text{rref}} \quad \begin{bmatrix} 1 & 0 & 3/4 & 0 \\ 0 & 1 & 2 & 0 \\ 0 & 0 & 0 & 0 \end{bmatrix}$$

Our solution, in vector form, is

$$\vec{x} = x_3 \begin{bmatrix} -3/4 \\ -2 \\ 1 \end{bmatrix}.$$

We can pick any nonzero value for x_3; a nice choice would get rid of the fractions. So we'll set $x_3 = 4$ and choose $\vec{x_1} = \begin{bmatrix} -3 \\ -8 \\ 4 \end{bmatrix}$ as our eigenvector.

For $\lambda_2 = 2$:

We need to solve the equation $(A - 2I)\vec{x} = \vec{0}$. To do this, we form the appropriate augmented matrix and put it into reduced row echelon form.

$$\begin{bmatrix} 0 & -1 & 1 & 0 \\ 0 & -1 & 6 & 0 \\ 0 & 3 & 2 & 0 \end{bmatrix} \quad \overrightarrow{\text{rref}} \quad \begin{bmatrix} 0 & 1 & 0 & 0 \\ 0 & 0 & 1 & 0 \\ 0 & 0 & 0 & 0 \end{bmatrix}$$

This looks funny, so we'll look remind ourselves how to solve this. The first two rows tell us that $x_2 = 0$ and $x_3 = 0$, respectively. Notice that no row/equation uses x_1; we conclude that it is free. Therefore, our solution in vector form is

$$\vec{x} = x_1 \begin{bmatrix} 1 \\ 0 \\ 0 \end{bmatrix}.$$

We can pick any nonzero value for x_1; an easy choice is $x_1 = 1$ and choose $\vec{x_2} = \begin{bmatrix} 1 \\ 0 \\ 0 \end{bmatrix}$ as our eigenvector.

For $\lambda_3 = 7$:

We need to solve the equation $(A - 7I)\vec{x} = \vec{0}$. To do this, we form the appropriate augmented matrix and put it into reduced row echelon form.

$$\begin{bmatrix} -5 & -1 & 1 & 0 \\ 0 & -6 & 6 & 0 \\ 0 & 3 & -3 & 0 \end{bmatrix} \quad \overrightarrow{\text{rref}} \quad \begin{bmatrix} 1 & 0 & 0 & 0 \\ 0 & 1 & -1 & 0 \\ 0 & 0 & 0 & 0 \end{bmatrix}$$

Our solution, in vector form, is (note that $x_1 = 0$):

$$\vec{x} = x_3 \begin{bmatrix} 0 \\ 1 \\ 1 \end{bmatrix}.$$

We can pick any nonzero value for x_3; an easy choice is $x_3 = 1$, so $\vec{x_3} = \begin{bmatrix} 0 \\ 1 \\ 1 \end{bmatrix}$ as our eigenvector.

To summarize, we have the following eigenvalue/eigenvector pairs:

$$\text{eigenvalue } \lambda_1 = -2 \text{ with eigenvector } \vec{x_1} = \begin{bmatrix} -3 \\ -8 \\ 4 \end{bmatrix}$$

$$\text{eigenvalue } \lambda_2 = 2 \text{ with eigenvector } \vec{x_2} = \begin{bmatrix} 1 \\ 0 \\ 0 \end{bmatrix}$$

$$\text{eigenvalue } \lambda_3 = 7 \text{ with eigenvector } \vec{x_3} = \begin{bmatrix} 0 \\ 1 \\ 1 \end{bmatrix}$$

In this section we have learned about a new concept: given a matrix A we can find certain values λ and vectors $\vec{x}$ where $A\vec{x} = \lambda\vec{x}$. In the next section we will continue to the pattern we have established in this text: after learning a new concept, we see how it interacts with other concepts we know about. That is, we'll look for connections between eigenvalues and eigenvectors and things like the inverse, determinants, the trace, the transpose, etc.

Exercises 4.1

In Exercises 1 – 6, a matrix A and one of its eigenvectors are given. Find the eigenvalue of A for the given eigenvector.

1. $A = \begin{bmatrix} 9 & 8 \\ -6 & -5 \end{bmatrix}$

   $\vec{x} = \begin{bmatrix} -4 \\ 3 \end{bmatrix}$

2. $A = \begin{bmatrix} 19 & -6 \\ 48 & -15 \end{bmatrix}$

   $\vec{x} = \begin{bmatrix} 1 \\ 3 \end{bmatrix}$

3. $A = \begin{bmatrix} 1 & -2 \\ -2 & 4 \end{bmatrix}$

   $\vec{x} = \begin{bmatrix} 2 \\ 1 \end{bmatrix}$

4. $A = \begin{bmatrix} -11 & -19 & 14 \\ -6 & -8 & 6 \\ -12 & -22 & 15 \end{bmatrix}$

   $\vec{x} = \begin{bmatrix} 3 \\ 2 \\ 4 \end{bmatrix}$

5. $A = \begin{bmatrix} -7 & 1 & 3 \\ 10 & 2 & -3 \\ -20 & -14 & 1 \end{bmatrix}$

   $\vec{x} = \begin{bmatrix} 1 \\ -2 \\ 4 \end{bmatrix}$

6. $A = \begin{bmatrix} -12 & -10 & 0 \\ 15 & 13 & 0 \\ 15 & 18 & -5 \end{bmatrix}$

   $\vec{x} = \begin{bmatrix} -1 \\ 1 \\ 1 \end{bmatrix}$

In Exercises 7 – 11, a matrix A and one of its eigenvalues are given. Find an eigenvector of A for the given eigenvalue.

7. $A = \begin{bmatrix} 16 & 6 \\ -18 & -5 \end{bmatrix}$

   $\lambda = 4$

8. $A = \begin{bmatrix} -2 & 6 \\ -9 & 13 \end{bmatrix}$

   $\lambda = 7$

9. $A = \begin{bmatrix} -16 & -28 & -19 \\ 42 & 69 & 46 \\ -42 & -72 & -49 \end{bmatrix}$

   $\lambda = 5$

10. $A = \begin{bmatrix} 7 & -5 & -10 \\ 6 & 2 & -6 \\ 2 & -5 & -5 \end{bmatrix}$

    $\lambda = -3$

11. $A = \begin{bmatrix} 4 & 5 & -3 \\ -7 & -8 & 3 \\ 1 & -5 & 8 \end{bmatrix}$

    $\lambda = 2$

In Exercises 12 – 28, find the eigenvalues of the given matrix. For each eigenvalue, give an eigenvector.

12. $\begin{bmatrix} -1 & -4 \\ -3 & -2 \end{bmatrix}$

13. $\begin{bmatrix} -4 & 72 \\ -1 & 13 \end{bmatrix}$

14. $\begin{bmatrix} 2 & -12 \\ 2 & -8 \end{bmatrix}$

15. $\begin{bmatrix} 3 & 12 \\ 1 & -1 \end{bmatrix}$

16. $\begin{bmatrix} 5 & 9 \\ -1 & -5 \end{bmatrix}$

17. $\begin{bmatrix} 3 & -1 \\ -1 & 3 \end{bmatrix}$

18. $\begin{bmatrix} 0 & 1 \\ 25 & 0 \end{bmatrix}$

19. $\begin{bmatrix} -3 & 1 \\ 0 & -1 \end{bmatrix}$

20. $\begin{bmatrix} 1 & -2 & -3 \\ 0 & 3 & 0 \\ 0 & -1 & -1 \end{bmatrix}$

21. $\begin{bmatrix} 5 & -2 & 3 \\ 0 & 4 & 0 \\ 0 & -1 & 3 \end{bmatrix}$

22. $\begin{bmatrix} 1 & 0 & 12 \\ 2 & -5 & 0 \\ 1 & 0 & 2 \end{bmatrix}$

23. $\begin{bmatrix} 1 & 0 & -18 \\ -4 & 3 & -1 \\ 1 & 0 & -8 \end{bmatrix}$

24. $\begin{bmatrix} -1 & 18 & 0 \\ 1 & 2 & 0 \\ 5 & -3 & -1 \end{bmatrix}$

25. $\begin{bmatrix} 5 & 0 & 0 \\ 1 & 1 & 0 \\ -1 & 5 & -2 \end{bmatrix}$

26. $\begin{bmatrix} 2 & -1 & 1 \\ 0 & 3 & 6 \\ 0 & 0 & 7 \end{bmatrix}$

27. $\begin{bmatrix} 3 & 5 & -5 \\ -2 & 3 & 2 \\ -2 & 5 & 0 \end{bmatrix}$

28. $\begin{bmatrix} 1 & 2 & 1 \\ 1 & 2 & 3 \\ 1 & 1 & 1 \end{bmatrix}$

4.2 Properties of Eigenvalues and Eigenvectors

AS YOU READ . . .

1. T/F: A and A^T have the same eigenvectors.
2. T/F: A and A^{-1} have the same eigenvalues.
3. T/F: Marie Ennemond Camille Jordan was a guy.
4. T/F: Matrices with a trace of 0 are important, although we haven't seen why.
5. T/F: A matrix A is invertible only if 1 is an eigenvalue of A.

In this section we'll explore how the eigenvalues and eigenvectors of a matrix relate to other properties of that matrix. This section is essentially a hodgepodge of interesting facts about eigenvalues; the goal here is not to memorize various facts about matrix algebra, but to again be amazed at the many connections between mathematical concepts.

We'll begin our investigations with an example that will give a foundation for other discoveries.

Example 90 Let $A = \begin{bmatrix} 1 & 2 & 3 \\ 0 & 4 & 5 \\ 0 & 0 & 6 \end{bmatrix}$. Find the eigenvalues of A.

Solution To find the eigenvalues, we compute $\det(A - \lambda I)$:

$$\det(A - \lambda I) = \begin{vmatrix} 1-\lambda & 2 & 3 \\ 0 & 4-\lambda & 5 \\ 0 & 0 & 6-\lambda \end{vmatrix}$$
$$= (1-\lambda)(4-\lambda)(6-\lambda)$$

Since our matrix is triangular, the determinant is easy to compute; it is just the product of the diagonal elements. Therefore, we found (and factored) our characteristic polynomial very easily, and we see that we have eigenvalues of $\lambda = 1, 4$, and 6.

This examples demonstrates a wonderful fact for us: the eigenvalues of a triangular matrix are simply the entries on the diagonal. Finding the corresponding eigenvectors still takes some work, but finding the eigenvalues is easy.

With that fact in the backs of our minds, let us proceed to the next example where we will come across some more interesting facts about eigenvalues and eigenvectors.

Example 91 Let $A = \begin{bmatrix} -3 & 15 \\ 3 & 9 \end{bmatrix}$ and let $B = \begin{bmatrix} -7 & -2 & 10 \\ -3 & 2 & 3 \\ -6 & -2 & 9 \end{bmatrix}$ (as used in Examples 86 and 88, respectively). Find the following:

1. eigenvalues and eigenvectors of A and B
2. eigenvalues and eigenvectors of A^{-1} and B^{-1}
3. eigenvalues and eigenvectors of A^T and B^T
4. The trace of A and B
5. The determinant of A and B

Solution We'll answer each in turn.

1. We already know the answer to these for we did this work in previous examples. Therefore we just list the answers.

   For A, we have eigenvalues $\lambda = -6$ and 12, with eigenvectors

   $$\vec{x} = x_2 \begin{bmatrix} -5 \\ 1 \end{bmatrix} \text{ and } x_2 \begin{bmatrix} 1 \\ 1 \end{bmatrix}, \text{ respectively.}$$

   For B, we have eigenvalues $\lambda = -1,\ 2$, and 3 with eigenvectors

   $$\vec{x} = x_3 \begin{bmatrix} 3 \\ 1 \\ 2 \end{bmatrix}, \ x_3 \begin{bmatrix} 2 \\ 1 \\ 2 \end{bmatrix} \text{ and } x_3 \begin{bmatrix} 1 \\ 0 \\ 1 \end{bmatrix}, \text{ respectively.}$$

2. We first compute the inverses of A and B. They are:

$$A^{-1} = \begin{bmatrix} -1/8 & 5/24 \\ 1/24 & 1/24 \end{bmatrix} \quad \text{and} \quad B^{-1} = \begin{bmatrix} -4 & 1/3 & 13/3 \\ -3/2 & 1/2 & 3/2 \\ -3 & 1/3 & 10/3 \end{bmatrix}.$$

Finding the eigenvalues and eigenvectors of these matrices is not terribly hard, but it is not "easy," either. Therefore, we omit showing the intermediate steps and go right to the conclusions.

For A^{-1}, we have eigenvalues $\lambda = -1/6$ and $1/12$, with eigenvectors

$$\vec{x} = x_2 \begin{bmatrix} -5 \\ 1 \end{bmatrix} \text{ and } x_2 \begin{bmatrix} 1 \\ 1 \end{bmatrix}, \text{ respectively.}$$

For B^{-1}, we have eigenvalues $\lambda = -1, 1/2$ and $1/3$ with eigenvectors

$$\vec{x} = x_3 \begin{bmatrix} 3 \\ 1 \\ 2 \end{bmatrix}, x_3 \begin{bmatrix} 2 \\ 1 \\ 2 \end{bmatrix} \text{ and } x_3 \begin{bmatrix} 1 \\ 0 \\ 1 \end{bmatrix}, \text{ respectively.}$$

3. Of course, computing the transpose of A and B is easy; computing their eigenvalues and eigenvectors takes more work. Again, we omit the intermediate steps.

For A^T, we have eigenvalues $\lambda = -6$ and 12 with eigenvectors

$$\vec{x} = x_2 \begin{bmatrix} -1 \\ 1 \end{bmatrix} \text{ and } x_2 \begin{bmatrix} 5 \\ 1 \end{bmatrix}, \text{ respectively.}$$

For B^T, we have eigenvalues $\lambda = -1,\ 2$ and 3 with eigenvectors

$$\vec{x} = x_3 \begin{bmatrix} -1 \\ 0 \\ 1 \end{bmatrix}, x_3 \begin{bmatrix} -1 \\ 1 \\ 1 \end{bmatrix} \text{ and } x_3 \begin{bmatrix} 0 \\ -2 \\ 1 \end{bmatrix}, \text{ respectively.}$$

4. The trace of A is 6; the trace of B is 4.

5. The determinant of A is -72; the determinant of B is -6.

Now that we have completed the "grunt work," let's analyze the results of the previous example. We are looking for any patterns or relationships that we can find.

The eigenvalues and eigenvectors of A and A^{-1}.

In our example, we found that the eigenvalues of A are -6 and 12; the eigenvalues of A^{-1} are $-1/6$ and $1/12$. Also, the eigenvalues of B are -1, 2 and 3, whereas the eigenvalues of B^{-1} are -1, $1/2$ and $1/3$. There is an obvious relationship here; it seems that if λ is an eigenvalue of A, then $1/\lambda$ will be an eigenvalue of A^{-1}. We can also note that the corresponding eigenvectors matched, too.

Why is this the case? Consider an invertible matrix A with eigenvalue λ and eigenvector $\vec{x}$. Then, by definition, we know that $A\vec{x} = \lambda\vec{x}$. Now multiply both sides by A^{-1}:

$$\begin{aligned} A\vec{x} &= \lambda\vec{x} \\ A^{-1}A\vec{x} &= A^{-1}\lambda\vec{x} \\ \vec{x} &= \lambda A^{-1}\vec{x} \\ \frac{1}{\lambda}\vec{x} &= A^{-1}\vec{x} \end{aligned}$$

We have just shown that $A^{-1}\vec{x} = 1/\lambda\vec{x}$; this, by definition, shows that $\vec{x}$ is an eigenvector of A^{-1} with eigenvalue $1/\lambda$. This explains the result we saw above.

The eigenvalues and eigenvectors of A and A^T.

Our example showed that A and A^T had the same eigenvalues but different (but somehow similar) eigenvectors; it also showed that B and B^T had the same eigenvalues but unrelated eigenvectors. Why is this?

We can answer the eigenvalue question relatively easily; it follows from the properties of the determinant and the transpose. Recall the following two facts:

1. $(A+B)^T = A^T + B^T$ (Theorem 11) and
2. $\det(A) = \det\left(A^T\right)$ (Theorem 16).

We find the eigenvalues of a matrix by computing the characteristic polynomial; that is, we find $\det(A - \lambda I)$. What is the characteristic polynomial of A^T? Consider:

$$\begin{aligned} \det\left(A^T - \lambda I\right) &= \det\left(A^T - \lambda I^T\right) && \text{since } I = I^T \\ &= \det\left((A - \lambda I)^T\right) && \text{Theorem 11} \\ &= \det(A - \lambda I) && \text{Theorem 16} \end{aligned}$$

So we see that the characteristic polynomial of A^T is the same as that for A. Therefore they have the same eigenvalues.

What about their respective eigenvectors? Is there any relationship? The simple answer is "No."[11]

[11] We have defined an eigenvector to be a column vector. Some mathematicians prefer to use row vectors instead; in that case, the typical eigenvalue/eigenvector equation looks like $\vec{x}A = \lambda\vec{x}$. It turns out that doing things this way will give you the same eigenvalues as our method. What is more, take the transpose of the above equation: you get $(\vec{x}A)^T = (\lambda\vec{x})^T$ which is also $A^T\vec{x}^T = \lambda\vec{x}^T$. The transpose of a row vector is a column vector, so this equation is actually the kind we are used to, and we can say that $\vec{x}^T$ is an eigenvector of A^T.

In short, what we find is that the eigenvectors of A^T are the "row" eigenvectors of A, and vice–versa.

The eigenvalues and eigenvectors of A and The Trace.

Note that the eigenvalues of A are -6 and 12, and the trace is 6; the eigenvalues of B are -1, 2 and 3, and the trace of B is 4. Do we notice any relationship?

It seems that the sum of the eigenvalues is the trace! Why is this the case?

The answer to this is a bit out of the scope of this text; we can justify part of this fact, and another part we'll just state as being true without justification.

First, recall from Theorem 13 that $\text{tr}(AB) = \text{tr}(BA)$. Secondly, we state without justification that given a square matrix A, we can find a square matrix P such that $P^{-1}AP$ is an upper triangular matrix with the eigenvalues of A on the diagonal.[12] Thus $\text{tr}(P^{-1}AP)$ is the sum of the eigenvalues; also, using our Theorem 13, we know that $\text{tr}(P^{-1}AP) = \text{tr}(P^{-1}PA) = \text{tr}(A)$. Thus the trace of A is the sum of the eigenvalues.

The eigenvalues and eigenvectors of A and The Determinant.

Again, the eigenvalues of A are -6 and 12, and the determinant of A is -72. The eigenvalues of B are -1, 2 and 3; the determinant of B is -6. It seems as though the product of the eigenvalues is the determinant.

This is indeed true; we defend this with our argument from above. We know that the determinant of a triangular matrix is the product of the diagonal elements. Therefore, given a matrix A, we can find P such that $P^{-1}AP$ is upper triangular with the eigenvalues of A on the diagonal. Thus $\det\left(P^{-1}AP\right)$ is the product of the eigenvalues. Using Theorem 16, we know that $\det\left(P^{-1}AP\right) = \det\left(P^{-1}PA\right) = \det\left(A\right)$. Thus the determinant of A is the product of the eigenvalues.

We summarize the results of our example with the following theorem.

[12]Who in the world thinks up this stuff? It seems that the answer is Marie Ennemond Camille Jordan, who, despite having at least two girl names, was a guy.

Theorem 19

Properties of Eigenvalues and Eigenvectors

Let A be an $n \times n$ invertible matrix. The following are true:

1. If A is triangular, then the diagonal elements of A are the eigenvalues of A.
2. If λ is an eigenvalue of A with eigenvector $\vec{x}$, then $\frac{1}{\lambda}$ is an eigenvalue of A^{-1} with eigenvector $\vec{x}$.
3. If λ is an eigenvalue of A then λ is an eigenvalue of A^T.
4. The sum of the eigenvalues of A is equal to tr(A), the trace of A.
5. The product of the eigenvalues of A is the equal to $\det(A)$, the determinant of A.

There is one more concept concerning eigenvalues and eigenvectors that we will explore. We do so in the context of an example.

Example 92 Find the eigenvalues and eigenvectors of the matrix $A = \begin{bmatrix} 1 & 2 \\ 1 & 2 \end{bmatrix}$.

SOLUTION To find the eigenvalues, we compute $\det(A - \lambda I)$:

$$\begin{aligned} \det(A - \lambda I) &= \begin{vmatrix} 1-\lambda & 2 \\ 1 & 2-\lambda \end{vmatrix} \\ &= (1-\lambda)(2-\lambda) - 2 \\ &= \lambda^2 - 3\lambda \\ &= \lambda(\lambda - 3) \end{aligned}$$

Our eigenvalues are therefore $\lambda = 0, 3$.
For $\lambda = 0$, we find the eigenvectors:

$$\begin{bmatrix} 1 & 2 & 0 \\ 1 & 2 & 0 \end{bmatrix} \quad \overrightarrow{\text{rref}} \quad \begin{bmatrix} 1 & 2 & 0 \\ 0 & 0 & 0 \end{bmatrix}$$

This shows that $x_1 = -2x_2$, and so our eigenvectors $\vec{x}$ are

$$\vec{x} = x_2 \begin{bmatrix} -2 \\ 1 \end{bmatrix}.$$

For $\lambda = 3$, we find the eigenvectors:

$$\begin{bmatrix} -2 & 2 & 0 \\ 1 & -1 & 0 \end{bmatrix} \xrightarrow{\text{rref}} \begin{bmatrix} 1 & -1 & 0 \\ 0 & 0 & 0 \end{bmatrix}$$

This shows that $x_1 = x_2$, and so our eigenvectors $\vec{x}$ are

$$\vec{x} = x_2 \begin{bmatrix} 1 \\ 1 \end{bmatrix}.$$

One interesting thing about the above example is that we see that 0 is an eigenvalue of A; we have not officially encountered this before. Does this mean anything significant?[13]

Think about what an eigenvalue of 0 means: there exists an nonzero vector $\vec{x}$ where $A\vec{x} = 0\vec{x} = \vec{0}$. That is, we have a nontrivial solution to $A\vec{x} = \vec{0}$. We know this only happens when A is not invertible.

So if A is invertible, there is no nontrivial solution to $A\vec{x} = \vec{0}$, and hence 0 *is not* an eigenvalue of A. If A is not invertible, then there is a nontrivial solution to $A\vec{x} = \vec{0}$, and hence 0 *is* an eigenvalue of A. This leads us to our final addition to the Invertible Matrix Theorem.

Theorem 20 **Invertible Matrix Theorem**

Let A be an $n \times n$ matrix. The following statements are equivalent.

(a) A is invertible.

(h) A does not have an eigenvalue of 0.

This section is about the properties of eigenvalues and eigenvectors. Of course, we have not investigated all of the numerous properties of eigenvalues and eigenvectors; we have just surveyed some of the most common (and most important) concepts. Here are four quick examples of the many things that still exist to be explored.

First, recall the matrix

$$A = \begin{bmatrix} 1 & 4 \\ 2 & 3 \end{bmatrix}$$

that we used in Example 84. It's characteristic polynomial is $p(\lambda) = \lambda^2 - 4\lambda - 5$. Compute $p(A)$; that is, compute $A^2 - 4A - 5I$. You should get something "interesting," and you should wonder "does this always work?"[14]

[13] Since 0 is a "special" number, we might think so – afterall, we found that having a determinant of 0 is important. Then again, a matrix with a trace of 0 isn't all that important. (Well, as far as we have seen; it actually *is*). So, having an eigenvalue of 0 may or may not be significant, but we would be doing well if we recognized the possibility of significance and decided to investigate further.

[14] Yes.

Second, in all of our examples, we have considered matrices where eigenvalues "appeared only once." Since we know that the eigenvalues of a triangular matrix appear on the diagonal, we know that the eigenvalues of

$$A = \begin{bmatrix} 1 & 1 \\ 0 & 1 \end{bmatrix}$$

are "1 and 1;" that is, the eigenvalue $\lambda = 1$ appears twice. What does that mean when we consider the eigenvectors of $\lambda = 1$? Compare the result of this to the matrix

$$A = \begin{bmatrix} 1 & 0 \\ 0 & 1 \end{bmatrix},$$

which also has the eigenvalue $\lambda = 1$ appearing twice.[15]

Third, consider the matrix

$$A = \begin{bmatrix} 0 & -1 \\ 1 & 0 \end{bmatrix}.$$

What are the eigenvalues?[16] We quickly compute the characteristic polynomial to be $p(\lambda) = \lambda^2 + 1$. Therefore the eigenvalues are $\pm\sqrt{-1} = \pm i$. What does this mean?

Finally, we have found the eigenvalues of matrices by finding the roots of the characteristic polynomial. We have limited our examples to quadratic and cubic polynomials; one would expect for larger sized matrices that a computer would be used to factor the characteristic polynomials. However, in general, this is *not* how the eigenvalues are found. Factoring high order polynomials is too unreliable, even with a computer – round off errors can cause unpredictable results. Also, to even compute the characteristic polynomial, one needs to compute the determinant, which is also expensive (as discussed in the previous chapter).

So how are eigenvalues found? There are *iterative* processes that can progressively transform a matrix A into another matrix that is *almost* an upper triangular matrix (the entries below the diagonal are almost zero) where the entries on the diagonal are the eigenvalues. The more iterations one performs, the better the approximation is.

These methods are so fast and reliable that some computer programs convert polynomial root finding problems into eigenvalue problems!

Most textbooks on Linear Algebra will provide direction on exploring the above topics and give further insight to what is going on. We have mentioned all the eigenvalue and eigenvector properties in this section for the same reasons we gave in the previous section. First, knowing these properties helps us solve numerous real world problems, and second, it is fascinating to see how rich and deep the theory of matrices is.

[15] To direct further study, it helps to know that mathematicians refer to this as the *duplicity* of an eigenvalue. In each of these two examples, A has the eigenvalue $\lambda = 1$ with duplicity of 2.

[16] Be careful; this matrix is *not* triangular.

Exercises 4.2

In Exercises 1 – 6, a matrix A is given. For each,

(a) Find the eigenvalues of A, and for each eigenvalue, find an eigenvector.

(b) Do the same for A^T.

(c) Do the same for A^{-1}.

(d) Find tr(A).

(e) Find det (A).

Use Theorem 19 to verify your results.

1. $\begin{bmatrix} 0 & 4 \\ -1 & 5 \end{bmatrix}$

2. $\begin{bmatrix} -2 & -14 \\ -1 & 3 \end{bmatrix}$

3. $\begin{bmatrix} 5 & 30 \\ -1 & -6 \end{bmatrix}$

4. $\begin{bmatrix} -4 & 72 \\ -1 & 13 \end{bmatrix}$

5. $\begin{bmatrix} 5 & -9 & 0 \\ 1 & -5 & 0 \\ 2 & 4 & 3 \end{bmatrix}$

6. $\begin{bmatrix} 0 & 25 & 0 \\ 1 & 0 & 0 \\ 1 & 1 & -3 \end{bmatrix}$

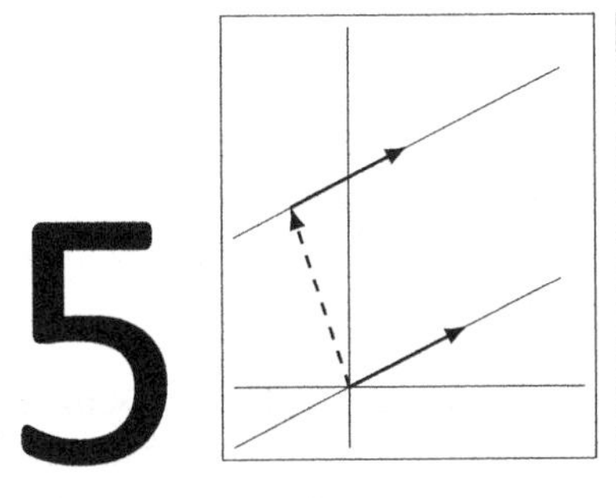
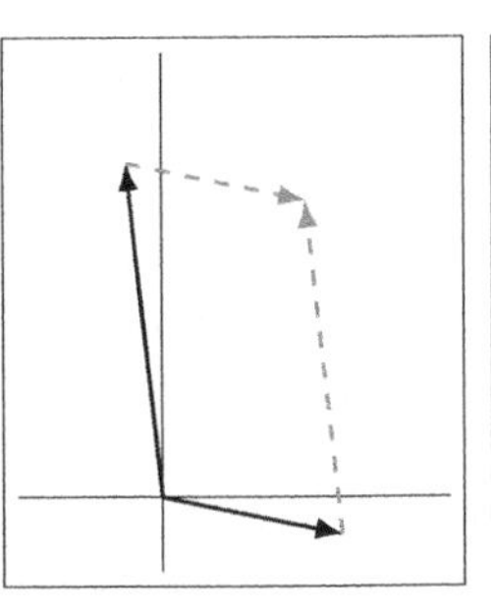
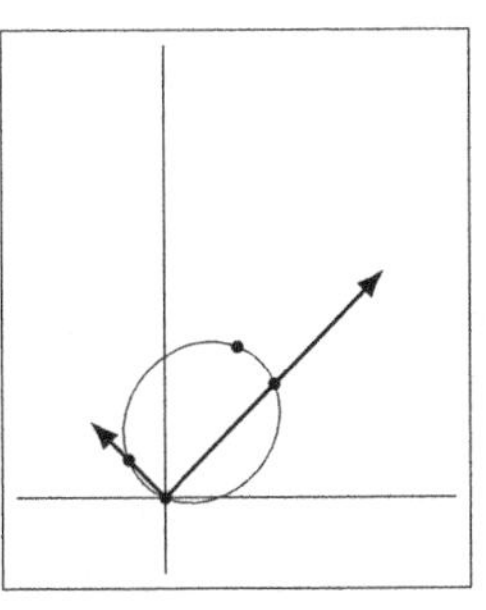

5 GRAPHICAL EXPLORATIONS OF VECTORS

We already looked at the basics of graphing vectors. In this chapter, we'll explore these ideas more fully. One often gains a better understanding of a concept by "seeing" it. For instance, one can study the function $f(x) = x^2$ and describe many properties of how the output relates to the input without producing a graph, but the graph can quickly bring meaning and insight to equations and formulae. Not only that, but the study of graphs of functions is in itself a wonderful mathematical world, worthy of exploration.

We've studied the graphing of vectors; in this chapter we'll take this a step further and study some fantastic graphical properties of vectors and matrix arithmetic. We mentioned earlier that these concepts form the basis of computer graphics; in this chapter, we'll see even better how that is true.

5.1 Transformations of the Cartesian Plane

1. To understand how the Cartesian plane is affected by multiplication by a matrix, it helps to study how what is affected?

2. Transforming the Cartesian plane through matrix multiplication transforms straight lines into what kind of lines?

3. T/F: If one draws a picture of a sheep on the Cartesian plane, then transformed the plane using the matrix

$$\begin{bmatrix} -1 & 0 \\ 0 & 1 \end{bmatrix},$$

one could say that the sheep was "sheared."

We studied in Section 2.3 how to visualize vectors and how certain matrix arithmetic operations can be graphically represented. We limited our visual understanding of matrix multiplication to graphing a vector, multiplying it by a matrix, then graphing the resulting vector. In this section we'll explore these multiplication ideas in greater depth. Instead of multiplying individual vectors by a matrix A, we'll study what happens when we multiply *every* vector in the Cartesian plans by A.[1]

Because of the Distributive Property as we saw demonstrated way back in Example 41, we can say that the Cartesian plane will be *transformed* in a very nice, predictable way. Straight lines will be transformed into other straight lines (and they won't become curvy, or jagged, or broken). Curved lines will be transformed into other curved lines (perhaps the curve will become "straight," but it won't become jagged or broken).

One way of studying how the whole Cartesian plane is affected by multiplication by a matrix A is to study how the *unit square* is affected. The unit square is the square with corners at the points $(0,0)$, $(1,0)$, $(1,1)$, and $(0,1)$. Each corner can be represented by the vector that points to it; multiply each of these vectors by A and we can get an idea of how A affects the whole Cartesian plane.

Let's try an example.

Example 93 Plot the vectors of the unit square before and after they have been multiplied by A, where

$$A = \begin{bmatrix} 1 & 4 \\ 2 & 3 \end{bmatrix}.$$

SOLUTION The four corners of the unit square can be represented by the vectors

$$\begin{bmatrix} 0 \\ 0 \end{bmatrix}, \quad \begin{bmatrix} 1 \\ 0 \end{bmatrix}, \quad \begin{bmatrix} 1 \\ 1 \end{bmatrix}, \quad \begin{bmatrix} 0 \\ 1 \end{bmatrix}.$$

Multiplying each by A gives the vectors

$$\begin{bmatrix} 0 \\ 0 \end{bmatrix}, \quad \begin{bmatrix} 1 \\ 2 \end{bmatrix}, \quad \begin{bmatrix} 5 \\ 5 \end{bmatrix}, \quad \begin{bmatrix} 4 \\ 3 \end{bmatrix},$$

respectively.

(Hint: one way of using your calculator to do this for you quickly is to make a 2×4 matrix whose columns are each of these vectors. In this case, create a matrix

$$B = \begin{bmatrix} 0 & 1 & 1 & 0 \\ 0 & 0 & 1 & 1 \end{bmatrix}.$$

Then multiply B by A and read off the transformed vectors from the respective columns:

$$AB = \begin{bmatrix} 0 & 1 & 5 & 4 \\ 0 & 2 & 5 & 3 \end{bmatrix}.$$

[1] No, we won't do them one by one.

This saves time, especially if you do a similar procedure for multiple matrices A. Of course, we can save more time by skipping the first column; since it is the column of zeros, it will stay the column of zeros after multiplication by A.)

The unit square and its transformation are graphed in Figure 5.1, where the shaped vertices correspond to each other across the two graphs. Note how the square got turned into some sort of quadrilateral (it's actually a parallelogram). A really interesting thing is how the triangular and square vertices seem to have changed places – it is as though the square, in addition to being stretched out of shape, was flipped.

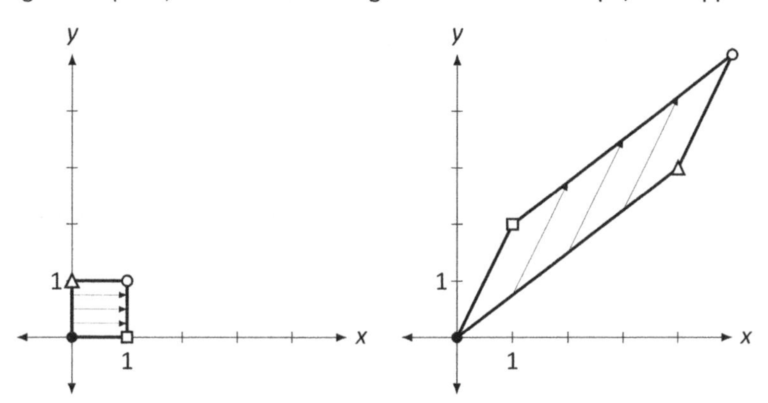

Figure 5.1: Transforming the unit square by matrix multiplication in Example 93.

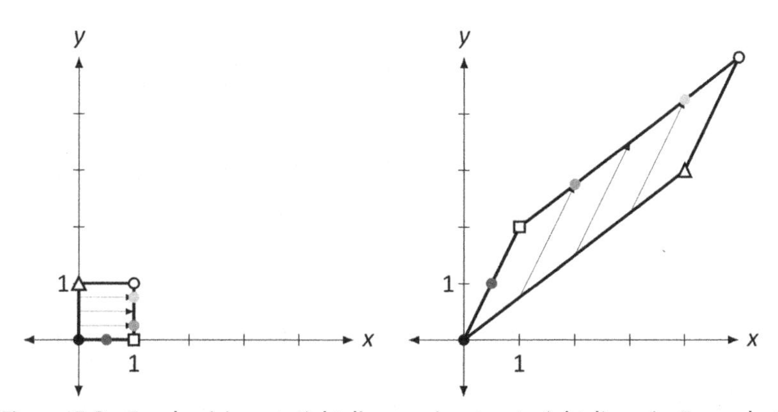

Figure 5.2: Emphasizing straight lines going to straight lines in Example 93.

To stress how "straight lines get transformed to straight lines," consider Figure 5.2. Here, the unit square has some additional points drawn on it which correspond to the shaded dots on the transformed parallelogram. Note how relative distances are also preserved; the dot halfway between the black and square dots is transformed to a position along the line, halfway between the black and square dots.

Much more can be said about this example. Before we delve into this, though, let's try one more example.

Example 94 Plot the transformed unit square after it has been transformed by A, where

$$A = \begin{bmatrix} 0 & -1 \\ 1 & 0 \end{bmatrix}.$$

SOLUTION We'll put the vectors that correspond to each corner in a matrix B as before and then multiply it on the left by A. Doing so gives:

$$\begin{aligned} AB &= \begin{bmatrix} 0 & -1 \\ 1 & 0 \end{bmatrix} \begin{bmatrix} 0 & 1 & 1 & 0 \\ 0 & 0 & 1 & 1 \end{bmatrix} \\ &= \begin{bmatrix} 0 & 0 & -1 & -1 \\ 0 & 1 & 1 & 0 \end{bmatrix} \end{aligned}$$

In Figure 5.3 the unit square is again drawn along with its transformation by A.

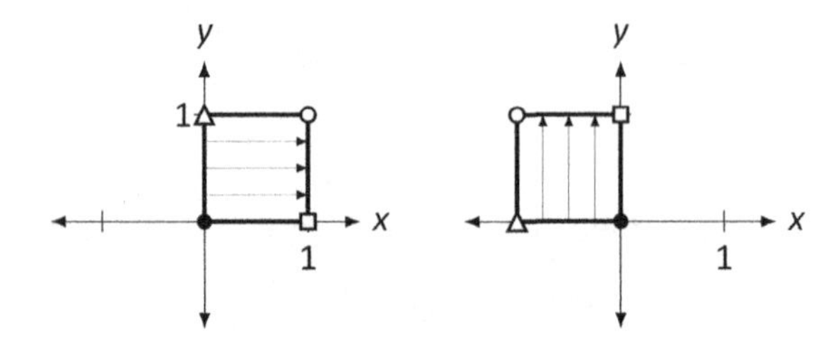

Figure 5.3: Transforming the unit square by matrix multiplication in Example 94.

Make note of how the square moved. It did not simply "slide" to the left;[2] nor did it "flip" across the y axis. Rather, it was *rotated* counterclockwise about the origin 90°. In a rotation, the shape of an object does not change; in our example, the square remained a square of the same size.

We have broached the topic of how the Cartesian plane can be transformed via multiplication by a 2×2 matrix A. We have seen two examples so far, and our intuition as to how the plane is changed has been informed only by seeing how the unit square changes. Let's explore this further by investigating two questions:

1. Suppose we want to transform the Cartesian plane in a known way (for instance, we may want to rotate the plane counterclockwise 180°). How do we find the matrix (if one even exists) which performs this transformation?

2. How does knowing how the unit square is transformed really help in understanding how the entire plane is transformed?

These questions are closely related, and as we answer one, we will help answer the other.

[2]mathematically, that is called a *translation*

To get started with the first question, look back at Examples 93 and 94 and consider again how the unit square was transformed. In particular, is there any correlation between where the vertices ended up and the matrix A?

If you are just reading on, and haven't actually gone back and looked at the examples, go back now and try to make some sort of connection. Otherwise – you may have noted some of the following things:

1. The zero vector ($\vec{0}$, the "black" corner) never moved. That makes sense, though; $A\vec{0} = \vec{0}$.

2. The "square" corner, i.e., the corner corresponding to the vector $\begin{bmatrix} 1 \\ 0 \end{bmatrix}$, is always transformed to the vector in the first column of A!

3. Likewise, the "triangular" corner, i.e., the corner corresponding to the vector $\begin{bmatrix} 0 \\ 1 \end{bmatrix}$, is always transformed to the vector in the second column of A![3]

4. The "white dot" corner is always transformed to the *sum* of the two column vectors of A.[4]

Let's now take the time to understand these four points. The first point should be clear; $\vec{0}$ will always be transformed to $\vec{0}$ via matrix multiplication. (Hence the hint in the middle of Example 93, where we are told that we can ignore entering in the column of zeros in the matrix B.)

We can understand the second and third points simultaneously. Let

$$A = \begin{bmatrix} a & b \\ c & d \end{bmatrix}, \quad \vec{e_1} = \begin{bmatrix} 1 \\ 0 \end{bmatrix} \quad \text{and} \quad \vec{e_2} = \begin{bmatrix} 0 \\ 1 \end{bmatrix}.$$

What are $A\vec{e_1}$ and $A\vec{e_2}$?

$$\begin{aligned} A\vec{e_1} &= \begin{bmatrix} a & b \\ c & d \end{bmatrix} \begin{bmatrix} 1 \\ 0 \end{bmatrix} \\ &= \begin{bmatrix} a \\ c \end{bmatrix} \end{aligned}$$

$$\begin{aligned} A\vec{e_2} &= \begin{bmatrix} a & b \\ c & d \end{bmatrix} \begin{bmatrix} 0 \\ 1 \end{bmatrix} \\ &= \begin{bmatrix} b \\ d \end{bmatrix} \end{aligned}$$

[3] Although this is less of a surprise, given the result of the previous point.

[4] This observation is a bit more obscure than the first three. It follows from the fact that this corner of the unit square is the "sum" of the other two nonzero corners.

So by mere mechanics of matrix multiplication, the square corner $\vec{e_1}$ is transformed to the first column of A, and the triangular corner $\vec{e_2}$ is transformed to the second column of A. A similar argument demonstrates why the white dot corner is transformed to the sum of the columns of A.[5]

Revisit now the question "How do we find the matrix that performs a given transformation on the Cartesian plane?" The answer follows from what we just did. Think about the given transformation and how it would transform the corners of the unit square. Make the first column of A the vector where $\vec{e_1}$ goes, and make the second column of A the vector where $\vec{e_2}$ goes.

Let's practice this in the context of an example.

Example 95 Find the matrix A that flips the Cartesian plane about the x axis and then stretches the plane horizontally by a factor of two.

Solution We first consider $\vec{e_1} = \begin{bmatrix} 1 \\ 0 \end{bmatrix}$. Where does this corner go to under the given transformation? Flipping the plane across the x axis does not change $\vec{e_1}$ at all; stretching the plane sends $\vec{e_1}$ to $\begin{bmatrix} 2 \\ 0 \end{bmatrix}$. Therefore, the first column of A is $\begin{bmatrix} 2 \\ 0 \end{bmatrix}$.

Now consider $\vec{e_2} = \begin{bmatrix} 0 \\ 1 \end{bmatrix}$. Flipping the plane about the x axis sends $\vec{e_2}$ to the vector $\begin{bmatrix} 0 \\ -1 \end{bmatrix}$; subsequently stretching the plane horizontally does not affect this vector. Therefore the second column of A is $\begin{bmatrix} 0 \\ -1 \end{bmatrix}$.

Putting this together gives

$$A = \begin{bmatrix} 2 & 0 \\ 0 & -1 \end{bmatrix}.$$

To help visualize this, consider Figure 5.4 where a shape is transformed under this matrix. Notice how it is turned upside down and is stretched horizontally by a factor of two. (The gridlines are given as a visual aid.)

[5] Another way of looking at all of this is to consider what $A \cdot I$ is: of course, it is just A. What are the columns of I? Just $\vec{e_1}$ and $\vec{e_2}$.

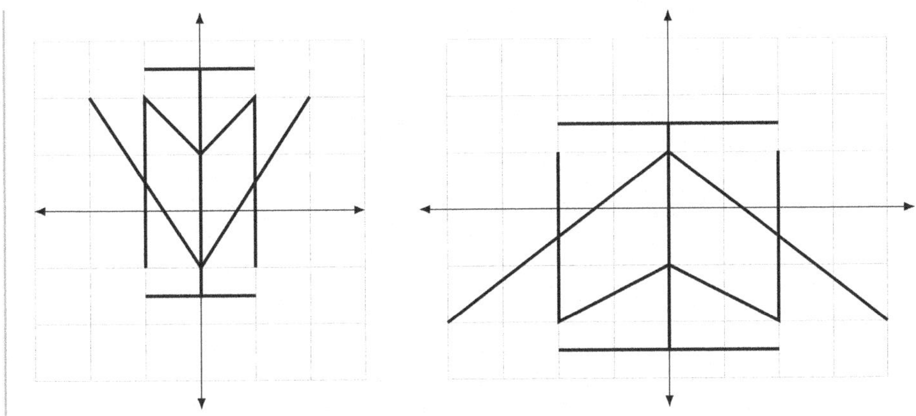

Figure 5.4: Transforming the Cartesian plane in Example 95

A while ago we asked two questions. The first was "How do we find the matrix that performs a given transformation?" We have just answered that question (although we will do more to explore it in the future). The second question was "How does knowing how the unit square is transformed really help us understand how the entire plane is transformed?"

Consider Figure 5.5 where the unit square (with vertices marked with shapes as before) is shown transformed under an unknown matrix. How does this help us understand how the whole Cartesian plane is transformed? For instance, how can we use this picture to figure out how the point $(2, 3)$ will be transformed?

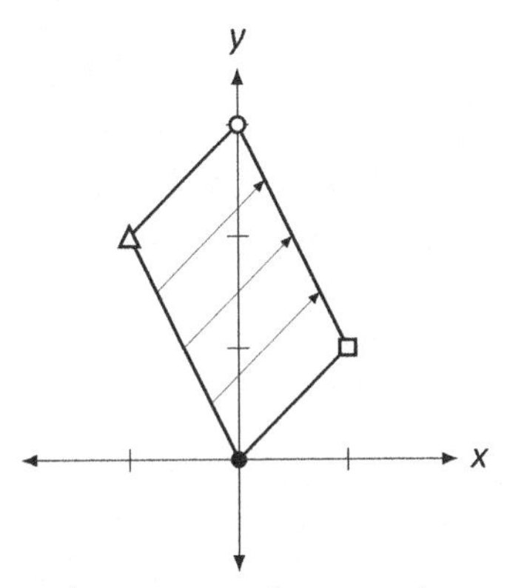

Figure 5.5: The unit square under an unknown transformation.

There are two ways to consider the solution to this question. First, we know now how to compute the transformation matrix; the new position of $\vec{e_1}$ is the first column of A, and the new position of $\vec{e_2}$ is the second column of A. Therefore, by looking at the figure, we can deduce that

$$A = \begin{bmatrix} 1 & -1 \\ 1 & 2 \end{bmatrix}.^6$$

[6] At least, A is close to that. The square corner could actually be at the point $(1.01, .99)$.

To find where the point $(2, 3)$ is sent, simply multiply

$$\begin{bmatrix} 1 & -1 \\ 1 & 2 \end{bmatrix} \begin{bmatrix} 2 \\ 3 \end{bmatrix} = \begin{bmatrix} -1 \\ 8 \end{bmatrix}.$$

There is another way of doing this which isn't as computational – it doesn't involve computing the transformation matrix. Consider the following equalities:

$$\begin{aligned} \begin{bmatrix} 2 \\ 3 \end{bmatrix} &= \begin{bmatrix} 2 \\ 0 \end{bmatrix} + \begin{bmatrix} 0 \\ 3 \end{bmatrix} \\ &= 2\begin{bmatrix} 1 \\ 0 \end{bmatrix} + 3\begin{bmatrix} 0 \\ 1 \end{bmatrix} \\ &= 2\vec{e_1} + 3\vec{e_2} \end{aligned}$$

This last equality states something that is somewhat obvious: to arrive at the vector $\begin{bmatrix} 2 \\ 3 \end{bmatrix}$, one needs to go 2 units in the $\vec{e_1}$ direction and 3 units in the $\vec{e_2}$ direction. To find where the point $(2, 3)$ is transformed, one needs to go 2 units in the *new* $\vec{e_1}$ direction and 3 units in the *new* $\vec{e_2}$ direction. This is demonstrated in Figure 5.6.

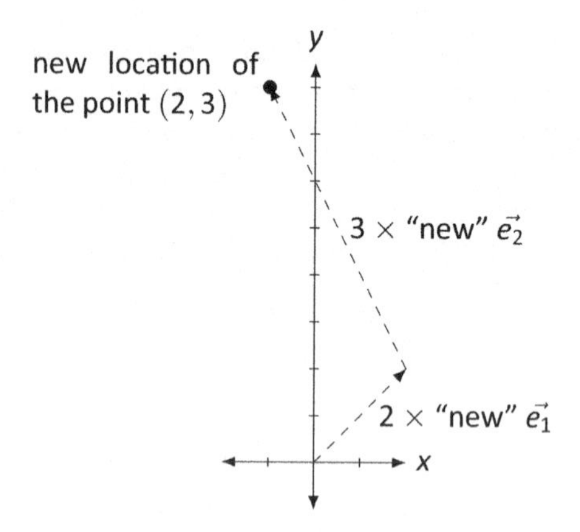

Figure 5.6: Finding the new location of the point $(2, 3)$.

We are coming to grips with how matrix transformations work. We asked two basic questions: "How do we find the matrix for a given transformation?" and "How do we understand the transformation without the matrix?", and we've answered each accompanied by one example. Let's do another example that demonstrates both techniques at once.

Example 96 First, find the matrix A that transforms the Cartesian plane by stretching it vertically by a factor of 1.5, then stretches it horizontally by a factor of 0.5, then rotates it clockwise about the origin 90°. Secondly, using the new locations of $\vec{e_1}$ and $\vec{e_2}$, find the transformed location of the point $(-1, 2)$.

Solution To find A, first consider the new location of $\vec{e_1}$. Stretching the plane vertically does not affect $\vec{e_1}$; stretching the plane horizontally by a factor of 0.5 changes $\vec{e_1}$ to $\begin{bmatrix} 1/2 \\ 0 \end{bmatrix}$, and then rotating it 90° about the origin moves it to $\begin{bmatrix} 0 \\ -1/2 \end{bmatrix}$. This is the first column of A.

Now consider the new location of $\vec{e_2}$. Stretching the plane vertically changes it to $\begin{bmatrix} 0 \\ 3/2 \end{bmatrix}$; stretching horizontally does not affect it, and rotating 90° moves it to $\begin{bmatrix} 3/2 \\ 0 \end{bmatrix}$. This is then the second column of A. This gives

$$A = \begin{bmatrix} 0 & 3/2 \\ -1/2 & 0 \end{bmatrix}.$$

Where does the point $(-1, 2)$ get sent to? The corresponding vector $\begin{bmatrix} -1 \\ 2 \end{bmatrix}$ is found by going -1 units in the $\vec{e_1}$ direction and 2 units in the $\vec{e_2}$ direction. Therefore, the transformation will send the vector to -1 units in the new $\vec{e_1}$ direction and 2 units in the new $\vec{e_2}$ direction. This is sketched in Figure 5.7, along with the transformed unit square. We can also check this multiplicatively:

$$\begin{bmatrix} 0 & 3/2 \\ -1/2 & 0 \end{bmatrix} \begin{bmatrix} -1 \\ 2 \end{bmatrix} = \begin{bmatrix} 3 \\ 1/2 \end{bmatrix}.$$

Figure 5.8 shows the effects of the transformation on another shape.

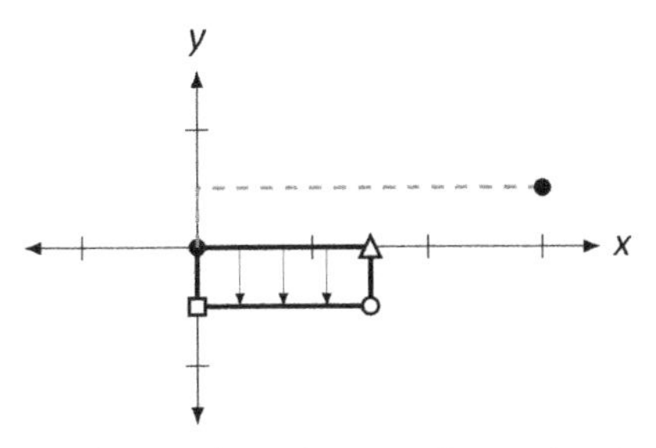

Figure 5.7: Understanding the transformation in Example 96.

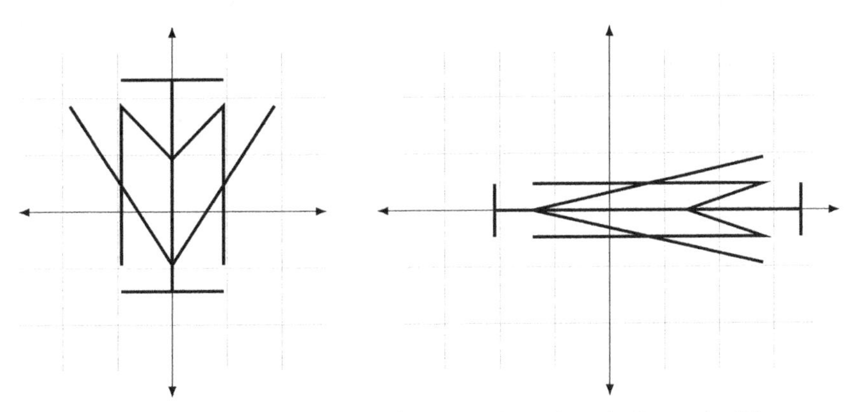

Figure 5.8: Transforming the Cartesian plane in Example 96

Right now we are focusing on transforming the Cartesian plane – we are making 2D transformations. Knowing how to do this provides a foundation for transforming 3D space,[7] which, among other things, is very important when producing 3D computer graphics. Basic shapes can be drawn and then rotated, stretched, and/or moved to other regions of space. This also allows for things like "moving the camera view."

What kinds of transformations are possible? We have already seen some of the things that are possible: rotations, stretches, and flips. We have also mentioned some things that are not possible. For instance, we stated that straight lines always get transformed to straight lines. Therefore, we cannot transform the unit square into a circle using a matrix.

Let's look at some common transformations of the Cartesian plane and the matrices that perform these operations. In the following figures, a transformation matrix will be given alongside a picture of the transformed unit square. (The original unit square is drawn lightly as well to serve as a reference.)

2D Matrix Transformations

Horizontal stretch by a factor of k.

$$\begin{bmatrix} k & 0 \\ 0 & 1 \end{bmatrix}$$

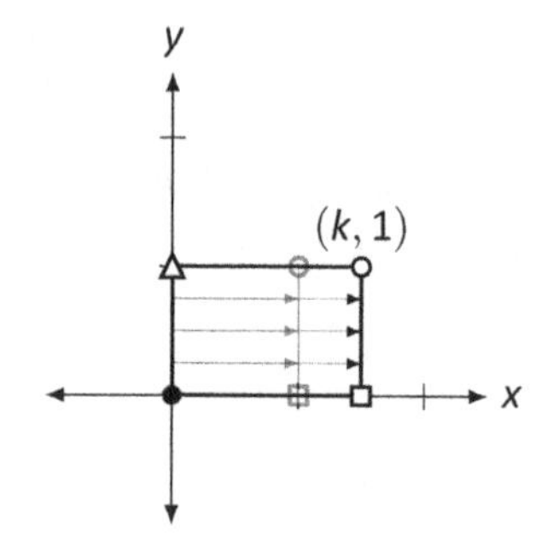

Vertical stretch by a factor of k.

$$\begin{bmatrix} 1 & 0 \\ 0 & k \end{bmatrix}$$

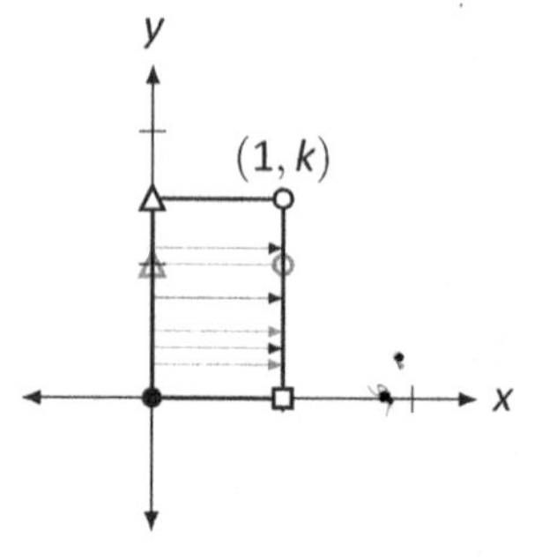

---

[7] Actually, it provides a foundation for doing it in 4D, 5D, . . ., 17D, etc. Those are just harder to visualize.

Horizontal shear by a factor of *k*.

$$\begin{bmatrix} 1 & k \\ 0 & 1 \end{bmatrix}$$

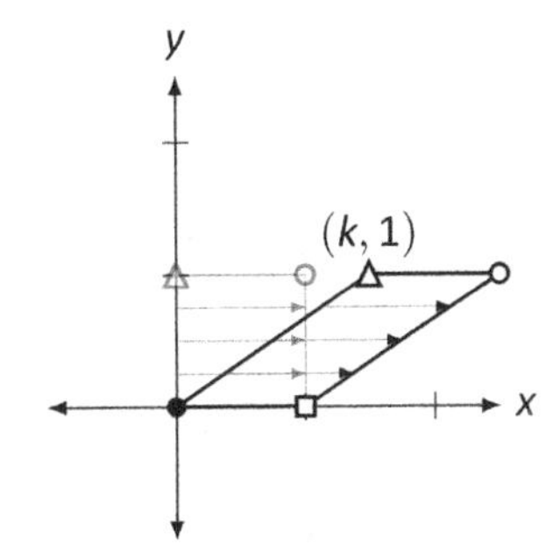

Vertical shear by a factor of *k*.

$$\begin{bmatrix} 1 & 0 \\ k & 1 \end{bmatrix}$$

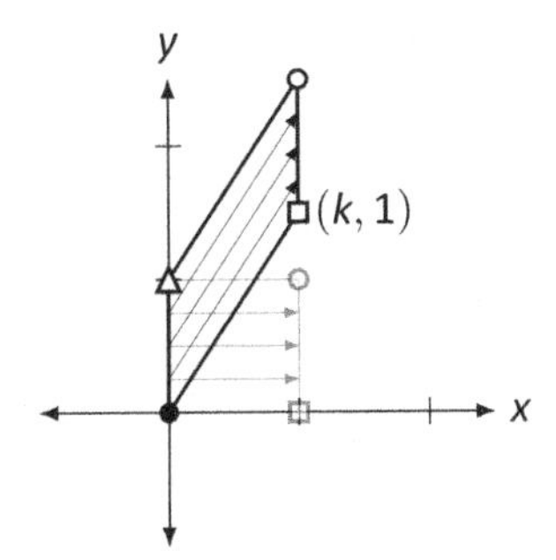

Horizontal reflection across the *y* axis.

$$\begin{bmatrix} -1 & 0 \\ 0 & 1 \end{bmatrix}$$

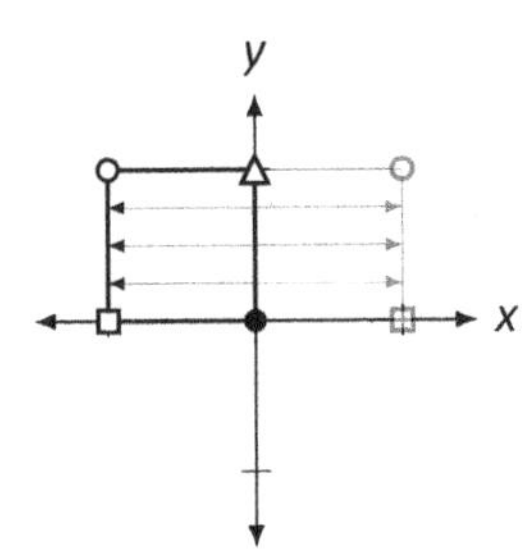

Vertical reflection across the *x* axis.

$$\begin{bmatrix} 1 & 0 \\ 0 & -1 \end{bmatrix}$$

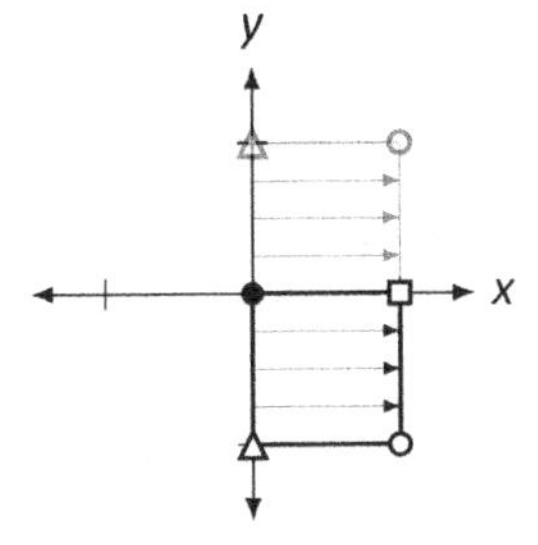

Diagonal reflection across the line $y = x$.

$$\begin{bmatrix} 0 & 1 \\ 1 & 0 \end{bmatrix}$$

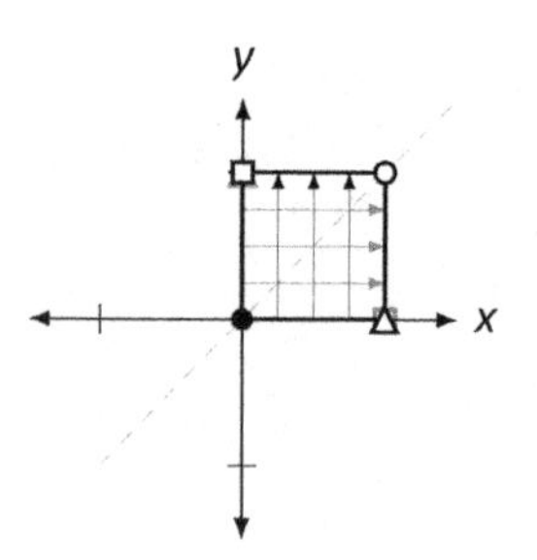

Rotation around the origin by an angle of θ.

$$\begin{bmatrix} \cos\theta & -\sin\theta \\ \sin\theta & \cos\theta \end{bmatrix}$$

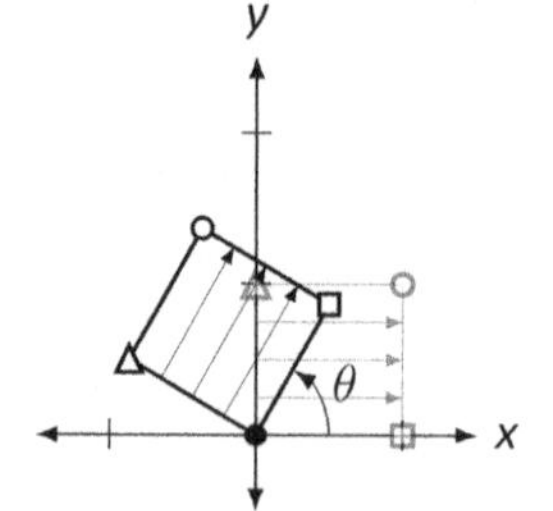

Projection onto the x axis.
(Note how the square is "squashed" down onto the x-axis.)

$$\begin{bmatrix} 1 & 0 \\ 0 & 0 \end{bmatrix}$$

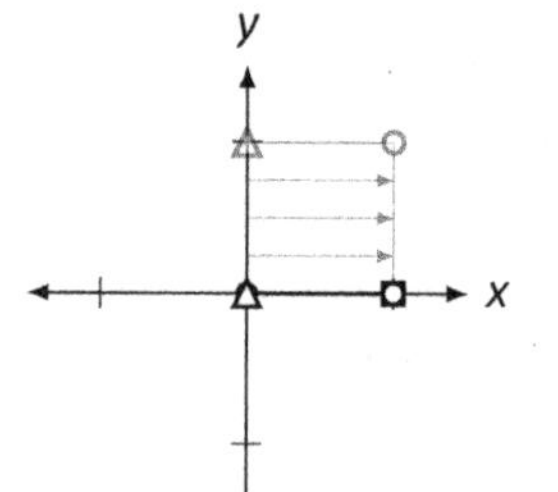

Projection onto the y axis.
(Note how the square is "squashed" over onto the y-axis.)

$$\begin{bmatrix} 0 & 0 \\ 0 & 1 \end{bmatrix}$$

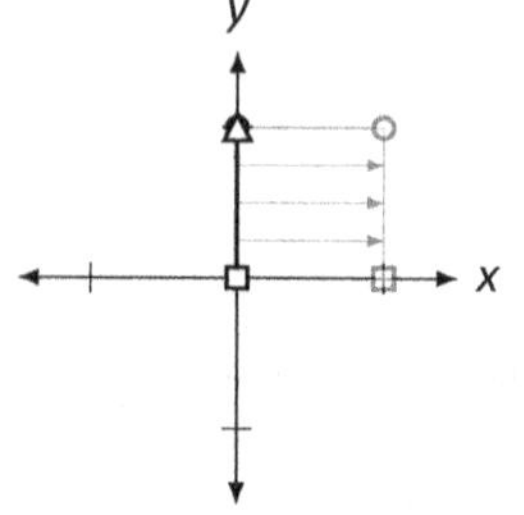

Now that we have seen a healthy list of transformations that we can perform on the Cartesian plane, let's practice a few more times creating the matrix that gives the desired transformation. In the following example, we develop our understanding one more critical step.

Example 97 Find the matrix A that transforms the Cartesian plane by performing the following operations in order:

1. Vertical shear by a factor of 0.5
2. Counterclockwise rotation about the origin by an angle of $\theta = 30^\circ$
3. Horizontal stretch by a factor of 2
4. Diagonal reflection across the line $y = x$

SOLUTION Wow! We already know how to do this – sort of. We know we can find the columns of A by tracing where $\vec{e_1}$ and $\vec{e_2}$ end up, but this also seems difficult. There is so much that is going on. Fortunately, we can accomplish what we need without much difficulty by being systematic.

First, let's perform the vertical shear. The matrix that performs this is

$$A_1 = \begin{bmatrix} 1 & 0 \\ 0.5 & 1 \end{bmatrix}.$$

After that, we want to rotate everything clockwise by 30°. To do this, we use

$$A_2 = \begin{bmatrix} \cos 30^\circ & -\sin 30^\circ \\ \sin 30^\circ & \cos 30^\circ \end{bmatrix} = \begin{bmatrix} \sqrt{3}/2 & -1/2 \\ 1/2 & \sqrt{3}/2 \end{bmatrix}.$$

In order to do both of these operations, in order, we multiply A_2A_1.[8]

To perform the final two operations, we note that

$$A_3 = \begin{bmatrix} 2 & 0 \\ 0 & 1 \end{bmatrix} \quad \text{and} \quad A_4 = \begin{bmatrix} 0 & 1 \\ 1 & 0 \end{bmatrix}$$

perform the horizontal stretch and diagonal reflection, respectively. Thus to perform all of the operations "at once," we need to multiply by

$$\begin{aligned} A &= A_4A_3A_2A_1 \\ &= \begin{bmatrix} 0 & 1 \\ 1 & 0 \end{bmatrix} \begin{bmatrix} 2 & 0 \\ 0 & 1 \end{bmatrix} \begin{bmatrix} \sqrt{3}/2 & -1/2 \\ 1/2 & \sqrt{3}/2 \end{bmatrix} \begin{bmatrix} 1 & 0 \\ 0.5 & 1 \end{bmatrix} \\ &= \begin{bmatrix} (\sqrt{3}+2)/4 & \sqrt{3}/2 \\ (2\sqrt{3}-1)/2 & -1 \end{bmatrix} \\ &\approx \begin{bmatrix} 0.933 & 0.866 \\ 1.232 & -1 \end{bmatrix}. \end{aligned}$$

[8]The reader might ask, "Is it important to do multiply these in that order? Could we have multiplied A_1A_2 instead?" Our answer starts with "Is matrix multiplication commutative?" The answer to our question is "No," so the answers to the reader's questions are "Yes" and "No," respectively.

Let's consider this closely. Suppose I want to know where a vector $\vec{x}$ ends up. We claim we can find the answer by multiplying $A\vec{x}$. Why does this work? Consider:

$$
\begin{aligned}
A\vec{x} &= A_4A_3A_2A_1\vec{x} \\
&= A_4A_3A_2(A_1\vec{x}) && \text{(performs the vertical shear)} \\
&= A_4A_3(A_2\vec{x}_1) && \text{(performs the rotation)} \\
&= A_4(A_3\vec{x}_2) && \text{(performs the horizontal stretch)} \\
&= A_4\vec{x}_3 && \text{(performs the diagonal reflection)} \\
&= \vec{x}_4 && \text{(the result of transforming } \vec{x}\text{)}
\end{aligned}
$$

Most readers are not able to visualize exactly what the given list of operations does to the Cartesian plane. In Figure 5.9 we sketch the transformed unit square; in Figure 5.10 we sketch a shape and its transformation.

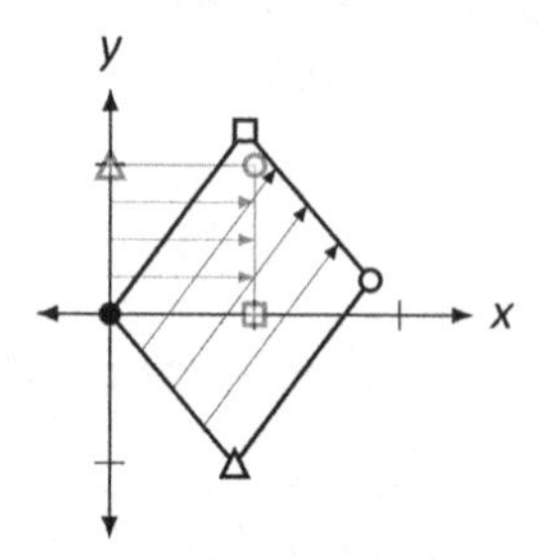

Figure 5.9: The transformed unit square in Example 97.

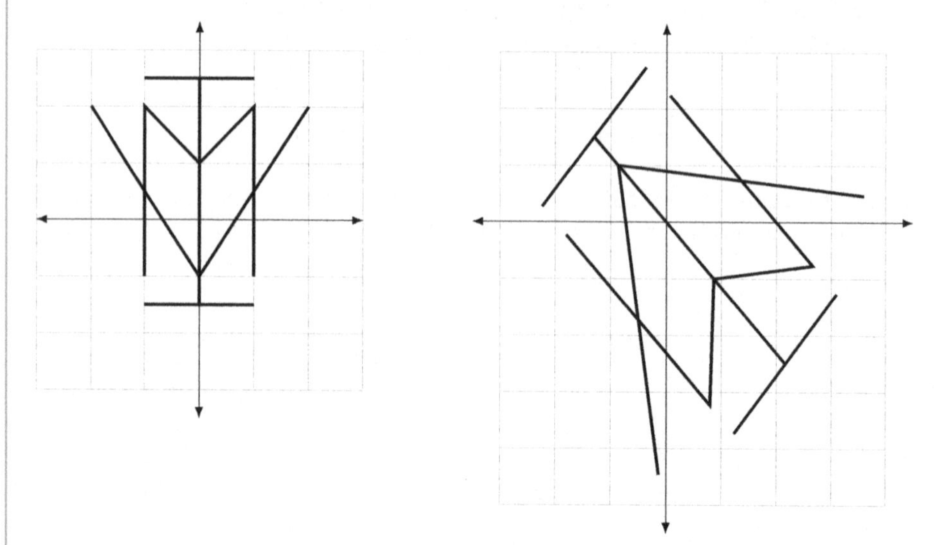

Figure 5.10: A transformed shape in Example 97.

Once we know what matrices perform the basic transformations,[9] performing complex transformations on the Cartesian plane really isn't that . . . complex. It boils down

[9]or know where to find them

to multiplying by a series of matrices.

We've shown many examples of transformations that we can do, and we've mentioned just a few that we can't – for instance, we can't turn a square into a circle. Why not? Why is it that straight lines get sent to straight lines? We spent a lot of time within this text looking at invertible matrices; what connections, if any,[10] are there between invertible matrices and their transformations on the Cartesian plane?

All these questions require us to think like mathematicians – we are being asked to study the *properties* of an object we just learned about and their connections to things we've already learned. We'll do all this (and more!) in the following section.

Exercises 5.1

In Exercises 1 – 4, a sketch of transformed unit square is given. Find the matrix A that performs this transformation.

1.
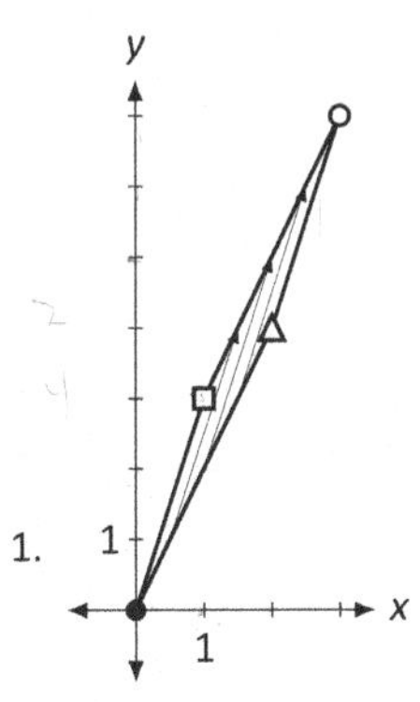

2.
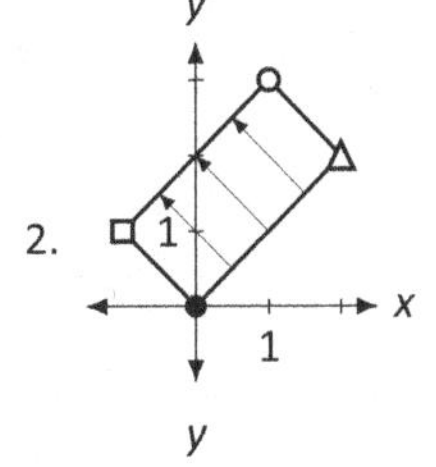

3.
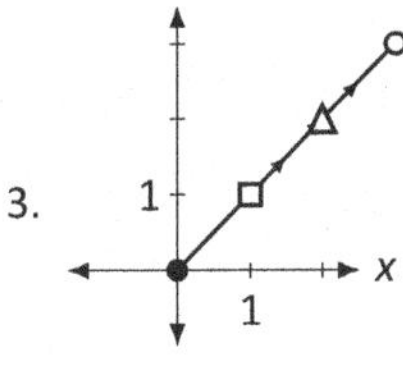

4.
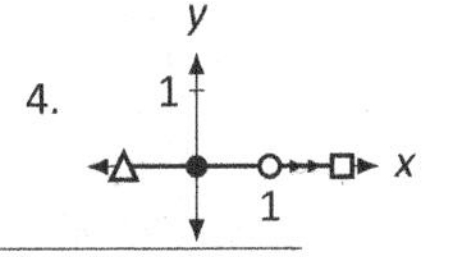

In Exercises 5 – 10, a list of transformations is given. Find the matrix A that performs those transformations, in order, on the Cartesian plane.

5. (a) vertical shear by a factor of 2
   (b) horizontal shear by a factor of 2

6. (a) horizontal shear by a factor of 2
   (b) vertical shear by a factor of 2

7. (a) horizontal stretch by a factor of 3
   (b) reflection across the line $y = x$

8. (a) counterclockwise rotation by an angle of 45°
   (b) vertical stretch by a factor of $1/2$

9. (a) clockwise rotation by an angle of 90°
   (b) horizontal reflection across the y axis
   (c) vertical shear by a factor of 1

10. (a) vertical reflection across the x axis
    (b) horizontal reflection across the y axis
    (c) diagonal reflection across the line $y = x$

In Exercises 11 – 14, two sets of transformations are given. Sketch the transformed unit square under each set of transformations. Are the transformations the same? Explain why/why not.

[10] By now, the reader should expect connections to exist.

11. (a) a horizontal reflection across the y axis, followed by a vertical reflection across the x axis, compared to

    (b) a counterclockise rotation of 180°

12. (a) a horizontal stretch by a factor of 2 followed by a reflection across the line $y = x$, compared to

    (b) a vertical stretch by a factor of 2

13. (a) a horizontal stretch by a factor of 1/2 followed by a vertical stretch by a factor of 3, compared to

    (b) the same operations but in opposite order

14. (a) a reflection across the line $y = x$ followed by a reflection across the x axis, compared to

    (b) a reflection across the the y axis, followed by a reflection across the line $y = x$.

5.2 Properties of Linear Transformations

1. T/F: Translating the Cartesian plane 2 units up is a linear transformation.
2. T/F: If T is a linear transformation, then $T(\vec{0}) = \vec{0}$.

In the previous section we discussed standard transformations of the Cartesian plane – rotations, reflections, etc. As a motivational example for this section's study, let's consider another transformation – let's find the matrix that moves the unit square one unit to the right (see Figure 5.11). This is called a *translation*.

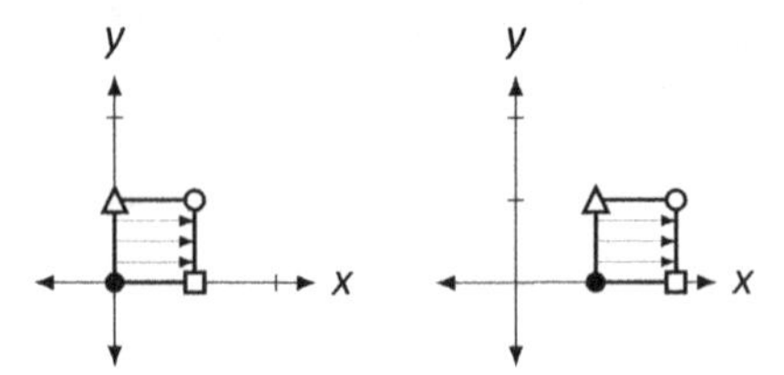

Figure 5.11: Translating the unit square one unit to the right.

Our work from the previous section allows us to find the matrix quickly. By looking at the picture, it is easy to see that $\vec{e_1}$ is moved to $\begin{bmatrix} 2 \\ 0 \end{bmatrix}$ and $\vec{e_2}$ is moved to $\begin{bmatrix} 1 \\ 1 \end{bmatrix}$. Therefore, the transformation matrix should be

$$A = \begin{bmatrix} 2 & 1 \\ 0 & 1 \end{bmatrix}.$$

However, look at Figure 5.12 where the unit square is drawn after being transformed by A. It is clear that we did not get the desired result; the unit square was not translated, but rather stretched/sheared in some way.

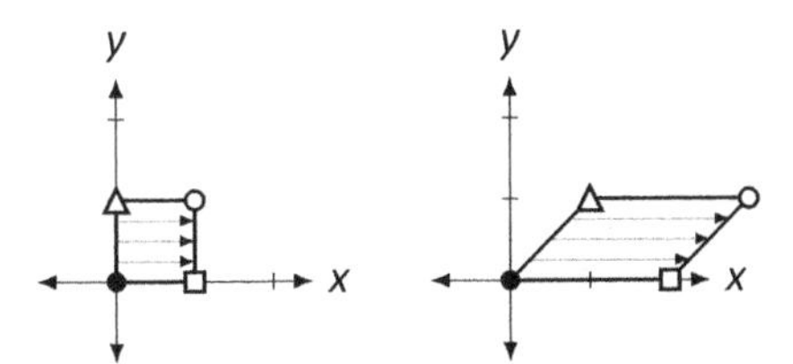

Figure 5.12: Actual transformation of the unit square by matrix A.

What did we do wrong? We will answer this question, but first we need to develop a few thoughts and vocabulary terms.

We've been using the term "transformation" to describe how we've changed vectors. In fact, "transformation" is synonymous to "function." We are used to functions like $f(x) = x^2$, where the input is a number and the output is another number. In the previous section, we learned about transformations (functions) where the input was a vector and the output was another vector. If A is a "transformation matrix," then we could create a function of the form $T(\vec{x}) = A\vec{x}$. That is, a vector $\vec{x}$ is the input, and the output is $\vec{x}$ multiplied by A.[11]

When we defined $f(x) = x^2$ above, we let the reader assume that the input was indeed a number. If we wanted to be complete, we should have stated

$$f : \mathbb{R} \to \mathbb{R} \quad \text{where} \quad f(x) = x^2.$$

The first part of that line told us that the input was a real number (that was the first $\mathbb{R}$) and the output was also a real number (the second $\mathbb{R}$).

To define a transformation where a 2D vector is transformed into another 2D vector via multiplication by a 2×2 matrix A, we should write

$$T : \mathbb{R}^2 \to \mathbb{R}^2 \quad \text{where} \quad T(\vec{x}) = A\vec{x}.$$

Here, the first $\mathbb{R}^2$ means that we are using 2D vectors as our input, and the second $\mathbb{R}^2$ means that a 2D vector is the output.

Consider a quick example:

$$T : \mathbb{R}^2 \to \mathbb{R}^3 \quad \text{where} \quad T\left(\begin{bmatrix} x_1 \\ x_2 \end{bmatrix}\right) = \begin{bmatrix} x_1^2 \\ 2x_1 \\ x_1x_2 \end{bmatrix}.$$

Notice that this takes 2D vectors as input and returns 3D vectors as output. For instance,

$$T\left(\begin{bmatrix} 3 \\ -2 \end{bmatrix}\right) = \begin{bmatrix} 9 \\ 6 \\ -6 \end{bmatrix}.$$

We now define a special type of transformation (function).

[11]We used T instead of f to define this function to help differentiate it from "regular" functions. "Normally" functions are defined using lower case letters when the input is a number; when the input is a vector, we use upper case letters.

Definition 29 **Linear Transformation**

A transformation $T : \mathbb{R}^n \to \mathbb{R}^m$ is a *linear transformation* if it satisfies the following two properties:

1. $T(\vec{x} + \vec{y}) = T(\vec{x}) + T(\vec{y})$ for all vectors $\vec{x}$ and $\vec{y}$, and
2. $T(k\vec{x}) = kT(\vec{x})$ for all vectors $\vec{x}$ and all scalars k.

If T is a linear transformation, it is often said that "T is *linear.*"

Let's learn about this definition through some examples.

Example 98 Determine whether or not the transformation $T : \mathbb{R}^2 \to \mathbb{R}^3$ is a linear transformation, where

$$T\left(\begin{bmatrix} x_1 \\ x_2 \end{bmatrix}\right) = \begin{bmatrix} x_1^2 \\ 2x_1 \\ x_1x_2 \end{bmatrix}.$$

SOLUTION We'll arbitrarily pick two vectors $\vec{x}$ and $\vec{y}$:

$$\vec{x} = \begin{bmatrix} 3 \\ -2 \end{bmatrix} \quad \text{and} \quad \vec{y} = \begin{bmatrix} 1 \\ 5 \end{bmatrix}.$$

Let's check to see if T is linear by using the definition.

1. Is $T(\vec{x} + \vec{y}) = T(\vec{x}) + T(\vec{y})$? First, compute $\vec{x} + \vec{y}$:

$$\vec{x} + \vec{y} = \begin{bmatrix} 3 \\ -2 \end{bmatrix} + \begin{bmatrix} 1 \\ 5 \end{bmatrix} = \begin{bmatrix} 4 \\ 3 \end{bmatrix}.$$

Now compute $T(\vec{x})$, $T(\vec{y})$, and $T(\vec{x} + \vec{y})$:

$$T(\vec{x}) = T\left(\begin{bmatrix} 3 \\ -2 \end{bmatrix}\right) = \begin{bmatrix} 9 \\ 6 \\ -6 \end{bmatrix} \qquad T(\vec{y}) = T\left(\begin{bmatrix} 1 \\ 5 \end{bmatrix}\right) = \begin{bmatrix} 1 \\ 2 \\ 5 \end{bmatrix} \qquad T(\vec{x} + \vec{y}) = T\left(\begin{bmatrix} 4 \\ 3 \end{bmatrix}\right) = \begin{bmatrix} 16 \\ 8 \\ 12 \end{bmatrix}$$

Is $T(\vec{x} + \vec{y}) = T(\vec{x}) + T(\vec{y})$?

$$\begin{bmatrix} 9 \\ 6 \\ -6 \end{bmatrix} + \begin{bmatrix} 1 \\ 2 \\ 5 \end{bmatrix} \overset{!}{\neq} \begin{bmatrix} 16 \\ 8 \\ 12 \end{bmatrix}.$$

Therefore, T is *not* a linear transformation.

So we have an example of something that *doesn't* work. Let's try an example where things *do* work.[12]

Example 99 Determine whether or not the transformation $T : \mathbb{R}^2 \to \mathbb{R}^2$ is a linear transformation, where $T(\vec{x}) = A\vec{x}$ and

$$A = \begin{bmatrix} 1 & 2 \\ 3 & 4 \end{bmatrix}.$$

SOLUTION Let's start by again considering arbitrary $\vec{x}$ and $\vec{y}$. Let's choose the same $\vec{x}$ and $\vec{y}$ from Example 98.

$$\vec{x} = \begin{bmatrix} 3 \\ -2 \end{bmatrix} \quad \text{and} \quad \vec{y} = \begin{bmatrix} 1 \\ 5 \end{bmatrix}.$$

If the lineararity properties hold for these vectors, then *maybe* it is actually linear (and we'll do more work).

1. Is $T(\vec{x} + \vec{y}) = T(\vec{x}) + T(\vec{y})$? Recall:

$$\vec{x} + \vec{y} = \begin{bmatrix} 4 \\ 3 \end{bmatrix}.$$

Now compute $T(\vec{x})$, $T(\vec{y})$, and $T(\vec{x}) + T(\vec{y})$:

$$T(\vec{x}) = T\left(\begin{bmatrix} 3 \\ -2 \end{bmatrix}\right) = \begin{bmatrix} -1 \\ 1 \end{bmatrix} \qquad T(\vec{y}) = T\left(\begin{bmatrix} 1 \\ 5 \end{bmatrix}\right) = \begin{bmatrix} 11 \\ 23 \end{bmatrix} \qquad T(\vec{x} + \vec{y}) = T\left(\begin{bmatrix} 4 \\ 3 \end{bmatrix}\right) = \begin{bmatrix} 10 \\ 24 \end{bmatrix}$$

Is $T(\vec{x} + \vec{y}) = T(\vec{x}) + T(\vec{y})$?

$$\begin{bmatrix} -1 \\ 1 \end{bmatrix} + \begin{bmatrix} 11 \\ 23 \end{bmatrix} \stackrel{!}{=} \begin{bmatrix} 10 \\ 24 \end{bmatrix}.$$

So far, so good: $T(\vec{x} + \vec{y})$ is equal to $T(\vec{x}) + T(\vec{y})$.

[12] Recall a principle of logic: to show that something doesn't work, we just need to show one case where it fails, which we did in Example 98. To show that something *always* works, we need to show it works for *all* cases – simply showing it works for a few cases isn't enough. However, doing so can be helpful in understanding the situation better.

2. Is $T(k\vec{x}) = kT(\vec{x})$? Let's arbitrarily pick $k = 7$, and use $\vec{x}$ as before.

$$\begin{aligned} T(7\vec{x}) &= T\left(\begin{bmatrix} 21 \\ -14 \end{bmatrix}\right) \\ &= \begin{bmatrix} -7 \\ 7 \end{bmatrix} \\ &= 7\begin{bmatrix} -1 \\ 1 \end{bmatrix} \\ &= 7 \cdot T(\vec{x}) \quad ! \end{aligned}$$

So far it *seems* that T is indeed linear, for it worked in one example with arbitrarily chosen vectors and scalar. Now we need to try to show it is always true.

Consider $T(\vec{x} + \vec{y})$. By the definition of T, we have

$$T(\vec{x} + \vec{y}) = A(\vec{x} + \vec{y}).$$

By Theorem 3, part 2 (on page 62) we state that the Distributive Property holds for matrix multiplication.[13] So $A(\vec{x} + \vec{y}) = A\vec{x} + A\vec{y}$. Recognize now that this last part is just $T(\vec{x}) + T(\vec{y})$! We repeat the above steps, all together:

$$\begin{aligned} T(\vec{x} + \vec{y}) &= A(\vec{x} + \vec{y}) && \text{(by the definition of } T \text{ in this example)} \\ &= A\vec{x} + A\vec{y} && \text{(by the Distributive Property)} \\ &= T(\vec{x}) + T(\vec{y}) && \text{(again, by the definition of } T\text{)} \end{aligned}$$

Therefore, no matter what $\vec{x}$ and $\vec{y}$ are chosen, $T(\vec{x} + \vec{y}) = T(\vec{x}) + T(\vec{y})$. Thus the first part of the lineararity definition is satisfied.

The second part is satisfied in a similar fashion. Let k be a scalar, and consider:

$$\begin{aligned} T(k\vec{x}) &= A(k\vec{x}) && \text{(by the definition of } T \text{ is this example)} \\ &= kA\vec{x} && \text{(by Theorem 3 part 3)} \\ &= kT(\vec{x}) && \text{(again, by the definition of } T\text{)} \end{aligned}$$

Since T satisfies both parts of the definition, we conclude that T is a linear transformation.

We have seen two examples of transformations so far, one which was not linear and one that was. One might wonder "Why is linearity important?", which we'll address shortly.

First, consider how we proved the transformation in Example 99 was linear. We defined T by matrix multiplication, that is, $T(\vec{x}) = A\vec{x}$. We proved T was linear using properties of matrix multiplication – *we never considered the specific values of A!* That is, we didn't just choose a good matrix for T; *any* matrix A would have worked. This

[13] Recall that a vector is just a special type of matrix, so this theorem applies to matrix–vector multiplication as well.

leads us to an important theorem. The first part we have essentially just proved; the second part we won't prove, although its truth is very powerful.

Theorem 21

Matrices and Linear Transformations

1. Define $T : \mathbb{R}^n \to \mathbb{R}^m$ by $T(\vec{x}) = A\vec{x}$, where A is an $m \times n$ matrix. Then T is a linear transformation.
2. Let $T : \mathbb{R}^n \to \mathbb{R}^m$ be any linear transformation. Then there exists an unique $m \times n$ matrix A such that $T(\vec{x}) = A\vec{x}$.

The second part of the theorem says that *all* linear transformations can be described using matrix multiplication. Given *any* linear transformation, there is a matrix that completely defines that transformation. This important matrix gets its own name.

Definition 30

Standard Matrix of a Linear Transformation

Let $T : \mathbb{R}^n \to \mathbb{R}^m$ be a linear transformation. By Theorem 21, there is a matrix A such that $T(\vec{x}) = A\vec{x}$. This matrix A is called the *standard matrix of the linear transformation T*, and is denoted $[\,T\,]$.[a]

[a] The matrix–like brackets around T suggest that the standard matrix A is a matrix "with T inside."

While exploring all of the ramifications of Theorem 21 is outside the scope of this text, let it suffice to say that since 1) linear transformations are very, very important in economics, science, engineering and mathematics, and 2) the theory of matrices is well developed and easy to implement by hand and on computers, then 3) it is great news that these two concepts go hand in hand.

We have already used the second part of this theorem in a small way. In the previous section we looked at transformations graphically and found the matrices that produced them. At the time, we didn't realize that these transformations were linear, but indeed they were.

This brings us back to the motivating example with which we started this section. We tried to find the matrix that translated the unit square one unit to the right. Our attempt failed, and we have yet to determine why. Given our link between matrices and linear transformations, the answer is likely "the translation transformation is not a linear transformation." While that is a true statement, it doesn't really explain things all that well. Is there some way we could have recognized that this transformation wasn't linear?[14]

Yes, there is. Consider the second part of the linear transformation definition. It states that $T(k\vec{x}) = kT(\vec{x})$ for all scalars k. If we let $k = 0$, we have $T(0\vec{x}) = 0 \cdot T(\vec{x})$, or more simply, $T(\vec{0}) = \vec{0}$. That is, if T is to be a linear transformation, it must send the zero vector to the zero vector.

This is a quick way to see that the translation transformation fails to be linear. By shifting the unit square to the right one unit, the corner at the point $(0, 0)$ was sent to the point $(1, 0)$, i.e.,

$$\text{the vector } \begin{bmatrix} 0 \\ 0 \end{bmatrix} \text{ was sent to the vector } \begin{bmatrix} 1 \\ 0 \end{bmatrix}.$$

This property relating to $\vec{0}$ is important, so we highlight it here.

Key Idea 15 **Linear Transformations and $\vec{0}$**

Let $T : \mathbb{R}^n \to \mathbb{R}^m$ be a linear transformation. Then:

$$T(\vec{0}_n) = \vec{0}_m.$$

That is, the zero vector in $\mathbb{R}^n$ gets sent to the zero vector in $\mathbb{R}^m$.

The interested reader may wish to read the footnote below.[15]

The Standard Matrix of a Linear Transformation

It is often the case that while one can describe a linear transformation, one doesn't know what matrix performs that transformation (i.e., one doesn't know the standard matrix of that linear transformation). How do we systematically find it? We'll need a new definition.

Definition 31 **Standard Unit Vectors**

In $\mathbb{R}^n$, the *standard unit vectors* $\vec{e_i}$ are the vectors with a 1 in the ith entry and 0s everywhere else.

---

[14] That is, apart from applying the definition directly?

[15] The idea that linear transformations "send zero to zero" has an interesting relation to terminology. The reader is likely familiar with functions like $f(x) = 2x + 3$ and would likely refer to this as a "linear function." However, $f(0) \neq 0$, so f is *not* "linear" by our new definition of linear. We erroneously call f "linear" since its graph produces a line, though we should be careful to instead state that "the graph of f is a line."

We've already seen these vectors in the previous section. In $\mathbb{R}^2$, we identified

$$\vec{e_1} = \begin{bmatrix} 1 \\ 0 \end{bmatrix} \quad \text{and} \quad \vec{e_2} = \begin{bmatrix} 0 \\ 1 \end{bmatrix}.$$

In $\mathbb{R}^4$, there are 4 standard unit vectors:

$$\vec{e_1} = \begin{bmatrix} 1 \\ 0 \\ 0 \\ 0 \end{bmatrix}, \quad \vec{e_2} = \begin{bmatrix} 0 \\ 1 \\ 0 \\ 0 \end{bmatrix}, \quad \vec{e_3} = \begin{bmatrix} 0 \\ 0 \\ 1 \\ 0 \end{bmatrix}, \quad \text{and} \quad \vec{e_4} = \begin{bmatrix} 0 \\ 0 \\ 0 \\ 1 \end{bmatrix}.$$

How do these vectors help us find the standard matrix of a linear transformation? Recall again our work in the previous section. There, we practiced looking at the transformed unit square and deducing the standard transformation matrix A. We did this by making the first column of A the vector where $\vec{e_1}$ ended up and making the second column of A the vector where $\vec{e_2}$ ended up. One could represent this with:

$$A = \begin{bmatrix} T(\vec{e_1}) & T(\vec{e_2}) \end{bmatrix} = [\,T\,].$$

That is, $T(\vec{e_1})$ is the vector where $\vec{e_1}$ ends up, and $T(\vec{e_2})$ is the vector where $\vec{e_2}$ ends up.

The same holds true in general. Given a linear transformation $T : \mathbb{R}^n \to \mathbb{R}^m$, the standard matrix of T is the matrix whose i^{th} column is the vector where $\vec{e_i}$ ends up. While we won't prove this is true, it is, and it is very useful. Therefore we'll state it again as a theorem.

Theorem 22 **The Standard Matrix of a Linear Transformation**

Let $T : \mathbb{R}^n \to \mathbb{R}^m$ be a linear transformation. Then $[\,T\,]$ is the $m \times n$ matrix:

$$[\,T\,] = \begin{bmatrix} T(\vec{e_1}) & T(\vec{e_2}) & \cdots & T(\vec{e_n}) \end{bmatrix}.$$

Let's practice this theorem in an example.

Example 100 Define $T : \mathbb{R}^3 \to \mathbb{R}^4$ to be the linear transformation where

$$T\left(\begin{bmatrix} x_1 \\ x_2 \\ x_3 \end{bmatrix}\right) = \begin{bmatrix} x_1 + x_2 \\ 3x_1 - x_3 \\ 2x_2 + 5x_3 \\ 4x_1 + 3x_2 + 2x_3 \end{bmatrix}.$$

Find $[\,T\,]$.

Solution T takes vectors from $\mathbb{R}^3$ into $\mathbb{R}^4$, so $[\,T\,]$ is going to be a 4×3 matrix. Note that

$$\vec{e_1} = \begin{bmatrix}1\\0\\0\end{bmatrix}, \quad \vec{e_2} = \begin{bmatrix}0\\1\\0\end{bmatrix} \quad \text{and} \quad \vec{e_3} = \begin{bmatrix}0\\0\\1\end{bmatrix}.$$

We find the columns of $[\,T\,]$ by finding where $\vec{e_1}$, $\vec{e_2}$ and $\vec{e_3}$ are sent, that is, we find $T(\vec{e_1})$, $T(\vec{e_2})$ and $T(\vec{e_3})$.

$$T(\vec{e_1}) = T\left(\begin{bmatrix}1\\0\\0\end{bmatrix}\right) = \begin{bmatrix}1\\3\\0\\4\end{bmatrix} \qquad T(\vec{e_2}) = T\left(\begin{bmatrix}0\\1\\0\end{bmatrix}\right) = \begin{bmatrix}1\\0\\2\\3\end{bmatrix} \qquad T(\vec{e_3}) = T\left(\begin{bmatrix}0\\0\\1\end{bmatrix}\right) = \begin{bmatrix}0\\-1\\5\\2\end{bmatrix}$$

Thus

$$[\,T\,] = A = \begin{bmatrix}1 & 1 & 0\\3 & 0 & -1\\0 & 2 & 5\\4 & 3 & 2\end{bmatrix}.$$

Let's check this. Consider the vector

$$\vec{x} = \begin{bmatrix}1\\2\\3\end{bmatrix}.$$

Strictly from the original definition, we can compute that

$$T(\vec{x}) = T\left(\begin{bmatrix}1\\2\\3\end{bmatrix}\right) = \begin{bmatrix}1+2\\3-3\\4+15\\4+6+6\end{bmatrix} = \begin{bmatrix}3\\0\\19\\16\end{bmatrix}.$$

Now compute $T(\vec{x})$ by computing $[\,T\,]\vec{x} = A\vec{x}$.

$$A\vec{x} = \begin{bmatrix}1 & 1 & 0\\3 & 0 & -1\\0 & 2 & 5\\4 & 3 & 2\end{bmatrix}\begin{bmatrix}1\\2\\3\end{bmatrix} = \begin{bmatrix}3\\0\\19\\16\end{bmatrix}.$$

They match![16]

Let's do another example, one that is more application oriented.

[16] Of course they do. That was the whole point.

Example 101 A baseball team manager has collected basic data concerning his hitters. He has the number of singles, doubles, triples, and home runs they have hit over the past year. For each player, he wants two more pieces of information: the total number of hits and the total number of bases.

Using the techniques developed in this section, devise a method for the manager to accomplish his goal.

SOLUTION If the manager only wants to compute this for a few players, then he could do it by hand fairly easily. After all:

total # hits = # of singles + # of doubles + # of triples + # of home runs,

and

total # bases = # of singles + 2×# of doubles + 3×# of triples + 4×# of home runs.

However, if he has a lot of players to do this for, he would likely want a way to automate the work. One way of approaching the problem starts with recognizing that he wants to input four numbers into a function (i.e., the number of singles, doubles, etc.) and he wants two numbers as output (i.e., number of hits and bases). Thus he wants a transformation $T : \mathbb{R}^4 \to \mathbb{R}^2$ where each vector in $\mathbb{R}^4$ can be interpreted as

$$\begin{bmatrix} \text{\# of singles} \\ \text{\# of doubles} \\ \text{\# of triples} \\ \text{\# of home runs} \end{bmatrix},$$

and each vector in $\mathbb{R}^2$ can be interpreted as

$$\begin{bmatrix} \text{\# of hits} \\ \text{\# of bases} \end{bmatrix}.$$

To find $[\,T\,]$, he computes $T(\vec{e_1})$, $T(\vec{e_2})$, $T(\vec{e_3})$ and $T(\vec{e_4})$.

$$T(\vec{e_1}) = T\left(\begin{bmatrix}1\\0\\0\\0\end{bmatrix}\right) = \begin{bmatrix}1\\1\end{bmatrix} \qquad T(\vec{e_2}) = T\left(\begin{bmatrix}0\\1\\0\\0\end{bmatrix}\right) = \begin{bmatrix}1\\2\end{bmatrix}$$

$$T(\vec{e_3}) = T\left(\begin{bmatrix}0\\0\\1\\0\end{bmatrix}\right) = \begin{bmatrix}1\\3\end{bmatrix} \qquad T(\vec{e_4}) = T\left(\begin{bmatrix}0\\0\\0\\1\end{bmatrix}\right) = \begin{bmatrix}1\\4\end{bmatrix}$$

(What do these calculations mean? For example, finding $T(\vec{e_3}) = \begin{bmatrix}1\\3\end{bmatrix}$ means that one triple counts as 1 hit and 3 bases.)

Thus our transformation matrix $[\,T\,]$is

$$[\,T\,] = A = \begin{bmatrix}1 & 1 & 1 & 1\\1 & 2 & 3 & 4\end{bmatrix}.$$

As an example, consider a player who had 102 singles, 30 doubles, 8 triples and 14 home runs. By using A, we find that

$$\begin{bmatrix}1 & 1 & 1 & 1\\1 & 2 & 3 & 4\end{bmatrix}\begin{bmatrix}102\\30\\8\\14\end{bmatrix} = \begin{bmatrix}154\\242\end{bmatrix},$$

meaning the player had 154 hits and 242 total bases.

A question that we should ask concerning the previous example is "How do we know that the function the manager used was actually a linear transformation? After all, we were wrong before – the translation example at the beginning of this section had us fooled at first."

This is a good point; the answer is fairly easy. Recall from Example 98 the transformation

$$T_{98}\left(\begin{bmatrix}x_1\\x_2\end{bmatrix}\right) = \begin{bmatrix}x_1^2\\2x_1\\x_1x_2\end{bmatrix}$$

and from Example 100

$$T_{100}\left(\begin{bmatrix} x_1 \\ x_2 \\ x_3 \end{bmatrix}\right) = \begin{bmatrix} x_1 + x_2 \\ 3x_1 - x_3 \\ 2x_2 + 5x_3 \\ 4x_1 + 3x_2 + 2x_3 \end{bmatrix},$$

where we use the subscripts for T to remind us which example they came from.

We found that T_{98} was not a linear transformation, but stated that T_{100} was (although we didn't prove this). What made the difference?

Look at the entries of $T_{98}(\vec{x})$ and $T_{100}(\vec{x})$. T_{98} contains entries where a variable is squared and where 2 variables are multiplied together – these prevent T_{98} from being linear. On the other hand, the entries of T_{100} are all of the form $a_1x_1 + \cdots + a_nx_n$; that is, they are just sums of the variables multiplied by coefficients. T is linear if and only if the entries of $T(\vec{x})$ are of this form. (Hence linear transformations are related to linear equations, as defined in Section 1.1.) This idea is important.

Key Idea 16 **Conditions on Linear Transformations**

Let $T : \mathbb{R}^n \to \mathbb{R}^m$ be a transformation and consider the entries of

$$T(\vec{x}) = T\left(\begin{bmatrix} x_1 \\ x_2 \\ \vdots \\ x_n \end{bmatrix}\right).$$

T is linear if and only if each entry of $T(\vec{x})$ is of the form $a_1x_1 + a_2x_2 + \cdots a_nx_n$.

Going back to our baseball example, the manager could have defined his transformation as

$$T\left(\begin{bmatrix} x_1 \\ x_2 \\ x_3 \\ x_4 \end{bmatrix}\right) = \begin{bmatrix} x_1 + x_2 + x_3 + x_4 \\ x_1 + 2x_2 + 3x_3 + 4x_4 \end{bmatrix}.$$

Since that fits the model shown in Key Idea 16, the transformation T is indeed linear and hence we can find a matrix $[\,T\,]$ that represents it.

Let's practice this concept further in an example.

Example 102 Using Key Idea 16, determine whether or not each of the following transformations is linear.

$$T_1\left(\begin{bmatrix} x_1 \\ x_2 \end{bmatrix}\right) = \begin{bmatrix} x_1 + 1 \\ x_2 \end{bmatrix} \qquad T_2\left(\begin{bmatrix} x_1 \\ x_2 \end{bmatrix}\right) = \begin{bmatrix} x_1/x_2 \\ \sqrt{x_2} \end{bmatrix}$$

$$T_3\left(\begin{bmatrix} x_1 \\ x_2 \end{bmatrix}\right) = \begin{bmatrix} \sqrt{7}x_1 - x_2 \\ \pi x_2 \end{bmatrix}$$

SOLUTION T_1 is *not* linear! This may come as a surprise, but we are not allowed to add constants to the variables. By thinking about this, we can see that this transformation is trying to accomplish the translation that got us started in this section – it adds 1 to all the x values and leaves the y values alone, shifting everything to the right one unit. However, this is not linear; again, notice how $\vec{0}$ does not get mapped to $\vec{0}$.

T_2 is also not linear. We cannot divide variables, nor can we put variabless inside the square root function (among other other things; again, see Section 1.1). This means that the baseball manager would not be able to use matrices to compute a batting average, which is (number of hits)/(number of at bats).

T_3 is linear. Recall that $\sqrt{7}$ and π are just numbers, just coefficients.

We've mentioned before that we can draw vectors other than 2D vectors, although the more dimensions one adds, the harder it gets to understand. In the next section we'll learn about graphing vectors in 3D – that is, how to draw on paper or a computer screen a 3D vector.

Exercises 5.2

In Exercises 1 – 5, a transformation T is given. Determine whether or not T is linear; if not, state why not.

1. $T\left(\begin{bmatrix} x_1 \\ x_2 \end{bmatrix}\right) = \begin{bmatrix} x_1 + x_2 \\ 3x_1 - x_2 \end{bmatrix}$

2. $T\left(\begin{bmatrix} x_1 \\ x_2 \end{bmatrix}\right) = \begin{bmatrix} x_1 + x_2^2 \\ x_1 - x_2 \end{bmatrix}$

3. $T\left(\begin{bmatrix} x_1 \\ x_2 \end{bmatrix}\right) = \begin{bmatrix} x_1 + 1 \\ x_2 + 1 \end{bmatrix}$

4. $T\left(\begin{bmatrix} x_1 \\ x_2 \end{bmatrix}\right) = \begin{bmatrix} 1 \\ 1 \end{bmatrix}$

5. $T\left(\begin{bmatrix} x_1 \\ x_2 \end{bmatrix}\right) = \begin{bmatrix} 0 \\ 0 \end{bmatrix}$

In Exercises 6 – 11, a linear transformation T is given. Find $[\,T\,]$.

6. $T\left(\begin{bmatrix} x_1 \\ x_2 \end{bmatrix}\right) = \begin{bmatrix} x_1 + x_2 \\ x_1 - x_2 \end{bmatrix}$

7. $T\left(\begin{bmatrix} x_1 \\ x_2 \end{bmatrix}\right) = \begin{bmatrix} x_1 + 2x_2 \\ 3x_1 - 5x_2 \\ 2x_2 \end{bmatrix}$

8. $T\left(\begin{bmatrix} x_1 \\ x_2 \\ x_3 \end{bmatrix}\right) = \begin{bmatrix} x_1 + 2x_2 - 3x_3 \\ 0 \\ x_1 + 4x_3 \\ 5x_2 + x_3 \end{bmatrix}$

9. $T\left(\begin{bmatrix} x_1 \\ x_2 \\ x_3 \end{bmatrix}\right) = \begin{bmatrix} x_1 + 3x_3 \\ x_1 - x_3 \\ x_1 + x_3 \end{bmatrix}$

10. $T\left(\begin{bmatrix} x_1 \\ x_2 \end{bmatrix}\right) = \begin{bmatrix} 0 \\ 0 \end{bmatrix}$

11. $T\left(\begin{bmatrix} x_1 \\ x_2 \\ x_3 \\ x_4 \end{bmatrix}\right) = \begin{bmatrix} x_1 + 2x_2 + 3x_3 + 4x_4 \end{bmatrix}$

5.3 Visualizing Vectors: Vectors in Three Dimensions

1. T/F: The viewpoint of the reader makes a difference in how vectors in 3D look.
2. T/F: If two vectors are not near each other, then they will not appear to be near each other when graphed.
3. T/F: The parallelogram law only applies to adding vectors in 2D.

We ended the last section by stating we could extend the ideas of drawing 2D vectors to drawing 3D vectors. Once we understand how to properly draw these vectors, addition and subtraction is relatively easy. We'll also discuss how to find the length of a vector in 3D.

We start with the basics of drawing a vector in 3D. Instead of having just the traditional x and y axes, we now add a third axis, the z axis. Without any additional vectors, a generic 3D coordinate system can be seen in Figure 5.13.

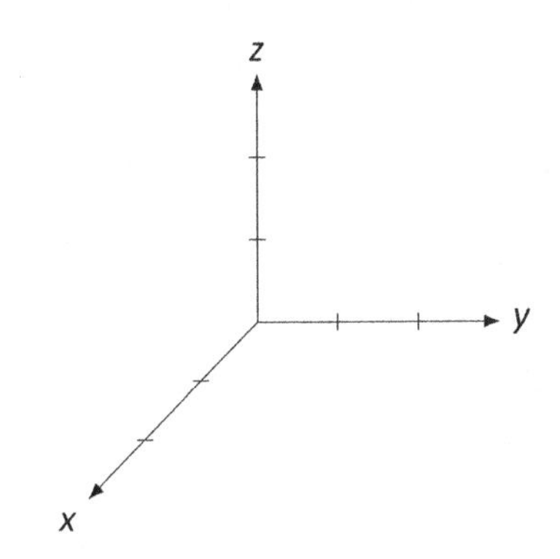

Figure 5.13: The 3D coordinate system

In 2D, the point $(2, 1)$ refers to going 2 units in the x direction followed by 1 unit in the y direction. In 3D, each point is referenced by 3 coordinates. The point $(4, 2, 3)$ is found by going 4 units in the x direction, 2 units in the y direction, and 3 units in the z direction.

How does one sketch a vector on this coordinate system? As one might expect, we can sketch the vector $\vec{v} = \begin{bmatrix} 1 \\ 2 \\ 3 \end{bmatrix}$ by drawing an arrow from the origin (the point (0,0,0)) to the point $(1, 2, 3)$.[17] The only "tricky" part comes from the fact that we are trying to represent three dimensional space on a two dimensional sheet of paper. However,

[17] Of course, we don't have to start at the origin; all that really matters is that the tip of the arrow is 1 unit in the x direction, 2 units in the y direction, and 3 units in the z direction from the origin of the arrow.

it isn't really hard. We'll discover a good way of approaching this in the context of an example.

Example 103 Sketch the following vectors with their origin at the origin.

$$\vec{v} = \begin{bmatrix} 2 \\ 1 \\ 3 \end{bmatrix} \quad \text{and} \quad \vec{u} = \begin{bmatrix} 1 \\ 3 \\ -1 \end{bmatrix}$$

Solution We'll start with $\vec{v}$ first. Starting at the origin, move 2 units in the x direction. This puts us at the point $(2, 0, 0)$ on the x axis. Then, move 1 unit in the y direction. (In our method of drawing, this means moving 1 unit directly to the right. Of course, we don't have a grid to follow, so we have to make a good approximation of this distance.) Finally, we move 3 units in the z direction. (Again, in our drawing, this means going straight "up" 3 units, and we must use our best judgment in a sketch to measure this.)

This allows us to locate the point $(2, 1, 3)$; now we draw an arrow from the origin to this point. In Figure 5.14 we have all 4 stages of this sketch. The dashed lines show us moving down the x axis in (a); in (b) we move over in the y direction; in (c) we move up in the z direction, and finally in (d) the arrow is drawn.

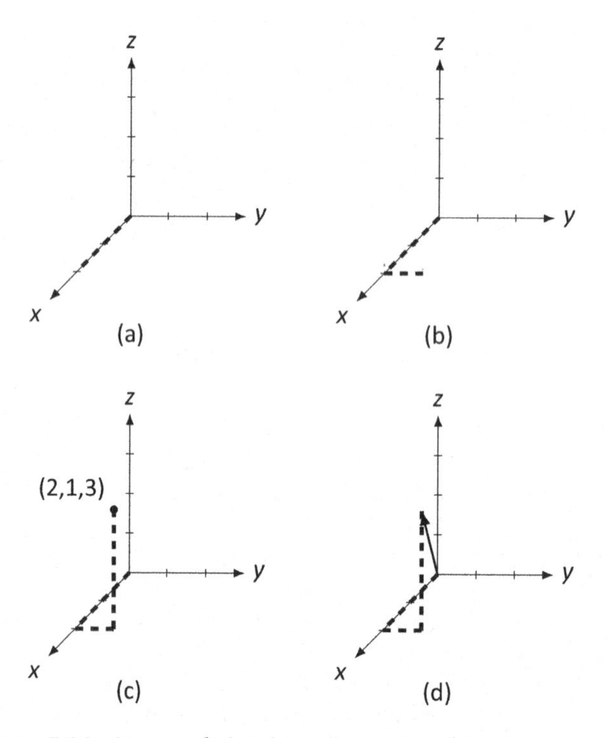

Figure 5.14: Stages of sketching the vector $\vec{v}$ for Example 103.

Drawing the dashed lines help us find our way in our representation of three dimensional space. Without them, it is hard to see how far in each direction the vector is supposed to have gone.

To draw $\vec{u}$, we follow the same procedure we used to draw $\vec{v}$. We first locate the point $(1, 3, -1)$, then draw the appropriate arrow. In Figure 5.15 we have $\vec{u}$ drawn along with $\vec{v}$. We have used different dashed and dotted lines for each vector to help distinguish them.

Notice that this time we had to go in the negative z direction; this just means we moved down one unit instead of up a unit.

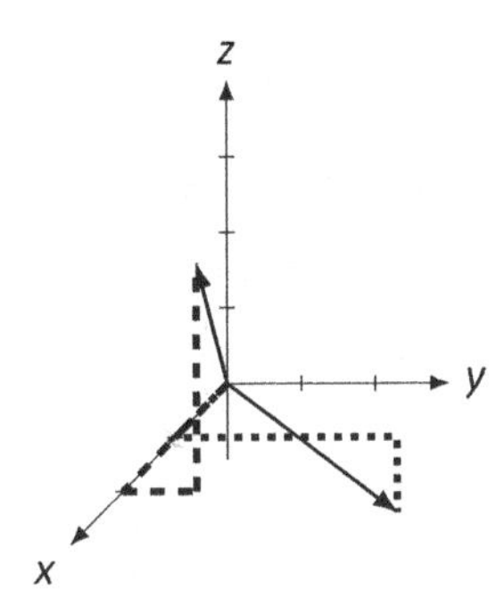

Figure 5.15: Vectors $\vec{v}$ and $\vec{u}$ in Example 103.

As in 2D, we don't usually draw the zero vector,

$$\vec{0} = \begin{bmatrix} 0 \\ 0 \\ 0 \end{bmatrix}.$$

It doesn't point anywhere. It is a conceptually important vector that does not have a terribly interesting visualization.

Our method of drawing 3D objects on a flat surface – a 2D surface – is pretty clever. It isn't perfect, though; visually, drawing vectors with negative components (especially negative x coordinates) can look a bit odd. Also, two very different vectors can point to the same place. We'll highlight this with our next two examples.

Example 104 Sketch the vector $\vec{v} = \begin{bmatrix} -3 \\ -1 \\ 2 \end{bmatrix}$.

Solution We use the same procedure we used in Example 103. Starting at the origin, we move in the negative x direction 3 units, then 1 unit in the negative y direction, and then finally up 2 units in the z direction to find the point $(-3, -1, 2)$. We follow by drawing an arrow. Our sketch is found in Figure 5.16; $\vec{v}$ is drawn in two coordinate systems, once with the helpful dashed lines, and once without. The second drawing makes it pretty clear that the dashed lines truly are helpful.

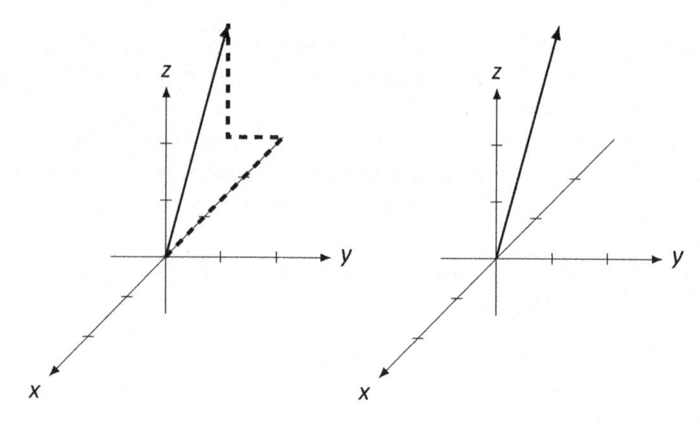

Figure 5.16: Vector $\vec{v}$ in Example 104.

Example 105 Draw the vectors $\vec{v} = \begin{bmatrix} 2 \\ 4 \\ 2 \end{bmatrix}$ and $\vec{u} = \begin{bmatrix} -2 \\ 1 \\ -1 \end{bmatrix}$ on the same coordinate system.

Solution We follow the steps we've taken before to sketch these vectors, shown in Figure 5.17. The dashed lines are aides for $\vec{v}$ and the dotted lines are aids for $\vec{u}$. We again include the vectors without the dashed and dotted lines; but without these, it is very difficult to tell which vector is which!

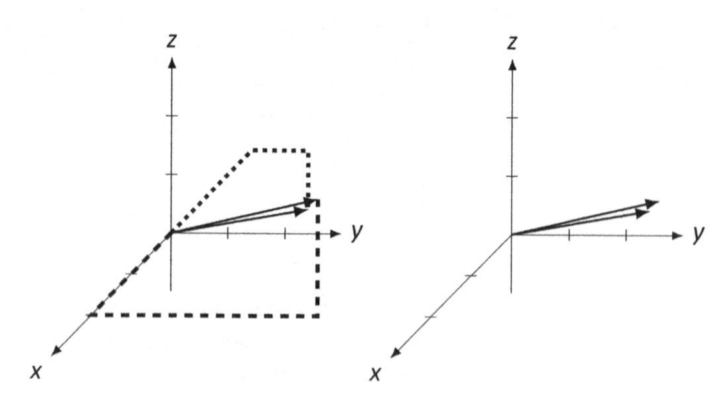

Figure 5.17: Vectors $\vec{v}$ and $\vec{u}$ in Example 105.

Our three examples have demonstrated that we have a pretty clever, albeit imperfect, method for drawing 3D vectors. The vectors in Example 105 look similar because of our *viewpoint*. In Figure 5.18 (a), we have rotated the coordinate axes, giving the vectors a different appearance. (Vector $\vec{v}$ now looks like it lies on the y axis.)

Another important factor in how things look is the scale we use for the x, y, and z axes. In 2D, it is easy to make the scale uniform for both axes; in 3D, it can be a bit tricky to make the scale the same on the axes that are "slanted." Figure 5.18 (b) again shows the same 2 vectors found in Example 105, but this time the scale of the x axis is a bit different. The end result is that again the vectors appear a bit different than they did before. These facts do not necessarily pose a big problem; we must merely be aware of these facts and not make judgments about 3D objects based on one 2D image.[18]

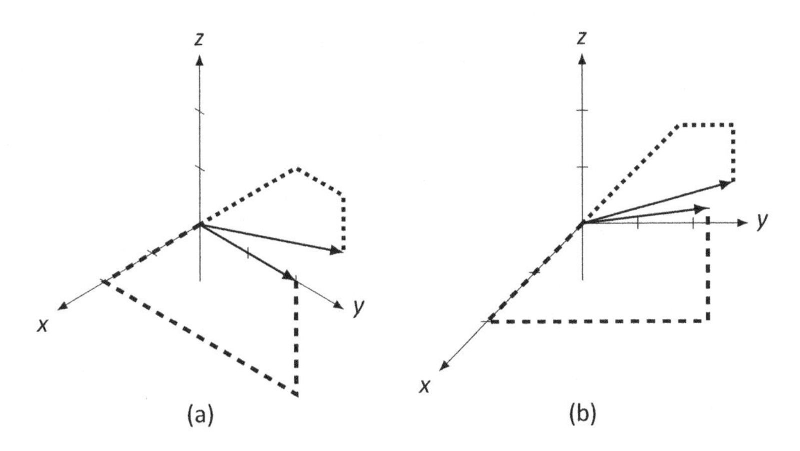

Figure 5.18: Vectors $\vec{v}$ and $\vec{u}$ in Example 105 with a different viewpoint (a) and x axis scale (b).

We now investigate properties of vector arithmetic: what happens (i.e., how do we draw) when we add 3D vectors and multiply by a scalar? How do we compute the length of a 3D vector?

Vector Addition and Subtraction

In 2D, we saw that we could add vectors together graphically using the Parallelogram Law. Does the same apply for adding vectors in 3D? We investigate in an example.

Example 106 Let $\vec{v} = \begin{bmatrix} 2 \\ 1 \\ 3 \end{bmatrix}$ and $\vec{u} = \begin{bmatrix} 1 \\ 3 \\ -1 \end{bmatrix}$. Sketch $\vec{v} + \vec{u}$.

SOLUTION We sketched each of these vectors previously in Example 103. We sketch them, along with $\vec{v} + \vec{u} = \begin{bmatrix} 3 \\ 4 \\ 2 \end{bmatrix}$, in Figure 5.19 (a). (We use loosely dashed lines for $\vec{v} + \vec{u}$.)

[18] The human brain uses both eyes to convey 3D, or depth, information. With one eye closed (or missing), we can have a very hard time with "depth perception." Two objects that are far apart can seem very close together. A simple example of this problem is this: close one eye, and place your index finger about a foot above this text, directly above this **WORD**. See if you were correct by dropping your finger straight down. Did you actually hit the proper spot? Try it again with both eyes, and you should see a noticable difference in your accuracy.

Looking at 3D objects on paper is a bit like viewing the world with one eye closed.

Does the Parallelogram Law still hold? In Figure 5.19 (b), we draw additional representations of $\vec{v}$ and $\vec{u}$ to form a parallelogram (without all the dotted lines), which seems to affirm the fact that the Parallelogram Law does indeed hold.

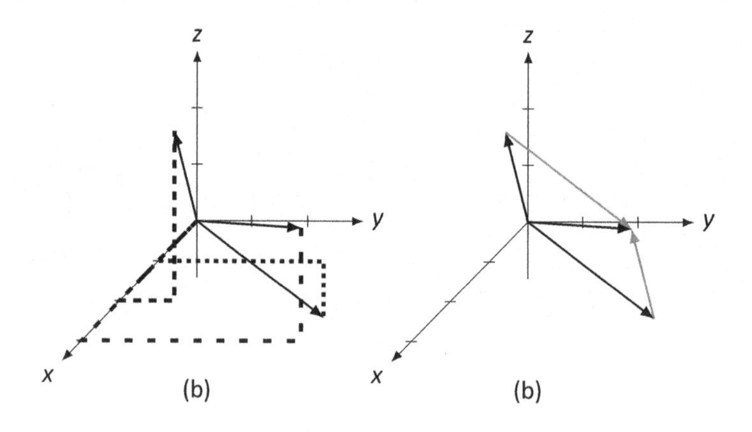

Figure 5.19: Vectors $\vec{v}$, $\vec{u}$, and $\vec{v} + \vec{u}$ Example 106.

We also learned that in 2D, we could subtract vectors by drawing a vector from the tip of one vector to the other.[19] Does this also work in 3D? We'll investigate again with an example, using the familiar vectors $\vec{v}$ and $\vec{u}$ from before.

Example 107 Let $\vec{v} = \begin{bmatrix} 2 \\ 1 \\ 3 \end{bmatrix}$ and $\vec{u} = \begin{bmatrix} 1 \\ 3 \\ -1 \end{bmatrix}$. Sketch $\vec{v} - \vec{u}$.

Solution It is simple to compute that $\vec{v} - \vec{u} = \begin{bmatrix} 1 \\ -2 \\ 4 \end{bmatrix}$. All three of these vectors are sketched in Figure 5.20 (a), where again $\vec{v}$ is guided by the dashed, $\vec{u}$ by the dotted, and $\vec{v} - \vec{u}$ by the loosely dashed lines.

Does the 2D subtraction rule still hold? That is, can we draw $\vec{v} - \vec{u}$ by drawing an arrow from the tip of $\vec{u}$ to the tip of $\vec{v}$? In Figure 5.20 (b), we translate the drawing of $\vec{v} - \vec{u}$ to the tip of $\vec{u}$, and sure enough, it looks like it works. (And in fact, it really does.)

[19] Recall that it is important which vector we used for the origin and which was used for the tip.

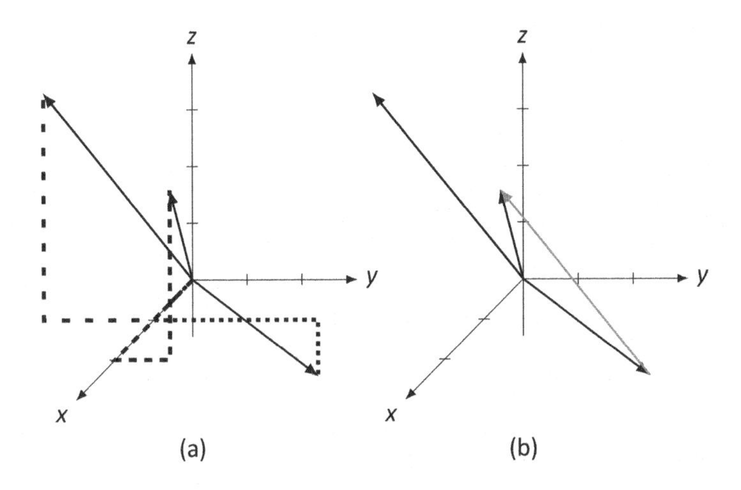

Figure 5.20: Vectors $\vec{v}$, $\vec{u}$, and $\vec{v} - \vec{u}$ from Example 107.

The previous two examples highlight the fact that even in 3D, we can sketch vectors without explicitly knowing what they are. We practice this one more time in the following example.

Example 108 Vectors $\vec{v}$ and $\vec{u}$ are drawn in Figure 5.21. Using this drawing, sketch the vectors $\vec{v} + \vec{u}$ and $\vec{v} - \vec{u}$.

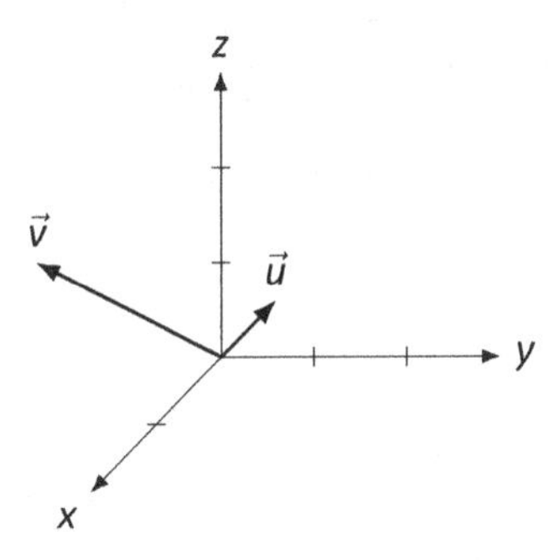

Figure 5.21: Vectors $\vec{v}$ and $\vec{u}$ for Example 108.

Solution Using the Parallelogram Law, we draw $\vec{v} + \vec{u}$ by first drawing a gray version of $\vec{u}$ coming from the tip of $\vec{v}$; $\vec{v} + \vec{u}$ is drawn dashed in Figure 5.22.

To draw $\vec{v} - \vec{u}$, we draw a dotted arrow from the tip of $\vec{u}$to the tip of $\vec{v}$.

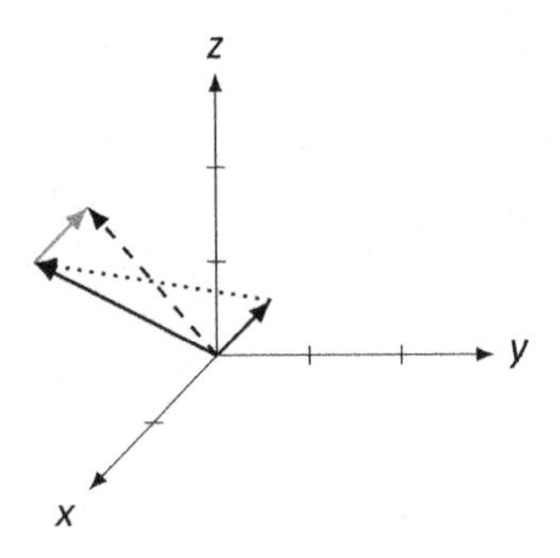

Figure 5.22: Vectors $\vec{v}$, $\vec{u}$, $\vec{v}+\vec{u}$ and $\vec{v}-\vec{u}$ for Example 108.

Scalar Multiplication

Given a vector $\vec{v}$ in 3D, what does the vector $2\vec{v}$ look like? How about $-\vec{v}$? After learning about vector addition and subtraction in 3D, we are probably gaining confidence in working in 3D and are tempted to say that $2\vec{v}$ is a vector twice as long as $\vec{v}$, pointing in the same direction, and $-\vec{v}$ is a vector of the same length as $\vec{v}$, pointing in the opposite direction. We would be right. We demonstrate this in the following example.

Example 109 Sketch $\vec{v}$, $2\vec{v}$, and $-\vec{v}$, where

$$\vec{v}=\begin{bmatrix}1\\2\\3\end{bmatrix}.$$

Solution

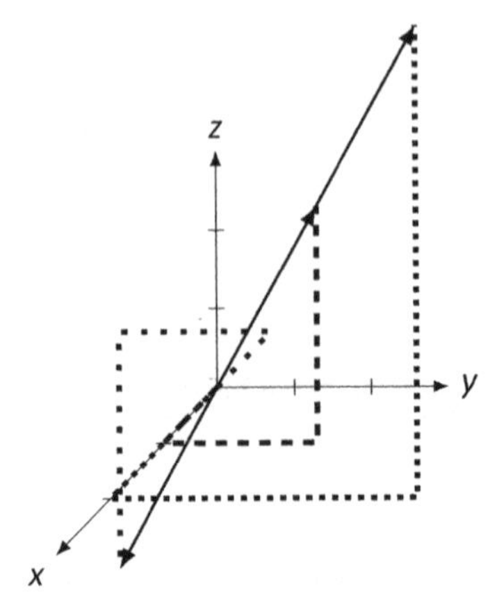

Figure 5.23: Sketching scalar multiples of $\vec{v}$ in Example 109.

It is easy to compute

$$2\vec{v} = \begin{bmatrix} 2 \\ 4 \\ 6 \end{bmatrix} \quad \text{and} \quad -\vec{v} = \begin{bmatrix} -1 \\ -2 \\ -3 \end{bmatrix}.$$

These are drawn in Figure 5.23. This figure is, in many ways, a mess, with all the dashed and dotted lines. They are useful though. Use them to see how each vector was formed, and note that $2\vec{v}$ at least looks twice as long as $\vec{v}$, and it looks like $-\vec{v}$ points in the opposite direction.[20]

Vector Length

How do we measure the length of a vector in 3D? In 2D, we were able to answer this question by using the Pythagorean Theorem. Does the Pythagorean Theorem apply in 3D? In a sense, it does.

Consider the vector $\vec{v} = \begin{bmatrix} 1 \\ 2 \\ 3 \end{bmatrix}$, as drawn in Figure 5.24 (a), with guiding dashed lines. Now look at part (b) of the same figure. Note how two lengths of the dashed lines have now been drawn gray, and another dotted line has been added.

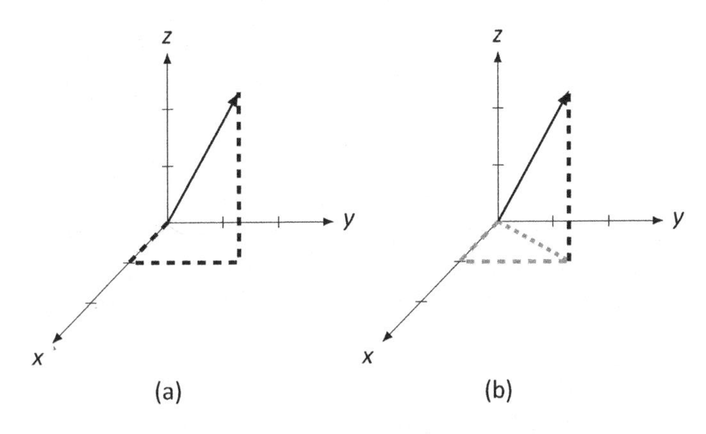

Figure 5.24: Computing the length of $\vec{v}$

These gray dashed and dotted lines form a right triangle with the dotted line forming the hypotenuse. We can find the length of the dotted line using the Pythagorean Theorem.

$$\text{length of the dotted line} = \sqrt{\text{sum of the squares of the dashed line lengths}}$$

That is, the length of the dotted line = $\sqrt{1^2 + 2^2} = \sqrt{5}$.

[20]Our previous work showed that looks can be deceiving, but it is indeed true in this case.

Now consider this: the vector $\vec{v}$ is the hypotenuse of another right triangle: the one formed by the dotted line and the vertical dashed line. Again, we employ the Pythagorean Theorem to find its length.

$$\text{length of } \vec{v} = \sqrt{(\text{length of dashed gray line})^2 + (\text{length of black dashed line})^2}$$

Thus, the length of $\vec{v}$ is (recall, we denote the length of $\vec{v}$ with $||\vec{v}||$):

$$\begin{aligned} ||\vec{v}|| &= \sqrt{(\text{length of gray line})^2 + (\text{length of black line})^2} \\ &= \sqrt{\sqrt{5}^2 + 3^2} \\ &= \sqrt{5 + 3^2} \end{aligned}$$

Let's stop for a moment and think: where did this 5 come from in the previous equation? It came from finding the length of the gray dashed line – it came from 1^2+2^2. Let's substitute that into the previous equation:

$$\begin{aligned} ||\vec{v}|| &= \sqrt{5 + 3^2} \\ &= \sqrt{1^2 + 2^2 + 3^2} \\ &= \sqrt{14} \end{aligned}$$

The key comes from the middle equation: $||\vec{v}|| = \sqrt{1^2 + 2^2 + 3^2}$. Do those numbers 1, 2, and 3 look familiar? They are the component values of $\vec{v}$! This is very similar to the definition of the length of a 2D vector. After formally defining this, we'll practice with an example.

Definition 32

3D Vector Length

Let

$$\vec{v} = \begin{bmatrix} x_1 \\ x_2 \\ x_3 \end{bmatrix}.$$

The *length* of $\vec{v}$, denoted $||\vec{v}||$, is

$$||\vec{v}|| = \sqrt{x_1^2 + x_2^2 + x_3^2}.$$

Example 110 Find the lengths of vectors $\vec{v}$ and $\vec{u}$, where

$$\vec{v} = \begin{bmatrix} 2 \\ -3 \\ 5 \end{bmatrix} \quad \text{and} \quad \vec{u} = \begin{bmatrix} -4 \\ 7 \\ 0 \end{bmatrix}.$$

SOLUTION We apply Definition 32 to each vector:

$$\begin{aligned} ||\vec{v}|| &= \sqrt{2^2 + (-3)^2 + 5^2} \\ &= \sqrt{4 + 9 + 25} \\ &= \sqrt{38} \end{aligned}$$

$$\begin{aligned} ||\vec{u}|| &= \sqrt{(-4)^2 + 7^2 + 0^2} \\ &= \sqrt{16 + 49} \\ &= \sqrt{65} \end{aligned}$$

Here we end our investigation into the world of graphing vectors. Extensions into graphing 4D vectors and beyond *can* be done, but they truly are confusing and not really done except for abstract purposes.

There are further things to explore, though. Just as in 2D, we can transform 3D space by matrix multiplication. Doing this properly – rotating, stretching, shearing, etc. – allows one to manipulate 3D space and create incredible computer graphics.

Exercises 5.3

In Exercises 1 – 4, vectors $\vec{x}$ and $\vec{y}$ are given. Sketch $\vec{x}$, $\vec{y}$, $\vec{x}+\vec{y}$, and $\vec{x}-\vec{y}$ on the same Cartesian axes.

1. $\vec{x} = \begin{bmatrix} 1 \\ -1 \\ 2 \end{bmatrix}, \vec{y} = \begin{bmatrix} 2 \\ 3 \\ 2 \end{bmatrix}$

2. $\vec{x} = \begin{bmatrix} 2 \\ 4 \\ -1 \end{bmatrix}, \vec{y} = \begin{bmatrix} -1 \\ -3 \\ -1 \end{bmatrix}$

3. $\vec{x} = \begin{bmatrix} 1 \\ 1 \\ 2 \end{bmatrix}, \vec{y} = \begin{bmatrix} 3 \\ 3 \\ 6 \end{bmatrix}$

4. $\vec{x} = \begin{bmatrix} 0 \\ 1 \\ 1 \end{bmatrix}, \vec{y} = \begin{bmatrix} 0 \\ -1 \\ 1 \end{bmatrix}$

In Exercises 5 – 8, vectors $\vec{x}$ and $\vec{y}$ are drawn. Sketch $2\vec{x}$, $-\vec{y}$, $\vec{x}+\vec{y}$, and $\vec{x}-\vec{y}$ on the same Cartesian axes.

5. 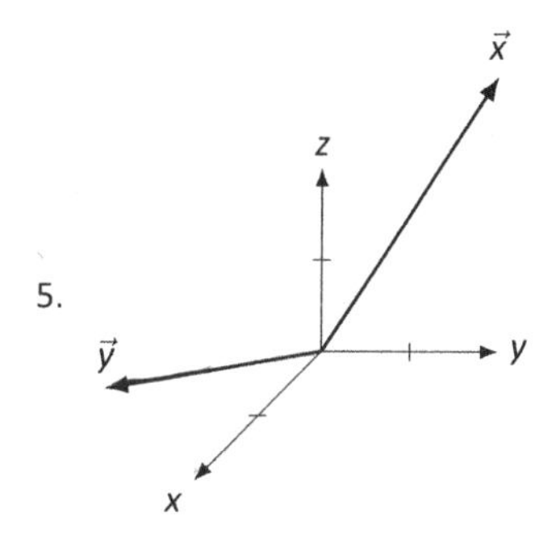

6. 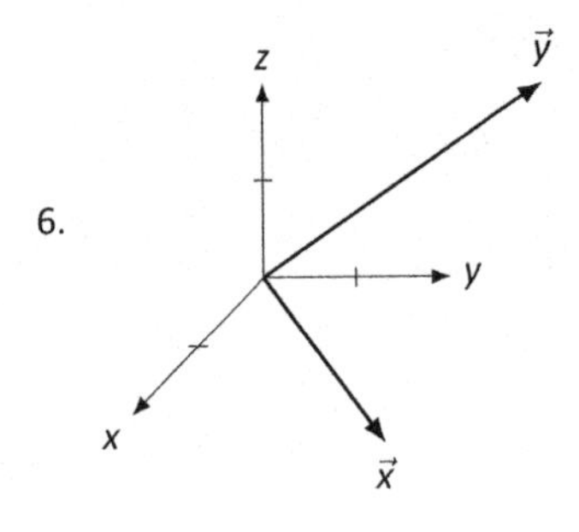

7. 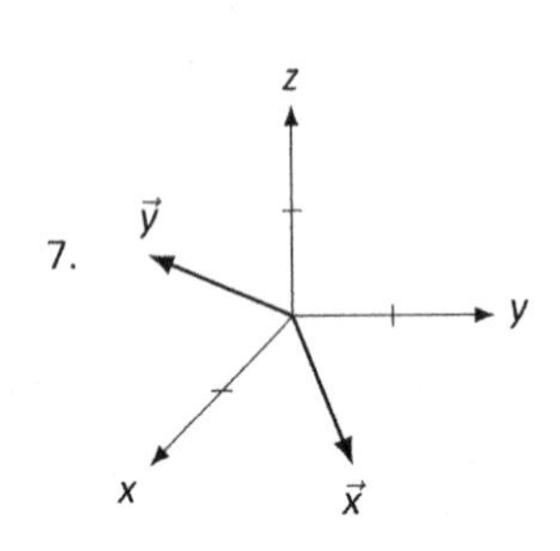

8. 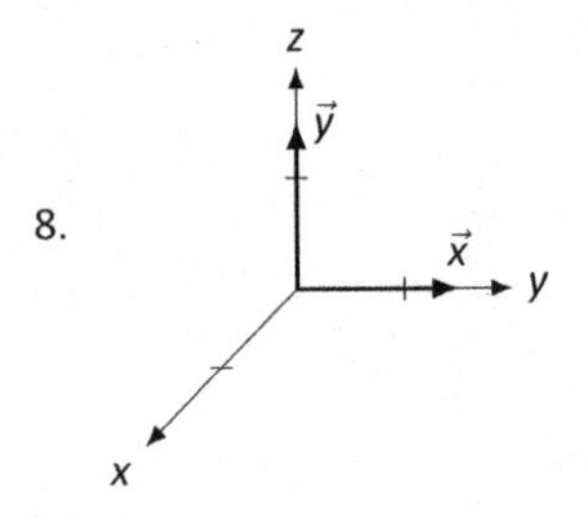

In Exercises 9 – 12, a vector $\vec{x}$ and a scalar a are given. Using Definition 32, compute the lengths of $\vec{x}$ and $a\vec{x}$, then compare these lengths.

9. $\vec{x} = \begin{bmatrix} 1 \\ -2 \\ 5 \end{bmatrix}, a = 2$

10. $\vec{x} = \begin{bmatrix} -3 \\ 4 \\ 3 \end{bmatrix}, a = -1$

11. $\vec{x} = \begin{bmatrix} 7 \\ 2 \\ 1 \end{bmatrix}, a = 5$

12. $\vec{x} = \begin{bmatrix} 1 \\ 2 \\ -2 \end{bmatrix}, a = 3$

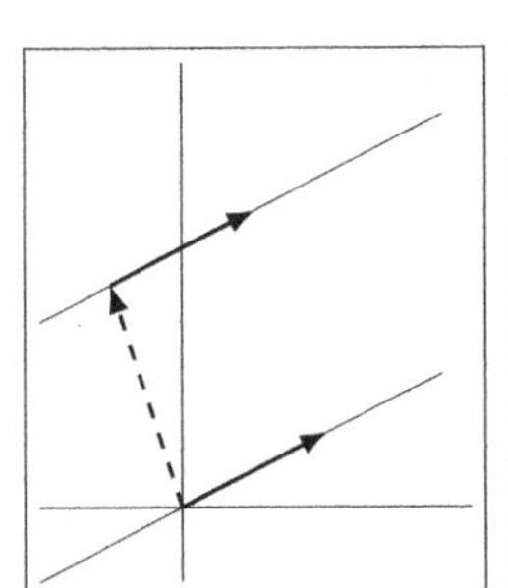
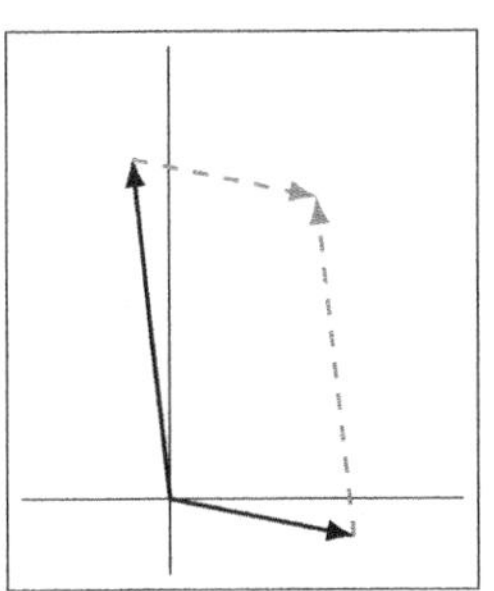
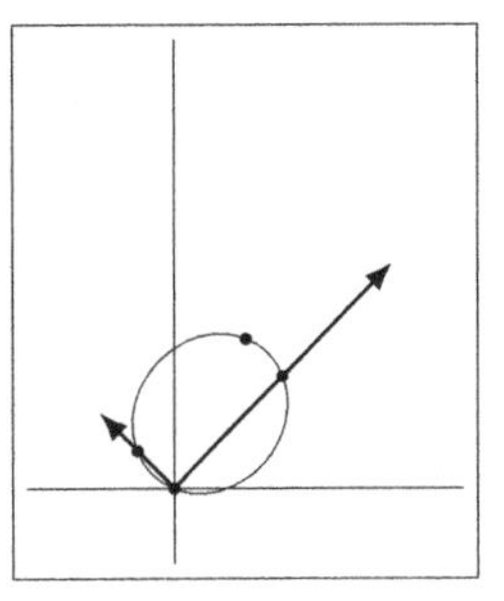

Solutions To Selected Problems

Chapter 1

Section 1.1

1. y
3. y
5. n
7. y
9. y
11. $x = 1, y = -2$
13. $x = -1, y = 0, z = 2$
15. 29 chickens and 33 pigs

Section 1.2

1. $\begin{bmatrix} 3 & 4 & 5 & 7 \\ -1 & 1 & -3 & 1 \\ 2 & -2 & 3 & 5 \end{bmatrix}$

3. $\begin{bmatrix} 1 & 3 & -4 & 5 & 17 \\ -1 & 0 & 4 & 8 & 1 \\ 2 & 3 & 4 & 5 & 6 \end{bmatrix}$

5. $\begin{aligned} x_1 + 2x_2 &= 3 \\ -x_1 + 3x_2 &= 9 \end{aligned}$

7. $\begin{aligned} x_1 + x_2 - x_3 - x_4 &= 2 \\ 2x_1 + x_2 + 3x_3 + 5x_4 &= 7 \end{aligned}$

9. $\begin{aligned} x_1 + x_3 + 7x_5 &= 2 \\ x_2 + 3x_3 + 2x_4 &= 5 \end{aligned}$

11. $\begin{bmatrix} 2 & -1 & 7 \\ 5 & 0 & 3 \\ 0 & 4 & -2 \end{bmatrix}$

13. $\begin{bmatrix} 2 & -1 & 7 \\ 0 & 4 & -2 \\ 5 & 8 & -1 \end{bmatrix}$

15. $\begin{bmatrix} 2 & -1 & 7 \\ 0 & 4 & -2 \\ 0 & 5/2 & -29/2 \end{bmatrix}$

17. $R_1 + R_2 \to R_2$
19. $R_1 \leftrightarrow R_2$
21. $x = 2, y = 1$
23. $x = -1, y = 0$
25. $x_1 = -2, x_2 = 1, x_3 = 2$

Section 1.3

1. (a) yes (b) no (c) no (d) yes

3. (a) no (b) yes (c) yes (d) yes

5. $\begin{bmatrix} 1 & 0 \\ 0 & 1 \end{bmatrix}$

7. $\begin{bmatrix} 1 & 3 \\ 0 & 0 \end{bmatrix}$

9. $\begin{bmatrix} 1 & 0 & 3 \\ 0 & 1 & 7 \end{bmatrix}$

11. $\begin{bmatrix} 1 & -1 & 2 \\ 0 & 0 & 0 \end{bmatrix}$

13. $\begin{bmatrix} 1 & 0 & 0 \\ 0 & 1 & 0 \\ 0 & 0 & 1 \end{bmatrix}$

15. $\begin{bmatrix} 1 & 0 & 0 \\ 0 & 1 & 0 \\ 0 & 0 & 1 \end{bmatrix}$

17. $\begin{bmatrix} 1 & 0 & 0 & 1 \\ 0 & 1 & 1 & 1 \\ 0 & 0 & 0 & 0 \end{bmatrix}$

19. $\begin{bmatrix} 1 & 0 & 1 & 3 \\ 0 & 1 & -2 & 4 \\ 0 & 0 & 0 & 0 \end{bmatrix}$

21. $\begin{bmatrix} 1 & 1 & 0 & 0 & 0 & 0 \\ 0 & 0 & 1 & 3 & 1 & 4 \end{bmatrix}$

Section 1.4

1. $x_1 = 1 - 2x_2$; x_2 is free. Possible solutions: $x_1 = 1, x_2 = 0$ and $x_1 = -1, x_2 = 1$.

3. $x_1 = 1$; $x_2 = 2$

5. No solution; the system is inconsistent.

7. $x_1 = -11 + 10x_3$; $x_2 = -4 + 4x_3$; x_3 is free. Possible solutions: $x_1 = -11$, $x_2 = -4, x_3 = 0$ and $x_1 = -1, x_2 = 0$ and $x_3 = 1$.

9. $x_1 = 1 - x_2 - x_4$; x_2 is free; $x_3 = 1 - 2x_4$; x_4 is free. Possible solutions: $x_1 = 1$, $x_2 = 0, x_3 = 1, x_4 = 0$ and $x_1 = -2$, $x_2 = 1, x_3 = -3, x_4 = 2$

11. No solution; the system is inconsistent.

13. $x_1 = \frac{1}{3} - \frac{4}{3}x_3$; $x_2 = \frac{1}{3} - \frac{1}{3}x_3$; x_3 is free. Possible solutions: $x_1 = \frac{1}{3}, x_2 = \frac{1}{3}, x_3 = 0$ and $x_1 = -1, x_2 = 0, x_3 = 1$

15. Never exactly 1 solution; infinite solutions if $k = 2$; no solution if $k \neq 2$.

17. Exactly 1 solution if $k \neq 2$; no solution if $k = 2$; never infinite solutions.

Section 1.5

1. 29 chickens and 33 pigs

3. 42 grande tables, 22 venti tables

5. 30 men, 15 women, 20 kids

7. $f(x) = -2x + 10$

9. $f(x) = \frac{1}{2}x^2 + 3x + 1$

11. $f(x) = 3$

13. $f(x) = x^3 + 1$

15. $f(x) = \frac{3}{2}x + 1$

17. The augmented matrix from this system is $\begin{bmatrix} 1 & 1 & 1 & 1 & 8 \\ 6 & 1 & 2 & 3 & 24 \\ 0 & 1 & -1 & 0 & 0 \end{bmatrix}$. From this we find the solution

$$t = \frac{8}{3} - \frac{1}{3}f$$
$$x = \frac{8}{3} - \frac{1}{3}f$$
$$w = \frac{8}{3} - \frac{1}{3}f.$$

The only time each of these variables are nonnegative integers is when $f = 2$ or $f = 8$. If $f = 2$, then we have 2 touchdowns, 2 extra points and 2 two point conversions (and 2 field goals); this doesn't make sense since the extra points and two point conversions follow touchdowns. If $f = 8$, then we have no touchdowns, extra points or two point conversions (just 8 field goals). This is the only solution; all points were scored from field goals.

19. Let x_1, x_2 and x_3 represent the number of free throws, 2 point and 3 point shots taken. The augmented matrix from this system is $\begin{bmatrix} 1 & 1 & 1 & 30 \\ 1 & 2 & 3 & 80 \end{bmatrix}$. From this we find the solution

$$x_1 = -20 + x_3$$
$$x_2 = 50 - 2x_3.$$

In order for x_1 and x_2 to be nonnegative, we need $20 \leq x_3 \leq 25$. Thus there are 6 different scenerios: the "first" is where 20 three point shots are taken, no free throws, and 10 two point shots; the "last" is where 25 three point shots are taken, 5 free throws, and no two point shots.

21. Let $y = ax + b$; all linear functions through (1,3) come in the form $y = (3 - b)x + b$. Examples: $b = 0$ yields $y = 3x$; $b = 2$ yields $y = x + 2$.

23. Let $y = ax^2 + bx + c$; we find that $a = -\frac{1}{2} + \frac{1}{2}c$ and $b = \frac{1}{2} - \frac{3}{2}c$. Examples: $c = 1$ yields $y = -x + 1$; $c = 3$ yields $y = x^2 - 4x + 3$.

Chapter 2

Section 2.1

1. $\begin{bmatrix} -2 & -1 \\ 12 & 13 \end{bmatrix}$

3. $\begin{bmatrix} 2 & -2 \\ 14 & 8 \end{bmatrix}$

5. $\begin{bmatrix} 9 & -7 \\ 11 & -6 \end{bmatrix}$

7. $\begin{bmatrix} -14 \\ 6 \end{bmatrix}$

9. $\begin{bmatrix} -15 \\ -25 \end{bmatrix}$

11. $X = \begin{bmatrix} -5 & 9 \\ -1 & -14 \end{bmatrix}$

13. $X = \begin{bmatrix} -5 & -2 \\ -9/2 & -19/2 \end{bmatrix}$

15. $a = 2, b = 1$

17. $a = 5/2 + 3/2b$

19. No solution.

21. No solution.

Section 2.2

1. -22

3. 0

5. 5

7. 15

9. -2

11. Not possible.

13. $AB = \begin{bmatrix} 8 & 3 \\ 10 & -9 \end{bmatrix}$

$BA = \begin{bmatrix} -3 & 24 \\ 4 & 2 \end{bmatrix}$

15. $AB = \begin{bmatrix} -1 & -2 & 12 \\ 10 & 4 & 32 \end{bmatrix}$

BA is not possible.

17. AB is not possible.

$BA = \begin{bmatrix} 27 & -33 & 39 \\ -27 & -3 & -15 \end{bmatrix}$

19. $AB = \begin{bmatrix} -32 & 34 & -24 \\ -32 & 38 & -8 \\ -16 & 21 & 4 \end{bmatrix}$

$BA = \begin{bmatrix} 22 & -14 \\ -4 & -12 \end{bmatrix}$

21. $AB = \begin{bmatrix} -56 & 2 & -36 \\ 20 & 19 & -30 \\ -50 & -13 & 0 \end{bmatrix}$

$BA = \begin{bmatrix} -46 & 40 \\ 72 & 9 \end{bmatrix}$

23. $AB = \begin{bmatrix} -15 & -22 & -21 & -1 \\ 16 & -53 & -59 & -31 \end{bmatrix}$

BA is not possible.

25. $AB = \begin{bmatrix} 0 & 0 & 4 \\ 6 & 4 & -2 \\ 2 & -4 & -6 \end{bmatrix}$

$BA = \begin{bmatrix} 2 & -2 & 6 \\ 2 & 2 & 4 \\ 4 & 0 & -6 \end{bmatrix}$

27. $AB = \begin{bmatrix} 21 & -17 & -5 \\ 19 & 5 & 19 \\ 5 & 9 & 4 \end{bmatrix}$

$BA = \begin{bmatrix} 19 & 5 & 23 \\ 5 & -7 & -1 \\ -14 & 6 & 18 \end{bmatrix}$

29. $DA = \begin{bmatrix} 4 & -6 \\ 4 & -6 \end{bmatrix}$

$AD = \begin{bmatrix} 4 & 8 \\ -3 & -6 \end{bmatrix}$

31. $DA = \begin{bmatrix} 2 & 2 & 2 \\ -6 & -6 & -6 \\ -15 & -15 & -15 \end{bmatrix}$

$AD = \begin{bmatrix} 2 & -3 & 5 \\ 4 & -6 & 10 \\ -6 & 9 & -15 \end{bmatrix}$

33. $DA = \begin{bmatrix} d_1a & d_1b & d_1c \\ d_2d & d_2e & d_2f \\ d_3g & d_3h & d_3i \end{bmatrix}$

$AD = \begin{bmatrix} d_1a & d_2b & d_3c \\ d_1d & d_2e & d_3f \\ d_1g & d_2h & d_3i \end{bmatrix}$

35. $A\vec{x} = \begin{bmatrix} -6 \\ 11 \end{bmatrix}$

37. $A\vec{x} = \begin{bmatrix} -5 \\ 5 \\ 21 \end{bmatrix}$

39. $A\vec{x} = \begin{bmatrix} x_1 + 2x_2 + 3x_3 \\ x_1 + 2x_3 \\ 2x_1 + 3x_2 + x_3 \end{bmatrix}$

41. $A^2 = \begin{bmatrix} 4 & 0 \\ 0 & 9 \end{bmatrix}; A^3 = \begin{bmatrix} 8 & 0 \\ 0 & 27 \end{bmatrix}$

43. $A^2 = \begin{bmatrix} 0 & 0 & 1 \\ 1 & 0 & 0 \\ 0 & 1 & 0 \end{bmatrix}; A^3 = \begin{bmatrix} 1 & 0 & 0 \\ 0 & 1 & 0 \\ 0 & 0 & 1 \end{bmatrix}$

45. (a) $\begin{bmatrix} 0 & -2 \\ -5 & -1 \end{bmatrix}$

(b) $\begin{bmatrix} 10 & 2 \\ 5 & 11 \end{bmatrix}$

(c) $\begin{bmatrix} -11 & -15 \\ 37 & 32 \end{bmatrix}$

(d) No.

(e) $(A+B)(A+B) = AA+AB+BA+BB = A^2 + AB + BA + B^2$.

Section 2.3

1. $\vec{x}+\vec{y}=\begin{bmatrix}-1\\4\end{bmatrix}, \vec{x}-\vec{y}=\begin{bmatrix}3\\-2\end{bmatrix}$

   Sketches will vary depending on choice of origin of each vector.

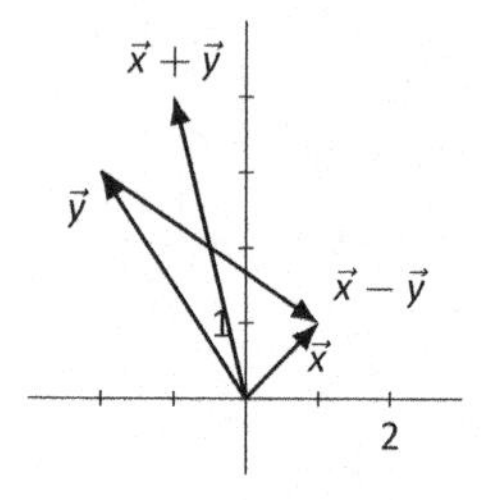

3. $\vec{x}+\vec{y}=\begin{bmatrix}-3\\3\end{bmatrix}, \vec{x}-\vec{y}=\begin{bmatrix}1\\-1\end{bmatrix}$

   Sketches will vary depending on choice of origin of each vector.

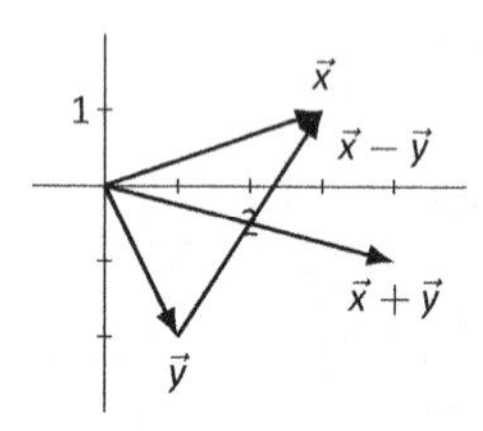

5. Sketches will vary depending on choice of origin of each vector.

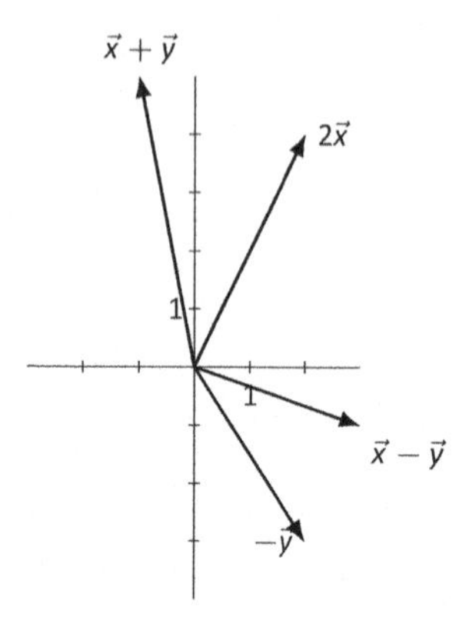

7. Sketches will vary depending on choice of origin of each vector.

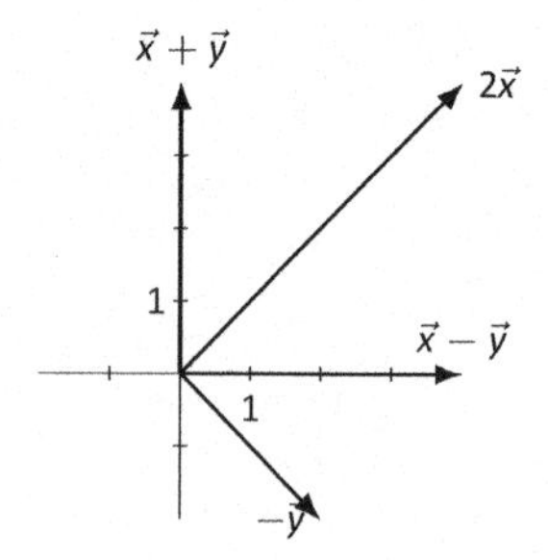

9. $||\vec{x}|| = \sqrt{5}$; $||a\vec{x}|| = \sqrt{45} = 3\sqrt{5}$. The vector $a\vec{x}$ is 3 times as long as $\vec{x}$.

11. $||\vec{x}|| = \sqrt{34}$; $||a\vec{x}|| = \sqrt{34}$. The vectors $a\vec{x}$ and $\vec{x}$ are the same length (they just point in opposite directions).

13. (a) $||\vec{x}|| = \sqrt{2}$; $||\vec{y}|| = \sqrt{13}$; $||\vec{x}+\vec{y}|| = 5$.

    (b) $||\vec{x}|| = \sqrt{5}$; $||\vec{y}|| = 3\sqrt{5}$; $||\vec{x}+\vec{y}|| = 4\sqrt{5}$.

    (c) $||\vec{x}|| = \sqrt{10}$; $||\vec{y}|| = \sqrt{29}$; $||\vec{x}+\vec{y}|| = \sqrt{65}$.

    (d) $||\vec{x}|| = \sqrt{5}$; $||\vec{y}|| = 2\sqrt{5}$; $||\vec{x}+\vec{y}|| = \sqrt{5}$.

    The equality holds sometimes; only when $\vec{x}$ and $\vec{y}$ point along the same line, in the same direction.

15. 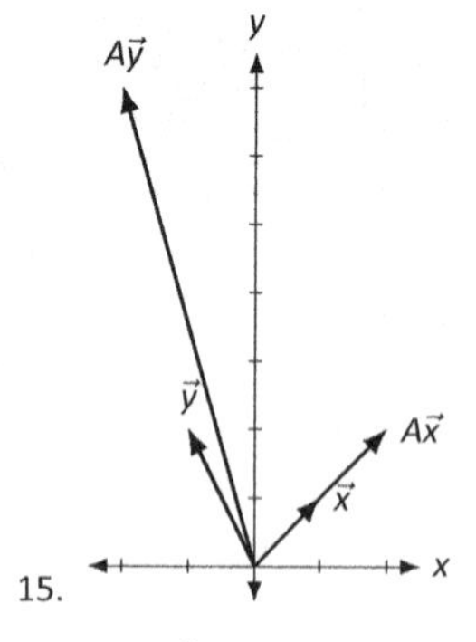

17. 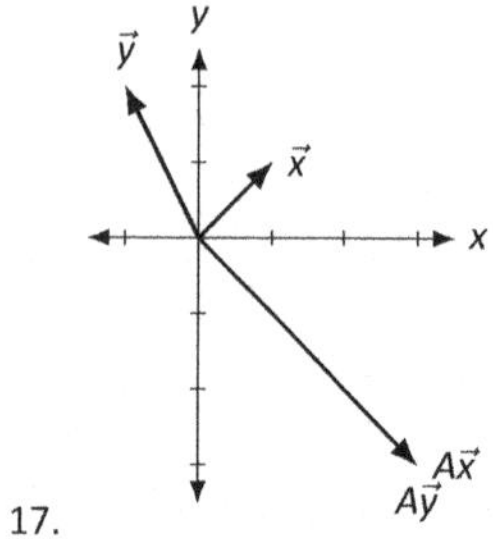

Section 2.4

1. Multiply $A\vec{u}$ and $A\vec{v}$ to verify.

3. Multiply $A\vec{u}$ and $A\vec{v}$ to verify.

5. Multiply $A\vec{u}$ and $A\vec{v}$ to verify.

7. Multiply $A\vec{u}$, $A\vec{v}$ and $A(\vec{u}+\vec{v})$ to verify.

9. Multiply $A\vec{u}$, $A\vec{v}$ and $A(\vec{u}+\vec{v})$ to verify.

11. (a) $\vec{x}=\begin{bmatrix}0\\0\end{bmatrix}$

(b) $\vec{x}=\begin{bmatrix}2/5\\-13/5\end{bmatrix}$

13. (a) $\vec{x}=\begin{bmatrix}0\\0\end{bmatrix}$

(b) $\vec{x}=\begin{bmatrix}-2\\-9/4\end{bmatrix}$

15. (a) $\vec{x}=x_3\begin{bmatrix}5/4\\1\\1\end{bmatrix}$

(b) $\vec{x}=\begin{bmatrix}1\\0\\0\end{bmatrix}+x_3\begin{bmatrix}5/4\\1\\1\end{bmatrix}$

17. (a) $\vec{x}=x_3\begin{bmatrix}14\\-10\\0\end{bmatrix}$

(b) $\vec{x}=\begin{bmatrix}-4\\2\end{bmatrix}+x_3\begin{bmatrix}14\\-10\\0\end{bmatrix}$

19. (a) $\vec{x}=x_3\begin{bmatrix}2\\2/5\\1\\0\end{bmatrix}+x_4\begin{bmatrix}-1\\2/5\\0\\1\end{bmatrix}$

(b) $\vec{x}=$
$\begin{bmatrix}-2\\2/5\\0\\0\end{bmatrix}+x_3\begin{bmatrix}2\\2/5\\1\\0\end{bmatrix}+x_4\begin{bmatrix}-1\\2/5\\0\\1\end{bmatrix}$

21. (a) $\vec{x}=x_2\begin{bmatrix}-1/2\\1\\0\\0\\0\end{bmatrix}+x_4\begin{bmatrix}1/2\\0\\-1/2\\1\\0\end{bmatrix}+$
$x_5\begin{bmatrix}13/2\\0\\-2\\0\\1\end{bmatrix}$

(b) $\vec{x}=\begin{bmatrix}-5\\0\\3/2\\0\\0\end{bmatrix}+x_2\begin{bmatrix}-1/2\\1\\0\\0\\0\end{bmatrix}+$
$x_4\begin{bmatrix}1/2\\0\\-1/2\\1\\0\end{bmatrix}+x_5\begin{bmatrix}13/2\\0\\-2\\0\\1\end{bmatrix}$

23. (a) $\vec{x}=x_4\begin{bmatrix}1\\13/9\\-1/3\\1\\0\end{bmatrix}+x_5\begin{bmatrix}0\\-1\\-1\\0\\1\end{bmatrix}$

(b) $\vec{x}=$
$\begin{bmatrix}1\\1/9\\5/3\\0\\0\end{bmatrix}+x_4\begin{bmatrix}1\\13/9\\-1/3\\1\\0\end{bmatrix}+x_5\begin{bmatrix}0\\-1\\-1\\0\\1\end{bmatrix}$

25. $\vec{x}=x_2\begin{bmatrix}-2\\1\end{bmatrix}=x_2\vec{v}$

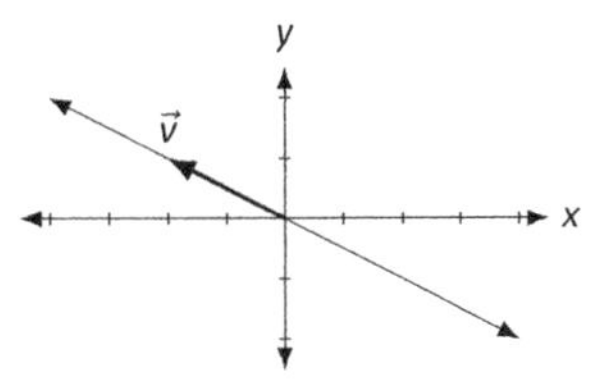

27. $\vec{x}=\begin{bmatrix}0.5\\0\end{bmatrix}+x_2\begin{bmatrix}2.5\\1\end{bmatrix}=\vec{x_p}+x_2\vec{v}$

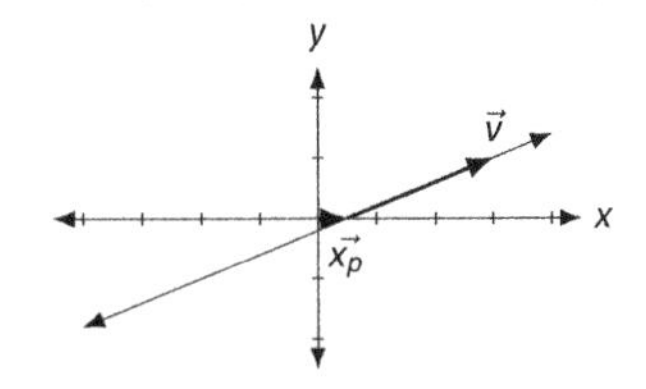

Section 2.5

1. $X=\begin{bmatrix}1&-9\\-4&-5\end{bmatrix}$

3. $X=\begin{bmatrix}-2&-7\\7&-6\end{bmatrix}$

5. $X=\begin{bmatrix}-5&2&-3\\-4&-3&-2\end{bmatrix}$

7. $X=\begin{bmatrix}1&0\\3&-1\end{bmatrix}$

9. $X=\begin{bmatrix}3&-3&3\\2&-2&-3\\-3&-1&-2\end{bmatrix}$

11. $X=\begin{bmatrix}5/3&2/3&1\\-1/3&1/6&0\\1/3&1/3&0\end{bmatrix}$

Section 2.6

1. $\begin{bmatrix} -24 & -5 \\ 5 & 1 \end{bmatrix}$

3. $\begin{bmatrix} 1/3 & 0 \\ 0 & 1/7 \end{bmatrix}$

5. A^{-1} does not exist.

7. $\begin{bmatrix} 1 & 0 \\ 0 & 1 \end{bmatrix}$

9. $\begin{bmatrix} -5/13 & 3/13 \\ 1/13 & 2/13 \end{bmatrix}$

11. $\begin{bmatrix} -2 & 1 \\ 3/2 & -1/2 \end{bmatrix}$

13. $\begin{bmatrix} 1 & 2 & -2 \\ 0 & 1 & -3 \\ 6 & 10 & -5 \end{bmatrix}$

15. $\begin{bmatrix} 1 & 0 & 0 \\ 52 & -48 & 7 \\ 8 & -7 & 1 \end{bmatrix}$

17. A^{-1} does not exist.

19. $\begin{bmatrix} 25 & 8 & 0 \\ 78 & 25 & 0 \\ -30 & -9 & 1 \end{bmatrix}$

21. $\begin{bmatrix} 0 & 1 & 0 \\ 0 & 0 & 1 \\ 1 & 0 & 0 \end{bmatrix}$

23. $\begin{bmatrix} 1 & 0 & 0 & 0 \\ -3 & -1 & 0 & -4 \\ -35 & -10 & 1 & -47 \\ -2 & -2 & 0 & -9 \end{bmatrix}$

25. $\begin{bmatrix} 28 & 18 & 3 & -19 \\ 5 & 1 & 0 & -5 \\ 4 & 5 & 1 & 0 \\ 52 & 60 & 12 & -15 \end{bmatrix}$

27. $\begin{bmatrix} 0 & 0 & 1 & 0 \\ 0 & 0 & 0 & 1 \\ 1 & 0 & 0 & 0 \\ 0 & 1 & 0 & 0 \end{bmatrix}$

29. $\vec{x} = \begin{bmatrix} 2 \\ 3 \end{bmatrix}$

31. $\vec{x} = \begin{bmatrix} -8 \\ 1 \end{bmatrix}$

33. $\vec{x} = \begin{bmatrix} -7 \\ 1 \\ -1 \end{bmatrix}$

35. $\vec{x} = \begin{bmatrix} 3 \\ -1 \\ -9 \end{bmatrix}$

Section 2.7

1. $(AB)^{-1} = \begin{bmatrix} -2 & 3 \\ 1 & -1.4 \end{bmatrix}$

3. $(AB)^{-1} = \begin{bmatrix} 29/5 & -18/5 \\ -11/5 & 7/5 \end{bmatrix}$

5. $A^{-1} = \begin{bmatrix} -2 & -5 \\ -1 & -3 \end{bmatrix}$,

   $(A^{-1})^{-1} = \begin{bmatrix} -3 & 5 \\ 1 & -2 \end{bmatrix}$

7. $A^{-1} = \begin{bmatrix} -3 & 7 \\ 1 & -2 \end{bmatrix}$,

   $(A^{-1})^{-1} = \begin{bmatrix} 2 & 7 \\ 1 & 3 \end{bmatrix}$

9. Solutions will vary.

11. Likely some entries that should be 0 will not be exactly 0, but rather very small values.

Chapter 3

Section 3.1

1. A is symmetric. $\begin{bmatrix} -7 & 4 \\ 4 & -6 \end{bmatrix}$

3. A is diagonal, as is A^T. $\begin{bmatrix} 1 & 0 \\ 0 & 9 \end{bmatrix}$

5. $\begin{bmatrix} -5 & 3 & -10 \\ -9 & 1 & -8 \end{bmatrix}$

7. $\begin{bmatrix} 4 & -9 \\ -7 & 6 \\ -4 & 3 \\ -9 & -9 \end{bmatrix}$

9. $\begin{bmatrix} -7 \\ -8 \\ 2 \\ -3 \end{bmatrix}$

11. $\begin{bmatrix} -9 & 6 & -8 \\ 4 & -3 & 1 \\ 10 & -7 & -1 \end{bmatrix}$

13. A is symmetric. $\begin{bmatrix} 4 & 0 & -2 \\ 0 & 2 & 3 \\ -2 & 3 & 6 \end{bmatrix}$

15. $\begin{bmatrix} 2 & 5 & 7 \\ -5 & 5 & -4 \\ -3 & -6 & -10 \end{bmatrix}$

17. $\begin{bmatrix} 4 & 5 & -6 \\ 2 & -4 & 6 \\ -9 & -10 & 9 \end{bmatrix}$

19. A is upper triangular; A^T is lower triangular. $\begin{bmatrix} -3 & 0 & 0 \\ -4 & -3 & 0 \\ -5 & 5 & -3 \end{bmatrix}$

21. A is diagonal, as is A^T. $\begin{bmatrix} 1 & 0 & 0 \\ 0 & 2 & 0 \\ 0 & 0 & -1 \end{bmatrix}$

23. A is skew symmetric. $\begin{bmatrix} 0 & -1 & 2 \\ 1 & 0 & -4 \\ -2 & 4 & 0 \end{bmatrix}$

Section 3.2

1. 6
3. 3
5. -9
7. 1
9. Not defined; the matrix must be square.
11. -23
13. 4
15. 0
17. (a) tr(A)=8; tr(B)=-2; tr($A+B$)=6
    (b) tr(AB) = 53 = tr(BA)
19. (a) tr(A)=-1; tr(B)=6; tr($A+B$)=5
    (b) tr(AB) = 201 = tr(BA)

Section 3.3

1. 34
3. -44
5. -44
7. 28
9. (a) The submatrices are $\begin{bmatrix} 7 & 6 \\ 6 & 10 \end{bmatrix}$, $\begin{bmatrix} 3 & 6 \\ 1 & 10 \end{bmatrix}$, and $\begin{bmatrix} 3 & 7 \\ 1 & 6 \end{bmatrix}$, respectively.
    (b) $C_{1,2} = 34$, $C_{1,2} = -24$, $C_{1,3} = 11$
11. (a) The submatrices are $\begin{bmatrix} 3 & 10 \\ 3 & 9 \end{bmatrix}$, $\begin{bmatrix} -3 & 10 \\ -9 & 9 \end{bmatrix}$, and $\begin{bmatrix} -3 & 3 \\ -9 & 3 \end{bmatrix}$, respectively.
    (b) $C_{1,2} = -3$, $C_{1,2} = -63$, $C_{1,3} = 18$
13. -59
15. 15
17. 3
19. 0
21. 0
23. -113
25. Hint: $C_{1,1} = d$.

Section 3.4

1. 84
3. 0
5. 10
7. 24
9. 175
11. -200
13. 34
15. (a) $\det(A) = 41$; $R_2 \leftrightarrow R_3$
    (b) $\det(B) = 164$; $-4R_3 \rightarrow R_3$
    (c) $\det(C) = -41$; $R_2 + R_1 \rightarrow R_1$
17. (a) $\det(A) = -16$; $R_1 \leftrightarrow R_2$ then $R_1 \leftrightarrow R_3$
    (b) $\det(B) = -16$; $-R_1 \rightarrow R_1$ and $-R_2 \rightarrow R_2$
    (c) $\det(C) = -432$; $C = 3 * M$
19. $\det(A) = 4$, $\det(B) = 4$, $\det(AB) = 16$
21. $\det(A) = -12$, $\det(B) = 29$, $\det(AB) = -348$
23. -59
25. 15
27. 3
29. 0

Section 3.5

1. (a) $\det(A) = 14$, $\det(A_1) = 70$, $\det(A_2) = 14$
   (b) $\vec{x} = \begin{bmatrix} 5 \\ 1 \end{bmatrix}$
3. (a) $\det(A) = 0$, $\det(A_1) = 0$, $\det(A_2) = 0$
   (b) Infinite solutions exist.
5. (a) $\det(A) = 16$, $\det(A_1) = -64$, $\det(A_2) = 80$
   (b) $\vec{x} = \begin{bmatrix} -4 \\ 5 \end{bmatrix}$
7. (a) $\det(A) = -123$, $\det(A_1) = -492$, $\det(A_2) = 123$, $\det(A_3) = 492$
   (b) $\vec{x} = \begin{bmatrix} 4 \\ -1 \\ -4 \end{bmatrix}$
9. (a) $\det(A) = 56$, $\det(A_1) = 224$, $\det(A_2) = 0$, $\det(A_3) = -112$

(b) $\vec{x} = \begin{bmatrix} 4 \\ 0 \\ -2 \end{bmatrix}$

11. (a) $\det(A) = 0$, $\det(A_1) = 147$, $\det(A_2) = -49$, $\det(A_3) = -49$

    (b) No solution exists.

Chapter 4

Section 4.1

1. $\lambda = 3$

3. $\lambda = 0$

5. $\lambda = 3$

7. $\vec{x} = \begin{bmatrix} -1 \\ 2 \end{bmatrix}$

9. $\vec{x} = \begin{bmatrix} 3 \\ -7 \\ 7 \end{bmatrix}$

11. $\vec{x} = \begin{bmatrix} -1 \\ 1 \\ 1 \end{bmatrix}$

13. $\lambda_1 = 4$ with $\vec{x_1} = \begin{bmatrix} 9 \\ 1 \end{bmatrix}$;

    $\lambda_2 = 5$ with $\vec{x_2} = \begin{bmatrix} 8 \\ 1 \end{bmatrix}$

15. $\lambda_1 = -3$ with $\vec{x_1} = \begin{bmatrix} -2 \\ 1 \end{bmatrix}$;

    $\lambda_2 = 5$ with $\vec{x_2} = \begin{bmatrix} 6 \\ 1 \end{bmatrix}$

17. $\lambda_1 = 2$ with $\vec{x_1} = \begin{bmatrix} 1 \\ 1 \end{bmatrix}$;

    $\lambda_2 = 4$ with $\vec{x_2} = \begin{bmatrix} -1 \\ 1 \end{bmatrix}$

19. $\lambda_1 = -1$ with $\vec{x_1} = \begin{bmatrix} 1 \\ 2 \end{bmatrix}$;

    $\lambda_2 = -3$ with $\vec{x_2} = \begin{bmatrix} 1 \\ 0 \end{bmatrix}$

21. $\lambda_1 = 3$ with $\vec{x_1} = \begin{bmatrix} -3 \\ 0 \\ 2 \end{bmatrix}$;

    $\lambda_2 = 4$ with $\vec{x_2} = \begin{bmatrix} -5 \\ -1 \\ 1 \end{bmatrix}$

    $\lambda_3 = 5$ with $\vec{x_3} = \begin{bmatrix} 1 \\ 0 \\ 0 \end{bmatrix}$

23. $\lambda_1 = -5$ with $\vec{x_1} = \begin{bmatrix} 24 \\ 13 \\ 8 \end{bmatrix}$;

    $\lambda_2 = -2$ with $\vec{x_2} = \begin{bmatrix} 6 \\ 5 \\ 1 \end{bmatrix}$

    $\lambda_3 = 3$ with $\vec{x_3} = \begin{bmatrix} 0 \\ 1 \\ 0 \end{bmatrix}$

25. $\lambda_1 = -2$ with $\vec{x_1} = \begin{bmatrix} 0 \\ 0 \\ 1 \end{bmatrix}$;

    $\lambda_2 = 1$ with $\vec{x_2} = \begin{bmatrix} 0 \\ 3 \\ 5 \end{bmatrix}$

    $\lambda_3 = 5$ with $\vec{x_3} = \begin{bmatrix} 28 \\ 7 \\ 1 \end{bmatrix}$

27. $\lambda_1 = -2$ with $\vec{x} = \begin{bmatrix} 1 \\ 0 \\ 1 \end{bmatrix}$;

    $\lambda_2 = 3$ with $\vec{x} = \begin{bmatrix} 1 \\ 1 \\ 1 \end{bmatrix}$;

    $\lambda_3 = 5$ with $\vec{x} = \begin{bmatrix} 0 \\ 1 \\ 1 \end{bmatrix}$

Section 4.2

1. (a) $\lambda_1 = 1$ with $\vec{x_1} = \begin{bmatrix} 4 \\ 1 \end{bmatrix}$;

       $\lambda_2 = 4$ with $\vec{x_2} = \begin{bmatrix} 1 \\ 1 \end{bmatrix}$

   (b) $\lambda_1 = 1$ with $\vec{x_1} = \begin{bmatrix} -1 \\ 1 \end{bmatrix}$;

       $\lambda_2 = 4$ with $\vec{x_2} = \begin{bmatrix} -1 \\ 4 \end{bmatrix}$

   (c) $\lambda_1 = 1/4$ with $\vec{x_1} = \begin{bmatrix} 1 \\ 1 \end{bmatrix}$;

       $\lambda_2 = 4$ with $\vec{x_2} = \begin{bmatrix} 4 \\ 1 \end{bmatrix}$

   (d) 5

   (e) 4

3. (a) $\lambda_1 = -1$ with $\vec{x_1} = \begin{bmatrix} -5 \\ 1 \end{bmatrix}$;

       $\lambda_2 = 0$ with $\vec{x_2} = \begin{bmatrix} -6 \\ 1 \end{bmatrix}$

   (b) $\lambda_1 = -1$ with $\vec{x_1} = \begin{bmatrix} 1 \\ 6 \end{bmatrix}$;

       $\lambda_2 = 0$ with $\vec{x_2} = \begin{bmatrix} 1 \\ 5 \end{bmatrix}$

   (c) A is not invertible.

(d) -1

(e) 0

5. (a) $\lambda_1 = -4$ with $\vec{x_1} = \begin{bmatrix} -7 \\ -7 \\ 6 \end{bmatrix}$;

$\lambda_2 = 3$ with $\vec{x_2} = \begin{bmatrix} 0 \\ 0 \\ 1 \end{bmatrix}$

$\lambda_3 = 4$ with $\vec{x_3} = \begin{bmatrix} 9 \\ 1 \\ 22 \end{bmatrix}$

(b) $\lambda_1 = -4$ with $\vec{x_1} = \begin{bmatrix} -1 \\ 9 \\ 0 \end{bmatrix}$;

$\lambda_2 = 3$ with $\vec{x_2} = \begin{bmatrix} -20 \\ 26 \\ 7 \end{bmatrix}$

$\lambda_3 = 4$ with $\vec{x_3} = \begin{bmatrix} -1 \\ 1 \\ 0 \end{bmatrix}$

(c) $\lambda_1 = -1/4$ with $\vec{x_1} = \begin{bmatrix} -7 \\ -7 \\ 6 \end{bmatrix}$;

$\lambda_2 = 1/3$ with $\vec{x_2} = \begin{bmatrix} 0 \\ 0 \\ 1 \end{bmatrix}$

$\lambda_3 = 1/4$ with $\vec{x_3} = \begin{bmatrix} 9 \\ 1 \\ 22 \end{bmatrix}$

(d) 3

(e) -48

Chapter 5

Section 5.1

1. $A = \begin{bmatrix} 1 & 2 \\ 3 & 4 \end{bmatrix}$

3. $A = \begin{bmatrix} 1 & 2 \\ 1 & 2 \end{bmatrix}$

5. $A = \begin{bmatrix} 5 & 2 \\ 2 & 1 \end{bmatrix}$

7. $A = \begin{bmatrix} 0 & 1 \\ 3 & 0 \end{bmatrix}$

9. $A = \begin{bmatrix} 0 & -1 \\ -1 & -1 \end{bmatrix}$

11. Yes, these are the same; the transformation matrix in each is $\begin{bmatrix} -1 & 0 \\ 0 & -1 \end{bmatrix}$.

13. Yes, these are the same. Each produces the transformation matrix $\begin{bmatrix} 1/2 & 0 \\ 0 & 3 \end{bmatrix}$.

Section 5.2

1. Yes

3. No; cannot add a constant.

5. Yes.

7. $[T] = \begin{bmatrix} 1 & 2 \\ 3 & -5 \\ 0 & 2 \end{bmatrix}$

9. $[T] = \begin{bmatrix} 1 & 0 & 3 \\ 1 & 0 & -1 \\ 1 & 0 & 1 \end{bmatrix}$

11. $[T] = \begin{bmatrix} 1 & 2 & 3 & 4 \end{bmatrix}$

Section 5.3

1. $\vec{x} + \vec{y} = \begin{bmatrix} 3 \\ 2 \\ 4 \end{bmatrix}$, $\vec{x} - \vec{y} = \begin{bmatrix} -1 \\ -4 \\ 0 \end{bmatrix}$

Sketches will vary slightly depending on orientation.

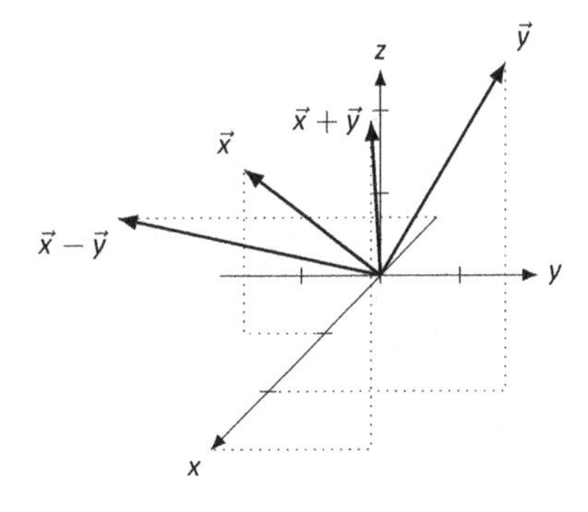

3. $\vec{x} + \vec{y} = \begin{bmatrix} 4 \\ 4 \\ 8 \end{bmatrix}$, $\vec{x} - \vec{y} = \begin{bmatrix} -2 \\ -2 \\ -4 \end{bmatrix}$

Sketches will vary slightly depending on orientation.

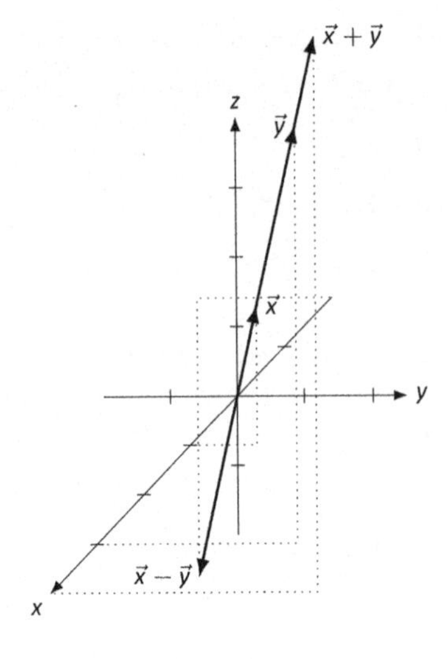

5. Sketches may vary slightly.

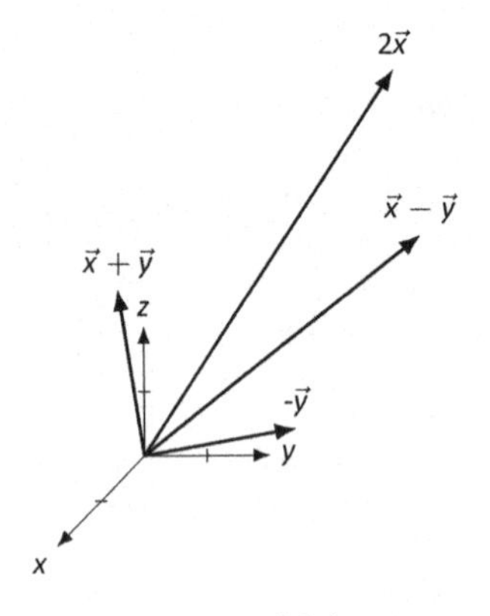

7. Sketches may vary slightly.

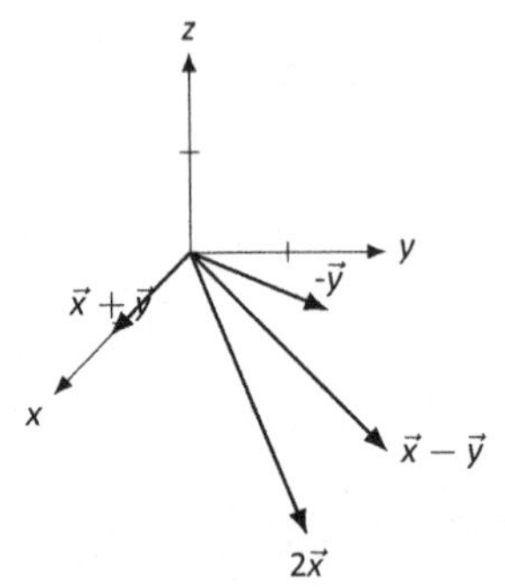

9. $||\vec{x}|| = \sqrt{30}$, $||a\vec{x}|| = \sqrt{120} = 2\sqrt{30}$

11. $||\vec{x}|| = \sqrt{54} = 3\sqrt{6}$,
$||a\vec{x}|| = \sqrt{270} = 15\sqrt{6}$

Index

Made in the USA
Middletown, DE
15 January 2019